WALTHER NERNST
Pioneer of Physics and of Chemistry

Hans-Georg Bartel
Humboldt University Berlin, Germany

Rudolf P Huebener
Eberhard-Karls-University Tübingen, Germany

NEW JERSEY · LONDON · SINGAPORE · BEIJING · SHANGHAI · HONG KONG · TAIPEI · CHENNAI

Published by

World Scientific Publishing Co. Pte. Ltd.
5 Toh Tuck Link, Singapore 596224
USA office: 27 Warren Street, Suite 401-402, Hackensack, NJ 07601
UK office: 57 Shelton Street, Covent Garden, London WC2H 9HE

British Library Cataloguing-in-Publication Data
A catalogue record for this book is available from the British Library.

WALTHER NERNST
Pioneer of Physics and of Chemistry

Copyright © 2007 by World Scientific Publishing Co. Pte. Ltd.

All rights reserved. This book, or parts thereof, may not be reproduced in any form or by any means, electronic or mechanical, including photocopying, recording or any information storage and retrieval system now known or to be invented, without written permission from the Publisher.

For photocopying of material in this volume, please pay a copying fee through the Copyright Clearance Center, Inc., 222 Rosewood Drive, Danvers, MA 01923, USA. In this case permission to photocopy is not required from the publisher.

ISBN-13 978-981-256-560-0
ISBN-10 981-256-560-4

Printed in Singapore by World Scientific Printers (S) Pte Ltd

Preface

WALTHER NERNST, physicist, chemist, Nobel laureate, cofounder of the field of physical chemistry, teacher, research manager, inventor, textbook author, farmer, and diplomat, covered an enormously wide range of activities. The discovery of his Thermal Law, subsequently referred to as the Third Law of Thermodynamics, for which in 1921 he was awarded the Nobel Prize in Chemistry (of the year 1920) represented perhaps his most outstanding and most widely recognized accomplishment. But there were many other important advances and actions connected with NERNST. We mention his work on the thermomagnetic effects in metals performed already during his thesis, his fundamental contributions to electrochemistry, his specific-heat measurements at low temperatures supporting the new quantum theory, his Presidency of the Physikalisch-Technische Reichsanstalt in Berlin, and his diplomatic initiatives for reaching an earlier ending of the First World War.

Already from this listing we can see that a description of the life and the activity of WALTHER NERNST is similarly rewarding as it is difficult, if one attempts to present at least their main aspects in a satisfactory way. In addition to many shorter and sometimes also longer articles devoted to the memory of NERNST, which are quite valuable in this connection, three authors have covered this subject in the past in the form of books.

Motivated by the memorial events on the occasion of the 100th birthday of NERNST in 1964, the low-temperature physicist KURT MENDELSSOHN first presented the highly noteworthy and valuable book on the great scholar and his times [Mendelssohn (1973)]. He had personally witnessed NERNST during his last creative period at the Physical Institute of the University of Berlin. Unfortunately, many times the fluent description suffered from incorrectness. EDITH VON ZANTHIER, the daughter of NERNST, who had supplied MENDELSSOHN with material, expressed her opinion on the book: *"In it there appear peculiar points, which really do not correspond to the nature of my father."* [Zanthier (1978)]. As a source of the history of science the book by MENDELSSOHN can be compared perhaps with the *"History of the Thirty Years War"* (1790) by FRIEDRICH SCHILLER in terms of its role as a secondary source.

In 1989 there appeared the shorter biography by HANS-GEORG BARTEL in the German language, in which an attempt was made to correct these inaccuracies at least regarding the most important points [Bartel (1989)]. As a volume within a series of biographies at the popular-science level it was limited to a relatively small size.

The works by DIANA K. BARKAN from the middle or the end of the 1990s are oriented more historically-epistemologically [Barkan (1995); (1999)]. Unfortunately, its great value is scattered in various directions. The historian of science DIETER HOFFMANN (Max Planck Institute for the History of Science, Berlin) hence remarked about this book in his review: *"The core of Barkan's study, where she deals with the growth of modern physics and Nernst's pivotal role in the process, is the best part of her analysis. However, her study is less detailed and insightful in its presentation of other parts of Nernst's scientific work. Indeed, those research topics in which he was engaged after the First World War are more or less neglected."* [Hoffmann (1999)]. Also the life of NERNST during the First World War is treated quite inadequately. Furthermore, there are several more incorrect points.

It is the goal of the present book to try and overcome the mentioned deficiencies of its predecessors. The book attempts to bridge the inseparable unity of biography and historical epistemology in the case of WALTHER NERNST, in many ways an exceptional human being and scientist, and his activities within the many fields mentioned above. In addition to the description of his life and the rise of his academic career up to reaching the very high peak in the second decade of the 20th century, special attention has been given to his activity during the First World

War and the subsequent time. Aside from some special treatments, up to now this period – from 1914 until his death in 1941 – has been covered only relatively briefly or in a summarizing way. This is in contrast to its importance. The period contains NERNST's actions during both World Wars, his attitude toward fascism, his occupation as Rector of the University of Berlin and as President of the Physikalisch-Technische Reichsanstalt, as well as his contributions to the electroacoustic musical instruments and to cosmology and astrophysics, which are frequently underrated in value.

The treatment in the book is supported by presenting quotations written by NERNST, his students, coworkers, colleagues, and members of his family. In this case sources have been cited, which up to now were used only rarely or not at all. An attempt to include only such statements the validity of which could hardly be questioned was made. Sometimes for the discussion of physical or physical-chemical facts, a mathematical presentation has been used. In this case, an opinion of the physical chemist HANS JAHN, having some importance in NERNST's biography, served as a good example: *"Because it is a fruitless endeavor ... to want to aim at a goal along bumpy, impassable secret paths, where the mathematical analysis has prepared already royal roads."* [Jahn (1895): IV].

In the attached name index to a large extent completeness has been observed regarding the (different) spellings, the first names, the years of the birth and the death, the titles of nobility, and any other information.

Due to the enormous amount of material, in this book too, some aspects could be treated only indirectly or marginally. In this case we have in mind NERNST's popularity as an academic teacher and his role as "father of the institute family", his attitude toward women, his love of motor cars and of traveling, in particular to Italy, but also the occupation with more special scientific problems. However, this did not noticeably affect the intended goal of a comprehensive treatment of the unique personality of WALTHER NERNST and his important achievements in science, technology, and society.

We hope that the present book has come reasonably close to this goal. Certainly, a total realization of it was not possible. Because of the large complexity of the subject "WALTHER NERNST", after one has treated it there always remains the fact that something had been left out which deserves to be discussed in the future.

We thank World Scientific for taking up the publication of our book, and we are especially grateful to Senior Editor Ms LAKSHMI NARAYANAN from the Singapore Office for handling capably and promptly our many questions. Prof. Dr. DIETER HOFFMANN and Dr. ULRICH SCHMITT (Institute of Physical Chemistry, University Göttingen) deserve thanks for their friendly advice and their support in connection with the science-historical questions. Last but not least our thanks go to Dr. SABINE BARTEL and Dipl.-Math. HANS-JOACHIM MUCHA for their support in the electronic traffic between Berlin and Tubingen.

Berlin and Tubingen, April 2007

Hans-Georg Bartel *Rudolf P. Huebener*

Contents

Preface   v

1. Development of Physics and Physical Chemistry from about 1800 until 1870   1
2. Youth and University Period (1864 – 1887)   9
   - 2.1 Ancestors and Parents   9
   - 2.2 Youth and High School in Graudenz   10
   - 2.3 University Studies in Zurich and Berlin   16
   - 2.4 Graz: The "Second Scientific Home"   19
     - 2.4.1 University and physics in Graz: Ludwig Boltzmann and Albert von Ettingshausen   19
     - 2.4.2 The Ettingshausen-Nernst effects and the Nernst effect   26
   - 2.5 Conclusion of the University Studies in Wurzburg   30
3. Habilitation in Leipzig (1887 – 1889)   35
   - 3.1 The Sciences at the University of Leipzig   36
   - 3.2 Wilhelm Ostwald   38

3.3 The Completion of the Thermodynamics of Electrochemistry: The Nernst Equation 40

3.4 The "Ionists" versus the "Anti-Ionists" 51

4. **The Göttingen Period: The Rise to World Fame (1890 – 1905)** 57

    4.1 The *Georgia Augusta* University in Göttingen 58

    4.2 Eduard Riecke, Felix Klein, and Mathematics in Göttingen 59

    4.3 Early Studies in Göttingen: The Nernst Distribution Law . 62

    4.4 Marriage with Emma Lohmeyer and the Walther Nernst Family 68

    4.5 The Textbook "Theoretical Chemistry from the Standpoint of Avogadro's Rule and Thermodynamics" 73

    4.6 The First Professorship and the Establishment of a Chair of Physical Chemistry 80

    4.7 The New Institute of Physical Chemistry and Electrochemistry 88

    4.8 Studies and Members in the New Institute 96

    4.9 The Nernst Lamp 104

    4.10 Nernst Law of Electrical Nerve Stimulus Threshold (*Reizschwellengesetz*) 114

    4.11 The Construction of Instruments 117

    4.12 Mathematics and Chemistry 126

5. **Professor of Physical Chemistry in Berlin (1905 – 1922)** 133

    5.1 The Friedrich-Wilhelm University and Other Academic Institutions in Berlin and Charlottenburg 133

    5.2 The Famous Year 1905 139

5.3	The Institute of Physical Chemistry at the University of Berlin	142
5.4	The First Lecture in Berlin – Announcement of a Fundamental Law of Nature	146
5.5	The Nernst Law of Heat or the Third Law of Thermodynamics	150
	5.5.1 Remarks on the First and Second Law of Thermodynamics	150
	5.5.2 The problem and its solution given by Nernst	154
	5.5.3 The calculation of chemical equilibria	165
	5.5.4 Specific heats and low-temperature physics	169
	5.5.5 Quantum Theory	177
	5.5.6 The impossibility of reaching the absolute zero of temperature	184
	5.5.7 Formulation of the Third Thermal Law by Max Planck	187
	5.5.8 Research between 1906 and 1916, the monograph, and the Nobel Prize in Chemistry	189
	5.5.9 Critique and priority conflict	195
5.6	Other Scientific Studies during this Period	198
5.7	Organization of Science	206
	5.7.1 Kaiser Wilhelm Institutes	206
	5.7.2 German Electrochemical Society	213
	5.7.3 Other developments	218
	5.7.4 Rector of the University and the German Institute for Foreigners	223
5.8	Managing a Country Estate, Hunting, and Fish Farming	229
5.9	The First World War	236
	5.9.1 War-related research: gas warfare, explosives, ballistics	237
	5.9.2 The effort on peace negotiations	248

	5.10	Political Activities	257
	5.11	Visits to the USA and to South America	260
6.	President of the Physikalisch-Technische Reichsanstalt (PTR) (1922 – 1924)		267
	6.1	Brief History of the PTR	267
	6.2	Activities of Nernst at the PTR	273
7.	Professor of Experimental Physics at the University of Berlin (1924 – 1933)		283
	7.1	Solutions of Strong Electrolytes	287
	7.2	Vibrating Strings and the Neo-Bechstein Grand Piano	292
	7.3	Studies in Cosmology and Astrophysics	306
8.	The Final Years (1933 – 1941)		327
	8.1	Attitude to the Fascism	327
	8.2	An Attempt to Participate in the War-Related Research during the Second World War	335
	8.3	The End in the Village of Zibelle	337
9.	Honors and Memorials		341

References 347

Name Index 373

Chapter 1

Development of Physics and Physical Chemistry from about 1800 until 1870

In his statement addressed to CASPAR VOGHT at the beginning of the 19th century, *"The sciences move forward not in a circle, but in a spiral – the same returns again, but higher and farther."* (quoted after [Dobel (1968)]), JOHANN WOLFGANG VON GOETHE had expressed his opinion on the development of the sciences, which at the same time implied the importance of the pre-occupation with the history of science. In the introduction to the historic part of his «*Farbenlehre*» GOETHE has remarked *"Indeed, a history of the sciences, as long as it is treated by human people, shows quite a different and highly educative meaning, than if discoveries and opinions are only arranged one after another."* [Goethe (1893): XII].

So it is necessary and rewarding at the same time to illuminate with a few words the historic development preceding the scientific activity and work of WALTHER NERNST, in which physical chemistry as a special discipline of physics occupies the central place.

The development of chemistry into an exact science happened around the turn of the 18th to the 19th century. Still in 1786 IMMANUEL KANT had denied chemistry this status: *"So as long as no concept, which can be constructed, is found regarding the 'chymischen' actions of matter upon one another, i.e., no law about getting closer or more apart of the parts can be formulated, according to which, say, in proportion to their densities ... their movements including their consequences can be made clear and be depicted a priory in space (a requirement which hardly ever will be fulfilled), so 'Chymie' can never become more than a systematic art, or an experimental doctrine, but never a proper science, since its principles are only empirical and do not allow a description a priory in a clear picture, hence, it cannot make the principles of 'chymischer' phenomena in the least understandable, because they are incapable of applying mathematics."* [Kant (1786): X]. Although at this time to a large extent this remark had lost already its validity, it still emphasizes the necessary connection of the chemistry with the mathematics and due to the latter with the physics as a requirement for its transformation into an exact science. In Section 4.12 a few aspects of the historic development of the relations between chemistry and mathematics or mechanics, at the time treated as part of this discipline, will be discussed. Here only a few points from the history of physics and chemistry from the end of the 18th until about the third quarter of the 19th century shall be mentioned, which were important for the rise of the classical physical chemistry, for the completion of which NERNST then played a decisive role.

The early history of these mutual relations extends to the time before 1600. During this period great scientists have demonstrated close relations between the chemical substances and their mutual transformations and measurable quantities such as weight (mass), temperature, and time.

ANDREAS LIBAVIUS, the founder of the quantitative chemical analysis and the discoverer of the *spiritus fumans Libavii* ($SnCl_4$), connected the relationship (affinity) between salts with their crystal size, mass, and magnetism, as well as smell and taste.

For the 17th century above all ROBERT BOYLE must be mentioned. His law of gases, which was discovered independent of him also by EDME MARIOTTE, and which, hence, is named after both scientists, quantitatively connects the pressure p, the volume V, the temperature T, and the amount n of the substance with one another: $p \sim V^{-1}$ for $T =$ const. and $n =$ const. BOYLE considered the qualitative and the quantitative determination of the compositions of the substances as the main task of the chemists. Also he had introduced the vacuum into the chemical experimental technique.

The natural scientist HERMAN BOERHAAVE from Leiden was similarly versatile and brilliant as the British BOYLE. His work *«Elementa chimiae»* of 1732 contains data on the freezing and boiling points, on the values of the heat of solution, mixture, and reaction, as well as on the thermal expansion of the materials. BOERHAAVE distinguished between purely chemical compounds and their mixtures.

The Russian universal scholar MIKHAIL LOMONOSOV advanced the mathematical and physical side of chemistry. He is credited with the discovery of the conservation of mass in the case of chemical reactions, one of the most fundamental laws of chemistry. The French scholar ANTOINE LAURENT LAVOISIER had also discovered this law a little later independent of his Russian colleague. It is said that LOMONOSOV had coined the word "Physical Chemistry" for the first time. He had emphasized: *"The chemistry of mine is mathematical."* (see also Section 4.12).

The activity of CARL FRIEDRICH WENZEL working in Saxony was highly important. In Freiberg he had recognized the influence of the amount of substance upon the result of a chemical reaction. The relationship between chemical substances he assumed to be inversely proportional to their time of solution in a given solvent.

These examples and names in place of many others may be sufficient in order to demonstrate how strongly chemistry was dominated by quantitative and thereby actually by physical ideas until the end of the 18th century. Of course, this was a great advance for its development. Now quantitative laws with far-reaching consequences could be discovered.

At the beginning of this development one can place the dissertation of JEREMIAS BENJAMIN RICHTER from 1789 of the University of Königsberg, which was devoted to the application of mathematics in chemistry (see Section 4.12). The law of the equivalent proportions (1792) also originates from RICHTER. In 1807 WILLIAM HYDE WOLLASTON formulated its general version: "Chemical elements always combine with each other following the ratio of certain compound masses (so-called equivalent masses) or integer multiples of these masses leading to chemical compounds." At this time also two additional fundamental laws of the mass balance of chemical reactions were found. We refer to the law of constant proportions by JOSEPH LOUIS PROUST from 1799 ("In a certain chemical compound elements always appear in the same mass ratio."), which JOHN DALTON extended in 1803 in terms of the law of the multiple proportions ("If two different chemical compounds contain the same elements, then there appear simple number ratios between the element masses."). Based on these laws in the beginning of the 19th century DALTON justified the model of the elementary nature of chemical compounds based on the assumption of atoms, which he treated in his book *"A New System of Chemical Philosophy"* published in 1808. Thereby the foundation was created for quantifying and mathematizing the chemistry. Therefore, more than LAVOISIER, DALTON is to be counted as the true founder of chemistry as an exact science, as has been noted already by the philosopher FRIEDRICH ENGELS: *"The new epoch starts in chemistry with the atomistics (so Dalton, not Lavoisier, the father of the newer chemistry) and correspondingly in physics with the molecular theory ..."* [Engels (1968): 552].

Essentially, the only common feature of the atomic theory of DALTON and the classical antique theory of DEMOCRITUS was the name of the treated objects, which goes back to the Greek ἄτομος ('indivisible'). Also this doctrine had been highly controversial. Prepared during the 16th century, perhaps during the 17th century the opinions collided with each other most strongly. For example, RENÉ DESCARTES represented the doctrine of the continuity; on the other hand his contemporary PIERRE GASSENDI followed the leading Greek atomist DEMOCRITUS. If the existence of atoms and, hence, a discontinuous structure of matter would be physically real, then between the atoms there should exist the empty space, i.e., the vacuum so strongly opposed by DESCARTES. The reality of the vacuum was demonstrated by the investigations by OTTO VON GUERICKE, in particular by his famous experiment performed in 1654 at the *Reichstag* in Regensburg using the Magdeburg hemispheres, by his construction of a vacuum pump, and also by the first proof of the chemical effect of the empty space by means of the extinction of a candle. In this way the validity of the atomic opinion had gained special weight. ROBERT BOYLE criticized the qualitative doctrines of the elements by

ARISTOTLE and by PARACELSUS, which were dominating during his time. He taught that the corpuscular or the atomic theory best explains the experimental experience and the existence of the vacuum. According to his opinion the atoms represent matter without any special quality, an idea which was contested, for example, by his contemporary NICOLAS LÉMERY.

AMADEO AVOGADRO from Piedmont worked out a molecular hypothesis, which is expressed in the law published in 1811 and named after him: "Under the same conditions (pressure, temperature) the same volumes of gases contain the same number of particles." Since initially AVOGADRO's paper remained unrecognized, in 1814 ANDRÉ MARIE AMPÈRE expressed the same hypothesis. Although the atomic or the corpuscular understanding of matter turned out to become an essential basis of chemistry and physical chemistry, even up to the times after 1900 one could find important scientists such as ERNST MACH and WILHELM OSTWALD, who considered the existence of atoms and molecules to be physically unreal, in contrast to, say, WALTHER NERNST.

In addition to the atomic theory, in particular thermodynamics and the theory of electricity represented further pillars upon which the structure of physical chemistry could be erected. The development of thermodynamics started from the interest to understand the physics of heat engines, in order to improve their efficiency. During the first half of the 19th century the most essential theoretical foundations have been created. Among the most important scientists who had contributed to these advances we must mention in particular SADI CARNOT, RUDOLF CLAUSIUS, JULIUS ROBERT MAYER, HERMANN VON HELMHOLTZ, JAMES PRESCOTT JOULE, and JOSIAH WILLARD GIBBS. In Section 5.5.1 this development will be treated in more detail. Together with the thermodynamics, the thermochemistry or the chemical thermodynamics developed into one of the most important disciplines of the physical chemistry.

The evolution of the theory of electricity and of the electrochemistry was closely connected with each other in this *"electric age"*, as the 19th century was called by FELIX PINNER [Pinner (1918)]. In 1780 in Bologna the medical doctor LUIGI GALVANI had discovered a new kind of electricity in the behavior of the thighs of dead frogs placed close to an electrostatic generator. This new kind could be counted in addition to the atmospheric and the frictional electricity. GALVANI interpreted his discovery in terms of a vitalistic point of view. On the other hand, during the years from 1792 until 1796 for the interpretation of the galvanism ALESSANDRO VOLTA created the concept of the contact electricity, which became highly important for the electrochemistry. In this work VOLTA had benefited from the previous studies of FRIEDRICH ALBRECHT CARL GREN and JOHANN CHRISTIAN REIL. In 1798 in Jena JOHANN WILHELM RITTER founded the electrochemistry by associating the galvanism with chemical effects [Ritter (1798)]. Only after 1810 the Englishman GEORGE JOHN SINGER coined the word "Electrochemistry". Prior to this the term galvanism had been used exclusively.

For the development of the electrochemistry during the first decade of the 19th century, in addition to the investigations by Sir HUMPHRY DAVY, JÖNS JAKOB BERZELIUS, and others, those by THEODOR GROTTHUSS are important. For example, already in 1805 the latter presented first ideas about the electrolytic dissociation, where, however, he considered the electric current to represent the cause of the dissociation. In 1857 RUDOLF CLAUSIUS interpreted this phenomenon statistically. The completion of the theory of the electrolytic dissociation fundamental to the natural science was achieved about 30 years later by the brilliant Swedish physicist SVANTE ARRHENIUS (see Section 3.4).

In the history of electrochemistry we must emphasize the activity of MICHAEL FARADAY also in this field. He created the basic terms electrode, electrolyte, anode, cathode, anion, and cation (see Section 3.4), and in 1832 he discovered the fundamental laws of electrolysis:

$$(1)\ n = (zF)^{-1} \cdot q, \quad (2)\ m = (zF)^{-1} \cdot qM,$$

where n denotes the amount of substance precipitated electrolytically, z the valency of the ion, q the electric charge, m the precipitated mass, M the mole mass, and $F = e_0 N_L$ FARADAY's constant (e_0 elementary charge, N_L LOSCHMIDT number).

For the theory of the electrolytes the investigations by WILHELM HITTORF on the transport of current within electrolytic solutions were highly important. Based on these studies in 1853 he discovered the transport law of the ions, which allowed a quantitative treatment of the mass transport in an electric field.

In particular GIBBS and HELMHOLTZ applied thermodynamics to electrochemical systems in equilibrium or in the absence of a current. So already in 1847 HELMHOLTZ had recognized the equivalence between the so-called electromotoric force (e.m.f, potential difference between the electrodes of a galvanic cell in the state of equilibrium) and the heat of reaction. In 1882 he found that the voltage of the cell can be particularly well described in terms of the free energy of the reaction. NERNST extended these scientific advances in his fundamental studies on electrochemistry.

In the 19th century for scientific experiments and as a current source the column developed by VOLTA and named after him played an important role. The discovery of FARADAY's laws led to new current sources, the galvanic elements, of which we mention those of JOHN FREDERIC DANIELL, WILLIAM GROVE, and ROBERT WILHELM BUNSEN. During his search for a new galvanic element, in 1854 the German physician JOSEPH SINSTEDEN invented the lead-acid storage battery, which five years later was improved by the French physicist GASTON PLANTÉ. Only in 1886 the inventor HENRI TUDOR from Luxemburg developed the first technically suitable lead accumulator.

The discovery of the dynamo-electric principle was extremely important for the generation of current. It had been recognized first in 1853 by the Hungarian physicist ÁNYOS JEDLIK, who initiated its technical application in 1861 with the construction of a dynamo engine. However, this principle became widely known only in 1866 when WERNER VON SIEMENS found it independently of JEDLIK and also built a dynamo engine. This then led to a rapid upturn of the electric technology. In the subsequent *"electric age"* the major innovation initially dealt with the electric light generation, to which also NERNST contributed significantly utilizing his electrochemical and physical experience (see Section 4.9).

Among the first scientists working in the field of the classical physical chemistry in the sense of today we can count ROBERT WILHELM BUNSEN and AUGUST FRIEDRICH HORSTMANN. Since 1859 the former worked on the spectrum analysis of the chemical elements together with GUSTAV ROBERT KIRCHHOFF. The paper by HORSTMANN on the vapor pressure and the heat of evaporation of ammonium chloride, which appeared in 1869, can be looked at as the first publication in the field of the chemical thermodynamics. Also HANS LANDOLT and HANS JAHN, which later will be mentioned several times, have achieved important results within the physical chemistry. Already in 1855 LANDOLT organized a physical-chemical colloquium at the Polytechnic Institute in Aachen.

On the completion of the classical physical chemistry the physical chemist WILHELM JOST from Göttingen has noted: *"Physical chemistry as a distinct field of chemistry, at least for practical purposes, originated in Germany during the last fifth of the 19th century. Who were the leading men of that period, i. e., born before 1860? They were Arrhenius, van't Hoff, and W. Ostwald, followed by Nernst – about ten years younger than the other three."* [Jost, W (1966): 1]. *" ... Physical chemistry in Germany started in Leipzig, and then spread by way of Göttingen to Berlin, there to experience an unprecedented flowering in a mutual exchange with all of Germany, and with the world."* [Jost, W (1966): 14]. In this last listing mainly the academic career and the development of the scientific impact of WALTHER NERNST is sketched.

Chapter 2

Youth and University Period (1864 – 1887)

2.1 Ancestors and Parents

ALBERT EINSTEIN had finished his obituary for his highly esteemed colleague WALTHER NERNST with the words *"He was an original personality; I have never met any one who resembled him in any essential way"* [Einstein (1942): 196], focusing on the human being and the scientist. NERNST occupied this unique position within human society in some sense also within his family. His ancestors have been craftsmen, tenants of an estate, ministers, and lawyers. Also his sons would not have pursued a career in science, even if they were not killed during the First World War.

The name 'Nernst' is the Low German version of 'Ernst', having the meaning of 'earnest', 'firmness', and 'fight'. This interpretation of his name fits WALTHER NERNST very well, and likely also in some sense many of his ancestors, which can be traced back into the 17th century. Around this time they lived in or near Prenzlau, a small village located in the Uckermark about 95 km north of Berlin. Originally, they worked as craftsmen in the village, as the cooper CHRISTIAN NERNST and his son, the carpenter JOHANNES CHRISTIAN. However, the great-grandfather of WALTHER NERNST, JOHANN DAVID was already a minister at the St. Mary's Church of this village. His grandfather PHILIPP NERNST, a second lieutenant of the cavalry, leased a farm in Potzlow, 13 km south of Prenzlau. He had distinguished himself as an officer in the Napoleonic Wars, similar to his brother HERMANN, who later became chief postmaster. This great-uncle of WALTHER had been given the honorable task by the Field Marshal BLÜCHER to deliver the news of the victory in the Battle of

Waterloo (June 18, 1815) to the Prussian King FRIEDRICH WILHELM III in Berlin.

WALTHER's father GUSTAV originated from the marriage between PHILIPP NERNST and ELISE MAGDALENA MITTERBACHER. He became a lawyer. He married OTTILIE NERGER, a daughter of the farmer KARL AUGUST NERGER and AUGUSTE SPERLING.

During his term of office as a judge in the small Western-Prussian town of Briesen (today Wąbrzeźno in Poland, Fig. 2.1) on June 25, 1864 WALTHER HERMANN NERNST was born as the third child to the married couple. There followed still two further children. Soon afterwards the family moved to Graudenz (today Grudziądz, Wojewod District Toruń/Poland, Fig. 2.1), located about 26 km north-west of Briesen, since the father had been promoted to county judge.

Fig. 2.1 The topography of Briesen and Graudenz (section of a modern map of Poland).

2.2 Youth and High School in Graudenz

WALTHER spent his childhood and youth in this small town located on the Vistula (Wisła). This town as well as the demesne Engelsburg located nearby (Fig. 2.2) had an important influence on his whole subsequent life. His uncle RUDOLF NERGER had leased this farming estate, where he lived together with his wife ANNA. Here WALTHER spent many days during the holidays and on weekends. Because of these visits and the associated experiences, he developed a life-long love of the country, the farming, and his only passionate sport activity, namely the hunting of small game. Since this youth period until the end of his life the open nature outside the cities played an important role for NERNST. In a later chapter we will mention the estates purchased by him, at which he spent a large part of his spare time and eventually the final period of his life, and where he died. It happened on the demesne Engelsburg near Graudenz, where in 1891 the lecturer Dr. NERNST from Göttingen received the news of his appointment as a Professor of Physical Chemistry. The importance, which NERNST attributed to his visits to Engelsburg, resulting in his love of the country life, can be seen from the fact, that after the death of his father he left his inheritance to his aunt ANNA, in order that she could keep up the demesne after the death of the uncle. Furthermore, he gave the name RUDOLF to his oldest son (Fig. 2.2), and he named his youngest daughter ANGELA, having a connection to Engelsburg because of the Latin word *angelus* meaning *Engel* in German.

The school education of WALTHER NERNST was provided by the Royal Protestant *Gymnasium* (high school) of Graudenz, which he entered in 1874. In order to illustrate the broad and even high teaching level of this institution, we list a selection of books mentioned by H. KRETSCHMANN, the Principal of the *Gymnasium*, to be added to the teacher's library during the period 1882 – 1883 [Kretschmann (1883): 39]: for history the third volume of *"World History"* by LEOPOLD VON RANKE, and the second volume of *"Germany in the 18th Century"* by KARL BIEDERMANN, for classical philology «*De agricultura*» by CATO, and «*Bibliotheca scriptorum classicorum*» by WILHELM ENGELMANN, for the area of art, ancient history, and archaeology *"The Ethics of the Greek"* by LEOPOLD SCHMIDT, the *"Handbook of the Roman Antiques"* by JOACHIM MARQUARDT and THEODOR MOMMSEN, the *"Handbook of the Biblical Archaeology"* by CARL FRIEDRICH KEIL, the *"Laokoon Studies"* by HUGO BLÜMNER, and *"Pictures of History: a Series of the outstanding Buildings of all Cultural Epoches"* by JOSEF LANGL, for theology *"The History of the Holy Scriptures"* by EDUARD REUSS, and in some sense on their own matters *"The Principal's Conferences of the State of Prussia"* by HEINRICH WILHELM ERLER.

Fig. 2.2 WALTHER NERNST (1) with the dog Tyras, the cousins EDITH (2) and FRIDA NERGER (3), his aunt ANNA NERGER (4), his wife EMMA (5), and his children GUSTAV (6), HILDE (7), and RUDOLF (8) at the demesne Engelsburg in 1904.

In addition to these areas of the humanities, also the fields of mathematics and the sciences were well represented. The second volume of the *"Textbook of Analysis"* by RUDOLF LIPSCHITZ treating differential and integral calculus has been purchased. Twelve years later the mathematics textbook by NERNST focused on these subjects. The Gymnasium also completed its stock of the works by CARL FRIEDRICH GAUSS published since 1863 by the Royal Society of the Sciences in Göttingen. This institution will play an important role in the life of NERNST. In addition, a book by FRIEDRICH AUGUST QUENSTEDT dealing with petrology is mentioned by KRETSCHMANN. Also we must mention the highly acclaimed *"Physics Demonstrations – Instructions for Experiments in Lessons at Gymnasiums, High Schools, and Business Schools"* by ADOLF FERDINAND WEINHOLD, who introduced the subject of Experimental Physics into the curriculum of the Technical Schools in Chemnitz, focusing in particular on Electricity and Electrical Engineering. Furthermore, he was the author of a number of well-received textbooks of Physics to be used for school instructions. The strong interest of the Gymnasium in well-founded instructions in the sciences can also be seen from the self-proclaimed goal of a *"continuity of the instructions in the description of nature"* [Kretschmann (1883): 37], but also from the high quality of the respective teachers. As an example we mention MAX BROSIG, who had studied and got his PhD in Breslau. Prior to his employment in Graudenz, he had worked as an assistant at the Mineralogy Museum of the University of Breslau and then as a teacher at the Agricultural School in Marienburg (Malbork).

Therefore, it is not surprising that the highly gifted and interested WALTHER NERNST utilized the available opportunities at school for gaining a broad and all-round knowledge in the humanities and the sciences, from which he profited during his whole life. Since his school period he demonstrated a strong inclination toward the Latin language. It is said that he read Latin texts to his father and that he delivered a brilliant graduation speech in this language at the Gymnasium of Graudenz.

The close relations of WALTHER to the theatre and to the dramatic art are likely a result of the humanistic instructions at the Gymnasium in literature and the languages. At any rate, they remain a characteristic feature of the later scientist. He had seen all stagings at the theatre of Graudenz, and in his youth he was so strongly attracted to the dramatic art and poetry, that for some time he desired to devote himself later to one or the other. Even without taking it up as a profession, the scientist NERNST was always a good actor. He was able in an excellent way to mimic the surprised or the naive person. Probably it was also one of his typical acting performances, with which in the beginning of August 1905 during his first lecture in Berlin he presented his greatest discovery, the Third Law of Thermodynamics, in such a way as if it had occurred to him just at this moment.

Sometimes NERNST presented himself even as a poet. It is said that already in 1899 as a Full Professor of Physical Chemistry in Göttingen he has tried to produce a play written by himself at a small theatre in Berlin. However, this project failed since prior to the first night the theatre went bankrupt (which was likely not due to the play written by NERNST). Unfortunately, title, text, and content of this play are lost. What has survived is a *"physical fairy-tale"* entitled *"Between Space and Time"*, which NERNST wrote in 1912 together with LOTTE WARBURG, a daughter of his colleague and friend in Berlin, EMIL WARBURG. Here we want to present the English translation of this tribute to the Special Relativity Theory of EINSTEIN ([Nernst and Warburg (1957)]):

"Once upon a time there lived a famous young scientist near the throne of a mighty king. What he was teaching was new and unique and could not be understood by the people. Therefore, they admired and revered him more than any other person. One day he came up with a revolutionary and challenging theory. A horrible tremor could be felt in the scientific community, since nature appeared to be shaken in its eternal laws. When the king heard about this, he called for the young scientist and said: 'If you are unable to prove in an experiment the validity of your theory, which has already severely disturbed the most gifted minds in my kingdom, you will have to sacrifice your head because of your theory.' The young scientist answered 'In this case have a light ball fabricated from a light metal. Then using the force of our electric machines, we will hurl it into space with such an enormous speed that it will race through space with light velocity suspended freely within the ether. My theory says that in the interior of this ball our time is brought to a standstill. If, for example, we fill it with flowers, they will continue to blossom with the same fragrance, since our time cannot affect them. And if during spring we hurl the ball into space at the right constellation of the planets, within a year it must return, and I can prove the validity of my theory.' Immediately, the king gave the order to prepare everything until spring including the last detail. All technical people of the kingdom were called upon to assist the scientist and to follow his instructions. The young queen, well known far and wide because of her beauty, had a magnificent physical laboratory built, which she gave as a present to the scientist, and in which she visited him every day. However, it happened that the young scientist and the beautiful queen fell in love with each other, and one day a servant of the palace caught the two while they were embracing each other. When this was reported to the king, he became very angry and had them brought to him in chains. Then he ordered that they be kept imprisoned in the interior of the ball. According to

his horrible plan of revenge, just at the moment when the fainting desire of the fettered lovers had reached its climax and when the metal ball was filled with the thrill of their hopeless passion, they would be sent into space in a never ending agony. So instead of being filled with flowers, the metal ball had turned into a prison for two unfortunate human beings. Now while the captain, who had to keep watch in front of the only window of the ball, went away for a short moment, the chained scientist succeeded to melt the iron by means of an electric short and to free himself and then also the queen. Nearly senseless from the ecstasy due to their recovered freedom, the two lovers embraced each other. At this moment the guard returned. Being frightened by what he had to see, he closed the switch, which released the electric force hurling away the ball. Within a fraction of a second the ball disappeared into the air with an immense noise, and after less than a few minutes it had reached light velocity. Hence, according to the theory of the scientist, time had come to a standstill, and simultaneously the lovers were united in an eternal kiss. In general, one believes that the ball follows a long, stretched trajectory through space similar to a comet, and that from time to time only for a short moment and invisible for the people on Earth because of the high velocity, it also comes close to our planet. In this case we feel its effect, when a human being is hit by the rays of an excessive feeling of love. According to the calculations of the scientist, until today the ball remains filled with a never-ending tenderness."

EINSTEIN himself has noted such artistic tendencies of NERNST, because in his obituary for his unique colleague he said: *"At the same time he was interested in literature and had such a sense of humor as is very seldom found with men who carry so heavy a load of work."* [Einstein (1942): 196].

On Easter 1883 WALTHER NERNST graduated from the Gymnasium of Graudenz ranking as *primus omnium*. The report of the principal said [Kretschmann (1883): 39]:

Nr.	first and family name	age	conf.	place of birth	profession of father	at the school	in highest gr.	study
69	Walther Nernst	18¾	protest.	Briesen, County Kulm	county judge in Graudenz	9½	2	medicine

In the case of four other of the altogether nine graduates, similar to NERNST, the plan to study medicine had been entered into the table. In fact, because of the excellent instructions at the Gymnasium and in particular due to the influence of his teachers, perhaps of Professor RÖHL, and last not least because of the laboratory installed in the basement of the house of the family, in the end WALTHER decided to study the sciences and mathematics.

As reported by ERIKA CREMER, regarding his excellent marks at graduation from the Gymnasium, later NERNST is said to have summarized: *"I graduated from the Gymnasium with distinction, I obtained my PhD with an average grade, I almost failed at the Habilitation, so it is only good, that I did not have to pass any further examinations."* [Cremer (1987): 185].

2.3 University Studies in Zurich and Berlin

The last statement attributed to NERNST must be qualified, if not revised. We must keep in mind that both with his dissertation as well as with his *Habilitation* (postdoctorate lecture qualification) within a period of only nine years NERNST has worked out laws of science the importance of which we can see from the fact that today they are directly connected with his name.

Before we start with this discussion in the following, we have to point out that his wish to study science already indicates an important feature of NERNST, which characterizes the activities of the student and even more those of the scientist during his whole life. We refer to his strong interest in everything which is modern and current in the exact sciences and in technology, in order to explore its technical use. If the new subjects were leading to questions having solutions with far reaching theoretical implications and promising practical applications, he pursued them vigorously.

During the decade, in which NERNST attended the gymnasium, a mentality had developed because of the rapid industrial growth, emphasizing scientific and technical knowledge and supporting this effort strongly. During this period sometimes even palaces were built for scientific research. Just at the time when NERNST graduated from the *Gymnasium* of Graudenz the opening of such a building was celebrated in Berlin. It was in this building where exactly 22 years later NERNST started his directorship of the Physical-Chemical Institute of the University of Berlin.

The strong change in the prestige of science during the last quarter of the 19th century can be illustrated, for example, by the fact that at the end of the second decade JUSTUS VON LIEBIG was drawing only laughter from his schoolmates and his teacher, when he said that he wanted to become a chemist. Even in 1869 JACOBUS HENRICUS VAN'T HOFF met complete lack of understanding by his parents when he told them a similar plan.

What a different situation was met by WALTHER NERNST when he signed up already in April 1883 as number 6676 in the list of the Philosophical Faculty of the University of Zurich, in order *"to devote himself to the study of physics, chemistry, and mathematics"* as we can read in his handwritten *curriculum vitae* [UAHUB: I, 3]. Perhaps he has chosen this place as his first University, since at the time its scientific departments were still very young and, hence, quite modern. The University had been founded in 1833, and the *Eidgenössische Technische Hochschule* (ETH) in 1854.

In his German *curriculum vitae* from 1890 NERNST mentioned as his academic teachers in Zurich only the chemist VICTOR MERZ and the mathematician ARNOLD MEYER, in the Latin «*Vita ... Lipsia 6/II 1890*» [UAG] in addition the physicist HEINRICH WEBER, whose perception of the properties of some chemical elements at low temperatures from the year 1875 will become important in connection with the Thermal Theorem of NERNST. In his textbook in connection with the deviations from the law of DULONG-PETIT, NERNST refers to this point: *"In the case of the most pronounced deviations, boron, carbon, silicon, H. F. Weber (1875) has shown, that for these elements the specific heat strongly increases with the temperature and approaches the value expected from the law of Dulong-Petit."* [Nernst (1909a): 174].

Having obtained a certificate for leaving from July 26, 1883, WALTHER NERNST left Zurich, and on October 29, 1883 he enrolled at the Friedrich-Wilhelms University in Berlin for studying the sciences, being registered under the number 902 during the term of office of the 74th rector, the classical philologist ADOLF JOHANN WILHELM KIRCHHOFF. This has been his first contact with the *alma mater*, at which more than a quarter of a century later he achieved important scientific successes as a researcher and a teacher, and where in 1921/22 he held the office of the 112th rector. However, at the time he left the University of Berlin already after one semester with a certificate for leaving from April 19, 1884 in order to continue his studies in Zurich again.

In his *curriculum vitae* mentioned above, as his academic teachers in Berlin he primarily refers to the physical chemist HANS LANDOLT (Fig. 2.3), who at the time as a Full Professor of Chemistry was still working at the Royal Agricultural Academy (*Landwirtschaftliche Hochschule*) in Berlin. In 1905 NERNST became his successor at this Chair at the University of Berlin. In addition, he also refers to HERMANN GEORG HETTNER, professor of mathematics at the University, in the German version and in the Latin version to the professor of physics at the Royal Agricultural Academy RICHARD BÖRNSTEIN. As we can see from a document of the University, *"Rector and Senate of the Royal Friedrich-Wilhelms University of Berlin attest by means of this certificate for leaving, that Mr. Walther Nernst"* has attended a laboratory course in the *"Chem. Laborat."* of the former at the Agricultural Academy, and a course on *"Differential Calculus and Introduction into Analysis"* as well as *"Problems of Differential Calculus"* of the latter. However, this document also indicates, that NERNST has attended three other courses: *"Introduction to the Theory of Telescopes"* by the astronomer WILHELM FÖRSTER, as well as *"Quantitative chemical Analysis"* and *"Qualitative chemical Analysis"* by the professors of chemistry FERDINAND TIEMANN and SIGISMUND GABRIEL, respectively (quoted from [Schultze (1992): 49– 50]).

Fig. 2.3   HANS HEINRICH LANDOLT (1905) drawn by his student WALTER ADOLF ROTH.

Apparently, the courses offered in organic chemistry by AUGUST WILHELM VON HOFMANN and KARL THEODOR LIEBERMANN did not interest NERNST. Possibly, he did not attend the courses by HERMANN VON HELMHOLTZ, because, as reported by MAX PLANCK, they *"did not mean any real benefit. Apparently, Helmholtz was never well prepared, he always spoke with a faltering voice, ... , in addition, he continuously had errors in his calculations at the blackboard, and we felt that he was bored during his lectures much the same as we were ourselves. As a result, his students gradually disappeared."* [Planck (1967): 8]. Perhaps, in the sequence of his studies NERNST followed a plan, which placed in particular theoretical physics at a later stage in Graz, where he realized it then in the fall of 1885. In Berlin NERNST could not attend lectures by GUSTAV ROBERT KIRCHHOFF, since he abstained from lecturing due to health reasons. Anyhow, according to PLANCK, his lecture *"seemed to be presented by heart in a dry and monotonous way"*, such that the students *"did admire the speaker, but not that what he did say."* [Planck (1967): 9].

The new matriculation at the Philosophical Faculty of the University of Zurich during the summer term in 1884 carried the number 6985. After one semester NERNST left this university with a certificate for leaving dated August 4, 1884.

In 1913 accompanied by MAX PLANCK, NERNST traveled again to Zurich in an important mission: ALBERT EINSTEIN, being Professor of Theoretical Physics at the ETH since August 1912, should be and was persuaded to come to Berlin.

2.4 Graz: The "Second Scientific Home"

2.4.1 *University and physics in Graz, Ludwig Boltzmann and Albert von Ettingshausen*

The University of Graz, where NERNST enrolled during the fall of 1895, was older than the two others, which he had attended before. In 1585/86 it was founded for the first time by Archduke KARL II and was given to the Jesuits in the spirit of the counter-reformation. Therefore, from 1763 until 1805 the Jesuit LEOPOLD BIWALD was the first Professor of Physics. It was credited to him, that the physics of NEWTON was known in Graz. In 1767 his work «*Physica generalis*» appeared. Prior to BIWALD, JOHANNES KEPLER had worked in Graz from 1594 until 1600 and had written his «*Mysterium cosmographicum*».

However, in 1782 Emperor JOSEPH II degraded the University to a Lyceum, and only in 1827 Emperor FRANZ I effected the second foundation of the University, which since then carries the name Karl-Franzens-University.

During the year of NERNST's birth Graz obtained another academic teaching institution: the Joanneum, founded in 1811 and serving for the agricultural-technical education, was raised to Technical Highschool, the Technical University of today. Ten years before NERNST took up his studies in Graz, the new building of the Physics Institute (Fig. 2.4) was completed. At this time AUGUST TOEPLER occupied the Chair of Physics.

Fig. 2.4 The Physics Institute of the Karl-Franzens-University of Graz.

In 1876 LUDWIG BOLTZMANN (Fig. 2.5) became his successor. Primarily because of him NERNST went to Graz, since in his works on the statistical foundation of thermodynamics this outstanding theoretical physicist had introduced a new concept in physics, which had been created by chemistry in the beginning of the 19th century: the atomic structure of matter.

BOLTZMANN was a student of the physicist JOSEF STEFAN from Vienna. To the latter we owe, among other things, important works on the theory of heat, electrodynamics, and the kinetic theory of gases. Already at the age of 25 years, in 1869 he became Full Professor of Mathematical Physics in Graz. From 1873 until 1876 he occupied the Chair of Mathematics in Vienna. Then followed his second period in Graz, this time as Director of the Physics Institute, which he developed into a world center of physics. In 1890 BOLTZMANN moved to Munich in order to occupy the Chair of Mathematical Physics. When in 1894 he became the successor of STEFAN in Vienna, NERNST was supposed to take over his Chair in Munich. Following an intermezzo from 1900 until 1902 at the University of Leipzig, BOLTZMANN returned to his still unoccupied Chair in Vienna, which he did not leave any more until his voluntary death in 1906.

However, the theoretician BOLTZMANN was also a good experimentalist. For example, during a research visit at the Physics Institute of HELMHOLTZ in Berlin, he was able to confirm experimentally some results derived theoretically from the electrodynamics of MAXWELL. During his second period in Graz BOLTZMANN achieved highly important and fruitful advances. Starting from a generalization of the MAXWELL distribution law of the molecular velocities of a gas, derived so far only for the equilibrium state, to nonequilibrium conditions, he had derived the so-called BOLTZMANN transport equation, interpreted mechanically for the first time concepts of the phenomenological thermodynamics, and formulated the H-theorem. In 1877 BOLTZMANN derived the logarithmic relation between the phenomenological entropy S, introduced by RUDOLF CLAUSIUS, and the so-called thermodynamic probability W: $S \sim \ln W$. The latter represents the number of the microscopic configurations of a particular state. In this way one obtained an access to thermodynamics, based on the probability calculus and mathematical statistics, representing the starting point of the field of statistical physics.

BOLTZMANN's atomistic concept of matter met strong opposition. As his opponents, in the first place we must mention ERNST MACH, one of the predecessors of BOLTZMANN in Graz and his colleague in Vienna. However, he was criticized also by WILHELM OSTWALD, with whom during his time in Leipzig he was privately on friendly terms, and who fluctuated in his opinion on the physical reality of the atoms.

NERNST has always accepted and defended the atomic concept. Its rejection by OSTWALD led to certain tensions between himself and his *"boss"* in Leipzig, although the latter had introduced him to physical chemistry which became for him so important.

The ingenious works by BOLTZMANN dealing with statistical thermodynamics have been extremely important, in addition to many other physicists, also for the later research by NERNST and in particular for that connected with the Third Thermal Law. PAUL GÜNTHER, a student of NERNST, remembers that his teacher told him, that prior to his discovery he had reflected much about this law completing the system of axioms of the phenomenological thermodynamics, in spite of the fact that BOLTZMANN had told him, that thermodynamics would be complete already [Günther (1951): 557].

NERNST also learned from BOLTZMANN how to illustrate complex subjects by means of intuitively clear models. This capability distinguished both scientists not only in their teaching style.

The law found empirically by STEFAN in 1879, the thermodynamic derivation of which by BOLTZMANN (1884) was called by HENDRIK ANTOON LORENTZ *"a true gem of theoretical physics"* and which states that the total density of radiation u is proportional to the fourth power of the temperature T $(u \sim T^4)$, played a conceptional role in the glow-lamp of NERNST.

NERNST's memories from 1930 express motivation, some disappointment, but also consolation: *"When (in the fall of 1885) I went to Graz, I did so with the intention to attend lectures in theoretical physics, which at the time were given at the University of Graz by the professors*

Ettingshausen, Heinrich and Franz Streintz, Klemenčič, and others in a larger number than at any other German university. ... My hope to work also with Boltzmann in theoretical physics could not be realized, since at the time Boltzmann only offered the lecture for beginners in experimental physics. However, in addition to the excellent lectures by the teachers mentioned above, I became richly compensated by the daily experimental collaboration with Prof. v. Ettingshausen, in particular since Boltzmann showed an increasing interest in our results, such that the genius of this great theoretical physicist often illuminated our experimental work during long scientific discussions." [Ne30: 279].

Fig. 2.5 LUDWIG BOLTZMANN (1) among his collaborators and students (Graz 1887): WALTHER NERNST (2), FRANZ STREINTZ (3), SVANTE ARRHENIUS (4), RICHARD HIECKE (5), EDUARD AULINGER (6), ALBERT VON ETTINGSHAUSEN (7), IGNAZ KLEMENČIČ (8), V. HAUSMANNINGER (9).

Still in 1927 also ALBERT VON ETTINGSHAUSEN reported to a colleague, perhaps HANS BENNDORF: *"Nernst came to Graz for the first time during the fall of 1885, and he remained there during the school year 1885/6 until about July; ... In Graz he was enrolled as a regular* *student, definitely at least with me, since I remember exactly that I have read his name as Ner<u>u</u>st on the list; most likely he has also attended experimental physics with Boltzmann and perhaps also some other course."* (quoted from the facsimile print in [Hohenester (1992): 13]).

As we can see from NERNST's memories, ALBERT VON ETTINGSHAUSEN (Fig. 2.5) in addition to BOLTZMANN became his true teacher in Graz. The research subjects of this physicist were primarily electricity and magnetism. In particular the investigation of the former became highly important in the "electric" 19th century. So it is not surprising, that also NERNST took up this subject in one form or another. In this field also other students of ETTINGSHAUSEN were successful. In this context we can mention, for example, FRANZ PICHLER, the founder of the famous Austrian electro-technical factories Elin in Weiz, OTTO NUSSBAUMER, who demonstrated in 1904 the first wireless transmission of sound, and also RICHARD HIECKE (Fig. 2.5).

ALBERT VON ETTINGSHAUSEN was a nephew of ANDREAS VON ETTINGSHAUSEN, who was a teacher of BOLTZMANN in addition to STEFAN and JOSEPH LOSCHMIDT, and whose son, CONSTANTIN, was teaching palaeo-botanics in Graz and was one of the founders of this scientific discipline in Austria. He himself had studied in Graz with AUGUST TOEPLER and got his PhD there in 1872. Before he left Graz, TOEPLER could arrange that in 1876 BOLTZMANN became his successor, and that ETTINGSHAUSEN was appointed at the same time, since in contrast to his student this outstanding scientist was known to have only little organizational talent. This combination of the two physicists was extremely beneficial for the Physics Institute and by no means hampered the scientific reputation of ETTINGSHAUSEN as a researcher and teacher. This is well indicated by the strong appreciation and gratitude WALTHER NERNST always showed for his teacher.

In 1888 ETTINGSHAUSEN was appointed *primo et unico loco* as Full Professor of General and Technical Physics at the newly established Physical Institute of the *Technische Hochschule* in Graz, together with a special teaching charge for Electrical Engineering. Here he worked as a teacher and researcher until 1920. It is likely that his departure from the University of Graz was one of the reasons causing BOLTZMANN to leave Graz in 1890.

"In the years 1885 and 1886 Professor v. Ettingshausen and myself performed in Graz an experimental investigation of the influence of magnetism on the flow of electricity and heat, the results of which are reported in three papers submitted to the Academy in Vienna.", NERNST wrote in his *curriculum vitae* from Göttingen in 1890 [UAHUB: I, 3]. In this case he had slightly underrated the number of publications, which had been prepared by the authors ETTINGSHAUSEN and NERNST on the indicated subject and then appeared between 1886 and 1888. The total number was seven [Ettingshausen and Nernst (1886)] (see also [Nernst (1887)]. The number of three in fact only refers to the publications in the Proceedings of the Imperial Academy of Science in Vienna.

The motivation for these studies had come from BOLTZMANN. NERNST later remembers, so *"we began with a quantitative study of Hall's phenomenon along the different directions; soon we noticed closely related phenomena, we uncovered the group of the thermomagnetic and galvanomagnetic effects, which already at the time generated a certain interest because of their curiosity."* This collaboration meant *"the richest and most fortunate opportunity for young beginners, to become acquainted with and admire Ettingshausen as a teacher and researcher. It is only rarely that a disciple of his science went to his laboratory within the magnificent building of the Physics Institute of Graz with such great joy and enthusiasm, as it was granted to me at the time."* [Nernst (1930): 279]. In the next section we will deal separately with the results obtained by the student and his teacher in Graz, which allowed NERNST to get his PhD in Wurzburg in 1887.

On December 19, 1886 ARRHENIUS mentioned to WILHELM OSTWALD in a letter from Wurzburg: *"By the way, there are ... good teachers (Boltzmann, v. Ettingshausen, and 4 lecturers, among others Jahn)."* [Körber (1969): 26]. The latter, HANS JAHN, worked on electrochemistry and thermodynamics as a physical chemist at the Chemical Institute of LEOPOLD VON PEBAL at the University. At the time he was one of the few people emphasizing mathematics in chemistry. His textbook on thermo-chemistry had appeared already in 1882 in Vienna. Since JAHN preferred modern subjects and concepts in his research and his teaching, which NERNST was interested in, he certainly had noticed also this teacher, in particular, since JAHN had been close to the people around BOLTZMANN and ETTINGSHAUSEN, as we can see from the letter of ARRHENIUS. Perhaps NERNST even had been introduced by JAHN to the problem, the solution of which he could find in the form of his Thermal Theorem. When NERNST formulated this law for the first time in his Institute in Berlin in 1905, JAHN was his department head (Fig. 5.6).

Later on, at least in his textbooks NERNST also contributed to the emphasis of mathematics in chemistry. In Graz he had studied originally mathematical physics with the teachers HEINRICH und FRANZ STREINITZ and IGNAZ KLEMENČIČ (Fig. 2.5), mentioned repeatedly in addition to BOLTZMANN and ETTINGSHAUSEN. He must have done this so emphatically, that in 1894 BOLTZMANN recommended NERNST to become his successor of the Chair of this physical discipline in Munich.

We can see that in Graz the teachers, the learning conditions, and the level of the scientific research were clearly excellent. Therefore, the remarks by NERNST in his speech on the Austrian Radio on February 17, 1936, *"that he came to Graz after five semesters at the Universities of Zurich and Berlin, and that he will never forget this town as his second scientific home"* (quoted from [Skrabal (1942): 199], see also [Hohenester (1992): 13]), came from his heart and are easily understood.

NERNST returned to Graz several times, so already during spring 1887, after he had gone to Wurzburg at the beginning of the academic year 1886/87.

2.4.2 The Ettingshausen-Nernst effects and the Nernst effect

The experimental studies performed by NERNST and VON ETTINGSHAUSEN in Graz during the years 1885 – 1887 were focused on the (transverse) galvanomagnetic and thermomagnetic effects.

The galvanomagnetic effects appear in an electric conductor in the presence of a magnetic field $\vec{H}$ and an electric field $\vec{E}$, causing an electric current of density $\vec{j}$ and magnitude I. In the case of a thermomagnetic effect the electric current is replaced by a heat current Φ. If the magnetic field and the electric or the heat current are oriented perpendicular to each other, one speaks of the transverse effects.

The HALL effect named after EDWIN HALL, who discovered it in 1879, belongs to the transverse galvanomagnetic effects. In this case an electric field $\vec{E}_{Hall} = A_{Hall} \vec{H} \times \vec{j}$ is generated, if an electric conductor is placed in a magnetic field $\vec{H}$ and carries an electric current of density $\vec{j}$ (A_{Hall}: HALL coefficient). In the case of a conducting strip of width b a potential difference, the HALL voltage $\Delta U = A_{Hall} b \,|\, \vec{H} \,\|\, \vec{j} \,|$ appears (for $\vec{H} \perp \vec{j}$). The HALL effect represents one of the simplest phenomena caused by moving electric charge carriers, when an external magnetic field is present.

The ETTINGSHAUSEN effect is analogous to the HALL effect, since for the same configuration of magnetic field and electric current instead of a potential difference a temperature difference is generated.

In the case of the 1st ETTINGSHAUSEN-NERNST effect a temperature gradient in the sample causes an electric field oriented perpendicular both to the temperature gradient and the magnetic field. Finally, the RIGHI-LEDUC effect of an applied temperature gradient represents the appearance of another temperature gradient perpendicular both to the applied temperature gradient and the magnetic field. A summary of the thermomagnetic effects is given by HERBERT B. CALLEN [Callen (1960): 305].

In the modern literature the 1st ETTINGSHAUSEN-NERNST effect is usually simply referred to as the NERNST effect [Ziman (1960): 497, Huebener (2001): 154]. The 2nd ETTINGSHAUSEN-NERNST effect is formally identical to the SEEBECK effect.

The ETTINGSHAUSEN effect can also be connected with the PELTIER effect (1834). The latter is caused by the fact, that an electric current always carries heat energy (PELTIER heat), in addition to the electric charges. Therefore, if an electric current passes through the contact between two different metals or semiconductors, heat energy is delivered or absorbed at the contact, depending on the current direction. In the case of the ETTINGSHAUSEN effect the electric current flow and, hence, the flow of heat energy experience a side-ways motion because of the LORENTZ force. Now we deal only with a single conductor, and the two edges of the conductor represent the locations, where the heat energy is delivered or absorbed, respectively.

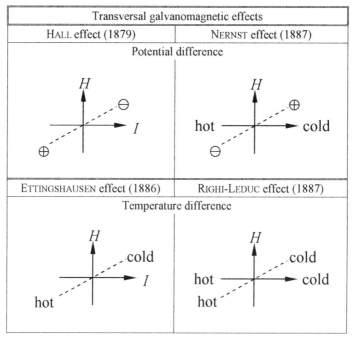

Fig. 2.6 The transverse galvanomagnetic and thermomagnetic effects. (Arranged underneath each other: name of the effect (year of publication) and the generated signal, schematically.)

Similarly, the 1st ETTINGSHAUSEN-NERNST effect can be connected with the SEEBECK effect. The thermal diffusion of the charge carriers in a temperature gradient results in an accumulation of charges at the contact between two different metals or semiconductors, leading to the longitudinal thermoelectric potential gradient (SEEBECK voltage). In the case of the 1st ETTINGSHAUSEN-NERNST effect this thermal diffusion also experiences a side-ways motion because of the LORENTZ force, and the accumulation and depletion of the charges at the two edges of the conductor, respectively, leads to a transverse potential gradient.

The 1st ETTINGSHAUSEN-NERNST effect and the ETTINGSHAUSEN effect (similarly as the SEEBECK effect and the PELTIER effect) are strongly coupled with each other by means of the ONSAGER reciprocity relations.

Remembering his time as a student in Graz, NERNST later summarized his activities dealing with the galvanomagnetic and thermomagnetic phenomena in his article celebrating the 80th birthday of ALBERT VON ETTINGSHAUSEN [Nernst (1930): 279]: *"Following a suggestion by Boltzmann, we began with a quantitative study of Hall's phenomenon along the different directions ... we uncovered the group of the thermomagnetic and galvanomagnetic phenomena, which already at the time generated a certain interest because of their curiosity, however, which should find a satisfactory theoretical explanation only after the theory of the metallic conduction would have been developed still further."* The latter development of the theory had to wait until the 1930s and later, when the quantum theory was successfully applied to the electrons in solids. NERNST may have had this visionary insight already soon after his stay in Graz and Wurzburg. Here we see for the first time the distinct mark characterizing the great scientist to terminate a research project, if the lack of the necessary theoretical knowledge imposes a limit for its meaningful continuation. So it is likely that it were not only the other fields of research, which caused NERNST to leave the subject of the behavior of metals in a magnetic field, if we exclude a joint paper with PAUL DRUDE [Drude and Nernst (1890)].

Today the effects discovered by NERNST and VON ETTINGSHAUSEN have gained a certain actuality in modern low temperature physics. On the one hand, this is associated with the quantum HALL effect discovered by KLAUS VON KLITZING, who received the Nobel Prize in Physics for this discovery in 1985. On the other hand, the thermomagnetic effects play a prominent role in the mixed state of type-II superconductors [Huebener (2001)].

If a type-II superconductor is placed in a magnetic field at low temperatures, magnetic flux can penetrate into the superconductor in the form of magnetic flux lines, each line carrying a single magnetic flux quantum. The magnetic flux lines arrange themselves forming an ordered lattice first predicted by ALEKSEĬ ALEKSEEVICH ABRIKOSOV, for which he received the Nobel Prize in Physics in 2003. Under the influence of an electric current (LORENTZ force) or of a temperature gradient (thermal force) the flux-line lattice can be set into motion, thereby generating an electric voltage. This results in a contribution to the SEEBECK and to the 1st ETTINGSHAUSEN-NERNST effect. The thermal energy transported by the moving flux-line lattice causes a contribution to the PELTIER and to the ETTINGSHAUSEN effect. In general, at low temperatures these galvanomagnetic and thermomagnetic effects due to the flux-line motion are much larger than the effects in the non-superconducting state [Huebener (2001)]. Recently, highly interesting measurements of the 1st ETTINGSHAUSEN-NERNST effect in high-temperature superconductors were reported, showing evidence for the appearance of a fluctuation regime of superconductivity as high as 50 – 100 K above the critical temperature T_c [Wang et al. (2001)]. This represents an important input for the theory of high-temperature superconductivity.

2.5 Conclusion of the University Studies in Wurzburg

The Royal Julius-Maximilians University of Wurzburg, at which NERNST studied during the winter term 1886/87, did exist already from 1402 until 1413 as a *"Hohe Schule"*. In 1582 it was newly founded again as a Catholic University by the Prince-Bishop JULIUS ECHTER VON MESPELBRUNN in the spirit of the Counter-Reformation. The Emperor MAXIMILIAN II had issued the privilege for this. Only in 1734 it became open also to non-Catholics, and it became secularized in the 19th century. Medicine, being added to theology and philosophy, was closely connected with the sciences, where famous scientists were working and teaching. To these belonged the physicist FRIEDRICH KOHLRAUSCH (Fig. 2.8), who caused NERNST to move to Wurzburg for concluding his studies primarily because of the pioneering research of KOHLRAUSCH on electrolytic solutions and the excellent modern research on electricity performed in his laboratory. At the time, the organic chemist EMIL FISCHER, who had been appointed in Wurzburg in 1885, may have been only of little interest for NERNST. However, two decades later in Berlin the two Full Professors were in close contact with each other because of questions dealing with the organization of science.

FRIEDRICH KOHLRAUSCH had been a student of WILHELM WEBER, from whom in Göttingen he had learned the delicate details of quantitative electric and magnetic measurements and from which he took over the atomic concept of electricity. In 1870 in Göttingen KOHLRAUSCH wrote the «*Leitfaden der praktischen Physik*» (*"Guide to practical Physics"*), which later became famous and widely accepted as the "Textbook of Practical Physics" or simply as the *"Kohlrausch"*. After holding appointments as Professor in Göttingen, Zurich, and Darmstadt, in 1875 he was appointed as Full Professor of Physics in Wurzburg. Here in 1878 he established a modern Physical Institute (Fig. 2.7). When in 1888 KOHLRAUSCH accepted the position as Full Professor in Strasbourg, WILHELM CONRAD RÖNTGEN became his successor, who later discovered the X-rays and investigated their nature at this Institute. In 1895 KOHLRAUSCH became President of the Physikalisch-Technische Reichsanstalt in Berlin and thereby one of the predecessors of NERNST in this position.

Fig. 2.7 The Physical Institute established by FRIEDRICH KOHLRAUSCH in Wurzburg.

Following the tradition of WILHELM WEBER and CARL FRIEDRICH GAUSS, KOHLRAUSCH occupied himself with the precise definition of the electric units and with the construction of the proper measuring instruments for this purpose. However, his special achievements concerned the investigation of the electrolytic solutions. In order to measure their conductivity he used alternating current. In 1870 together with AUGUST NIP-

POLDT he was able to demonstrate, that OHM's Law is valid also for electrolytes. The concept of the molar conductivity goes back to KOHLRAUSCH. He found the law of the independent velocity of the ion motion (1885), and, based on his own measurements, the Square Root Law named after him (1900), according to which the conductance λ of an electrolytic solution is connected with the concentration c according to $\lambda = \lambda_\infty - A\sqrt{c}$ (λ_∞ is the conductance at infinite dilution; A is a constant). Together with ADOLF HEYDWEILLER, KOHLRAUSCH determined the dissociation constant of water.

In addition to VON ETTINGSHAUSEN and BOLTZMANN, KOHLRAUSCH belonged to the true academic teachers of NERNST. For a long time the research and the findings by KOHLRAUSCH became extremely important for the studies in electrochemistry started soon by NERNST. He had also learned from KOHLRAUSCH to concentrate on the discovery of the fundamental ideas behind the ruling laws of nature, to limit himself to their foundation and to the instrumental and technical possibilities connected with this, and to leave the precise execution and the final tests to his students and coworkers.

It is interesting, that in both of his *curricula vitae* of 1890 among his teachers in chemistry, in addition to VICTOR MERZ and HANS LANDOLT, NERNST mentions only JOHANNES WISLICENUS, which, however, in 1885 had already accepted an offer from Leipzig. In contrast to his successor in Wurzburg, EMIL FISCHER, the former had also dealt with theoretical questions within organic chemistry, and in particular with the arrangement of the atoms in organic molecules, such that he can be looked at as a forerunner of stereochemistry. Probably, NERNST included his time in Leipzig, where certainly he had met WISLICENUS, within the time he spent as a student. The modern view of chemistry, accepted by WISLICENUS in contrast to many of his colleagues, will have caused NERNST to list him among his teachers.

In Wurzburg NERNST completed his studies of mathematics with FRIEDRICH EMIL PRYM and EDUARD SELLING, and possibly he attended lectures on mathematical physics by CARL ADOLF JOSEPH KRAZER.

At the time in the laboratory of KOHLRAUSCH, NERNST was joined by a number of young scientists, who had come from different countries to the famous physicist in Wurzburg in order to learn from him similarly as NERNST. The group of people working with KOHLRAUSCH in 1887 is shown in Fig. 2.8. In addition to the later Nobel Laureates NERNST and ARRHENIUS, also all the other young physicists did occupy honorable positions in the future: since 1908 ADOLF HEYDWEILLER was Professor of Physics at the University of Rostock, the electrical engineer GUSTAV RASCH occupied a Chair at the Technical University of Aix-la-Chapelle since 1905, the Italian LUIGI PALAZZO was head of the Meteorological and Geodynamic Central Office in Rome since 1901, and the American SAMUEL SHELDON was Professor of Physics at the Polytechnic Institute of Brooklyn since 1889.

Fig. 2.8 FRIEDRICH KOHLRAUSCH (6) in 1887 among his co-workers: ADOLF HEYDWEILLER (1), GUSTAV RASCH (2), SVANTE ARRHENIUS (3), WALTHER NERNST (4), LUIGI PALAZZO (5), SAMUEL SHELDON (7).

Also SVANTE ARRHENIUS on his visiting trip to famous physicists and physical chemists made possible by a scholarship worked in the laboratory of KOHLRAUSCH in Wurzburg during the winter term 1886/87, following a visit to WILHELM OSTWALD in Riga. ERNST HERMANN RIESENFELD commented this event: *"The accidental event, that at the same time also Nernst worked at the Institute of Kohlrausch, and that the two so extremely gifted and enthusiastic young scientists became close friends and supported themselves in terms of mutually motivating their work, had a decisive influence on the further development of physical chemistry."* [Riesenfeld (1930): 10].

In his German *curriculum vitae* of 1890 NERNST summarized: *"I studied at the Universities of Zurich, Berlin, Graz, Wurzburg, at the last of which I obtained my PhD on May 11, 1887."* [UAHUB: I, 3]. The PhD was graded as *summa cum laude*. The title of his dissertation was «*Über die elektromotorischen Kräfte, welche durch den Magnetismus in von einem Wärmestrome durchflossenen Metallplatten geweckt werden*» (*"On the electromotoric forces generated by the magnetism within metal plates through which a heat current is flowing"*) [Nernst (1887)] (Fig. 2.9). Of course, the thesis was based on the experimental results, which NERNST had obtained with VON ETTINGSHAUSEN in Graz and which were discussed above.

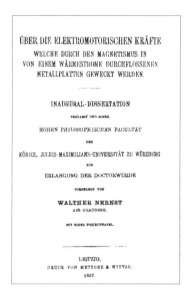

Fig. 2.9 Title page of the dissertation of WALTHER NERNST.

Because of his PhD obtained with KOHLRAUSCH, NERNST became an important part of the "WILHELM WEBER Tree", from which the following sequence can be listed:

Name		PhD		Nobel Prize in Chemistry
WILHELM	WEBER	Halle/S.	1826	
FRIEDRICH	KOHLRAUSCH	Göttingen	1863	
WALTHER	NERNST	Wurzburg	1887	1920
ARNOLD	EUCKEN	Berlin	1906	
MANFRED	EIGEN	Göttingen	1951	1967

Chapter 3

Habilitation in Leipzig (1887 – 1889)

In Wurzburg NERNST could persuade his friend ARRHENIUS to go to Graz together with him during the summer term in 1887. The young man from Sweden followed this suggestion with great pleasure, in order to meet there in addition to BOLTZMANN, VON ETTINGSHAUSEN, and other prominent University Lecturers *"a number of lively, highly motivated young men working along the same lines"* and to enjoy together with NERNST and these people *"the excitement generated by the ingenious teacher Boltzmann and the natural beauty of this southernmost German university town"* [Riesenfeld (1930): 12]. During this time the picture shown in Fig. 2.5 was taken.

During this term WILHELM OSTWALD also arrived in Graz on a visiting journey to many German universities. ARRHENIUS, who had visited him in Riga during the summer term in 1886, arranged the acquaintance between OSTWALD and NERNST, which should have important consequences. Because, after OSTWALD still in 1887 had been offered the Chair of Physical Chemistry in Leipzig, there *"was no candidate for the Physical-Chemical Section. Then I remembered Dr. Walter Nernst, whom I just had met in Graz and about whom Arrhenius had a very high opinion regarding his capabilities and knowledge. Since he had intended anyway to work together with me in Riga, I offered him a position in Leipzig, which he accepted immediately."* [Ostwald (1927): 36]. *"Therefore, Nernst was induced to turn his attention to problems of physical chemistry, the extraordinary success of which for the advancement of this science is well known."* [Riesenfeld (1930): 13], as it was properly stated by ERNST HERMANN RIESENFELD.

3.1 The Sciences at the University of Leipzig

In 1409, promoted by the invention of the letter-press printing technique, the *"Hohe Schule"* in Leipzig was founded as a State University of the Wettinic Saxony by the margrave of Meissen FRIEDRICH I and, therefore, it is one of the oldest Universities in Europe. In the beginning of the 16th century, when ULRICH VON HUTTEN studied here for a short time during his *«peregninatio academica»* (*"academic travels"*) the dominating academic subjects were Theology and Law. Around the middle of this century there existed already the official designation 'Professor of Physics', indicating, however, scholars who were dealing with the natural-philosophical writings of ARISTOTLE and of other antique authors. When JOHANN WOLFGANG VON GOETHE studied in Leipzig during 1765 to 1768, this title had obtained already its present-day meaning since more than half a century.

In 1710 the Professor of Medicine CHRISTIAN LEHMANN had been appointed as Professor of Physics. One of his successors, the experimental physicist JOHANN HEINRICH WINKLER, who occupied positions as Professor of Philosophy as well as of Greek and Latin Language before he was appointed to the Chair of Physics in 1750, partly together with the mathematician CHRISTIAN AUGUST HAUSEN started a line of tradition of Physics in Leipzig, namely the investigation of electric phenomena in the most general sense, which was also taken up by OSTWALD and NERNST.

An important milestone in the history of physics in Leipzig is the opening of the First Physical Institute in the year 1835. Its establishment had been initiated by HEINRICH WILHELM BRANDES. However, his successor GUSTAV THEODOR FECHNER became its first Director. When he had to leave his teaching position in 1839 at a very young age because of health reasons, he was suc-

Physical Institute		
Name		Director
GUSTAV THEODOR	FECHNER	1834–1839
WILHELM	WEBER	1843–1849
WILHELM	HANKEL	1849–1887
GUSTAV	WIEDEMANN	1887–1899
OTTO	WIENER	1899–1927
PETER	DEBYE	1927–1936
GERHARD	HOFFMANN	1937–1945

ceeded as Director of the Physical Institute by a number of important physicists, who are listed under the lifetime of NERNST in the overview.

This list is supplemented by the list of the Directors of the Theoretical-Physical Institute, among whom we find the teacher of NERNST, LUDWIG BOLTZMANN, and his friend THEODOR DES COUDRES, who before had been Professor in Wurzburg. In Göttingen NERNST had collaborated with PAUL DRUDE when the latter worked there as a lecturer. Later for a short time both were colleagues in Berlin in neighboring institutes.

Theoretical-Physical Institute		
Name		Director
HERMANN	EBERT	1894
PAUL	DRUDE	1894–1900
LUDWIG	BOLTZMANN	1900–1902
THEODOR	DES COUDRES	1902–1926
WERNER	HEISENBERG	1927–1941
FRIEDRICH	HUND	1942–1946

From the chemists in Leipzig we had mentioned already JOHANNES WISLICENUS, whom NERNST had mentioned as one of his teachers. In the following we name a selected number of scientists, who WILHELM OSTWALD counted among his *"Circle of Leipzig"* [Ostwald (1927): 80–110]. To this circle belongs CARL LUDWIG, one of the most famous representatives of experimental physiology. WILHELM WUNDT founded the modern, laboratory-based psychology, and RUDOLF LEUCKART the scientific zoology and parasitology. In the same year as OSTWALD the botanist WILHELM PFEFFER arrived in Leipzig, the studies of which on osmosis were most important for the theory of solutions by VAN'T HOFF. Furthermore, we must mention the astronomer HEINRICH BRUNS, the geographer FRIEDRICH RATZEL, the mineralogist and petrologist FERDINAND ZIRKEL, as well as the mathematicians ADOLPH MAYER and SOPHUS LIE. MAYER had been able to attract his famous colleague from Norway to Leipzig.

GUSTAV WIEDEMANN working in particular in the field of electricity since 1878 had edited the journal «*Annalen der Physik*», being important also for the young physical chemistry. Before he succeeded in 1887 to WILHELM HANKEL as Director of the Physical Institute, he had been Professor of Physical Chemistry in Leipzig since 1871. As his successor WILHELM OSTWALD (Fig. 3.1) had been appointed as Chair.

3.2 Wilhelm Ostwald

Already during his studies of chemistry in Dorpat (in Estonian: Tartu) with CARL SCHMIDT and JOHANN LEMBERG, OSTWALD had turned to problems of physical chemistry. In 1875 he became Assistant Professor at the Physical Institute of ARTHUR VON OETTINGEN. He obtained his PhD in 1878 with a thesis dealing with the theory of relationships in chemistry. In 1882 he obtained an offer from the Polytechnic Institute of Riga, where he continued his research on the application of the mass-action law he had started in Dorpat. After he had become acquainted with the dissociation theory of ARRHENIUS in 1884, he admired this theory so much that until 1897 his research concentrated on the electrolytic conductance and the theory of the ions. This period concerns in particular the first decade of his activities in Leipzig, from where in the fall of 1887 OSTWALD had obtained an offer, as mentioned above, much to his surprise, since still in the summer of that year *"I had given up all hope, ... to get there."* [Ostwald (1927): 92].

Fig. 3.1 WILHELM OSTWALD (left) in a discussion with SVANTE ARRHENIUS.

The Institute of OSTWALD in Leipzig became the nucleus and subsequently a world center of physical chemistry, from which about 60 professors in this field originated. In the new Institute completed in 1897 the research on problems of catalysis was continued. In addition to the modern definition of catalysis, the concept of autocatalysis was created. For this work and for the studies of reaction rates and chemical equilibria WILHELM OSTWALD was awarded the Nobel Prize in Chemistry for the year 1909. Already in 1906 he had retired early, in order to devote himself to philosophy, history of science, and his theory of colors while living on his estate Grossboten near Leipzig.

To WILHELM OSTWALD, who in addition to VAN'T HOFF, ARRHENIUS supported by him, and his student WALTHER NERNST belongs to the most important founders and organizers of the classical physical chemistry, we owe a number of laws of nature carrying his name. Among these we count the dilution law (1888), which represents a formulation of the mass action law for the dissociation of weak electrolytes in terms of the conductance at finite and infinite dilution. His step rule (1897) describes the stepwise generation of more stable out of less stable forms arising first in the case of chemical reactions. The OSTWALD ripening refers to the growth of larger particles at the expense of smaller ones. With the OSTWALD process patented in 1902, in which ammonia is burned catalytically using oxygen from the air and yielding nitric oxides important for the production of nitric acid, later OSTWALD came in contact with the research of NERNST dealing with the ammonia synthesis.

In 1895 the first textbook on Physical Chemistry written by WILHELM OSTWALD appeared. Later NERNST underlined its importance for this scientific discipline by connecting the year of publication with the birth of Physical Chemistry.

In contrast to NERNST, who always defended the kinetic atom theory, OSTWALD changed his mind several times regarding this point. Originally being a follower of this theory, after the 1890s he became its fighting opponent. However, in 1908 he returned to this theory again constructively such that he became a leading member of the Atomic Weights Commission.

OSTWALD was also an active historian of science. In addition to the many papers by him, this is also demonstrated by the original publications of important scientific works founded by him, which are continued until today, and which are now referred to as «*Ostwalds Klassiker der exakten Wissenschaften*» (*"Ostwald's Classics of the exact Sciences"*).

When OSTWALD accepted the offer from Leipzig in 1887, he was confronted with the task to direct the "Second Chemical Laboratory", which had been called "Physical-Chemical Institute" under GUSTAV WIEDEMANN. He took over the necessary space from the agricultural chemist WILHELM KNOP. The Laboratory consisted of three sections corresponding to the teaching program. JULIUS WAGNER was in charge of the analytical section, which he headed already under WIEDEMANN. Until then a pharmaceutical section belonged to the Laboratory of WISLICENUS, who transferred this section including its head ERNST BECKMANN to OSTWALD. The latter physical chemist, who became known among others because of the development of precision thermometry (BECKMANN thermometer), on which he worked in Leipzig in connection with the determination of the molar mass, was referred to by OSTWALD as one of his *"best and most successful coworkers"* and as his *"most loyal fellow worker"* [Ostwald (1927): 39].

As mentioned above, WALTHER NERNST obtained the third position as assistant professor. Although this position was connected with being in charge of the physical-chemical section, in 1890 NERNST writes in his *curriculum vitae* for his application in Göttingen, likely because there Physical Chemistry was part of Physics and not of Chemistry as in Leipzig: *"In October of 1887 I accepted an offer of Professor Ostwald, Director of the 2nd Chemical Laboratory of the University of Leipzig, to work in his Institute as Assistant Professor of Physics."* [UAHUB: I, 4].

3.3 The Completion of the Thermodynamics of Electrochemistry: The Nernst Equation

The year 1887, in the autumn of which NERNST entered the Laboratory of WILHELM OSTWALD in Leipzig, was called by the latter a *"turning point of science"*, *"a critical year of first order, in the sense of an exceptional fruitfulness"* [Ostwald (1927):19] and a *"year of general welfare"*

[Ostwald (1927): 26]. Indeed, with the offer of the position as Professor in Leipzig to OSTWALD for the first time a true academic teaching institution for physical chemistry had been created. All other locations, including that headed by GUSTAV WIEDEMANN, already carried this teaching subject in their notation, however, they did not have the rank and the importance of the institution established by OSTWALD in Leipzig. Furthermore, shortly before, on February 15, 1887, the first issue of the «*Zeitschrift für physikalische Chemie, Stöchiometrie und Verwandtschaftslehre*», initiated by OSTWALD and edited by him and VAN'T HOFF, was published by the Company founded in 1811 by WILHELM ENGELMANN in Leipzig. With this university chair and this publication channel, the *"Zeitschrift"*, as it was called later by NERNST and others, the two necessary prerequisites were created, which justified to accept Physical Chemistry now as an independent scientific discipline.

Also the perception of ARRHENIUS derived from his studies of the conductance of the electrolytes, that in general the electric conduction is caused only by one part of the electrolytes whereas the other part contributes nothing, experienced a dramatic breakthrough in 1887. Up to then it remained unclear what causes this difference. In this year ARRHENIUS saw clearly, that the part of the electrolyte responsible for the conductance is completely disintegrated into electrically charged particles, the ions, whereas the noncontributing part is not dissociated and, hence, is electrically neutral. Actually, this insight was not only a milestone for electrochemistry, but for the whole of chemistry. WILHELM OSTWALD commented: *"Also originating from this point there rapidly developed a large and new field of science."* [Ostwald (1927): 20].

This field now became interesting also to WALTHER NERNST. His first study in this field was initiated by deviations found in the case of mercury compounds between a thermodynamic theory of galvanic cells created by HERMANN VON HELMHOLTZ, connecting the reaction heat of the chemical reaction with the temperature dependence of the electromotoric force, and measurements by SIEGFRIED CZAPSKI [Czapski (1884)]. Here it had been assumed that the thermochemical data determined by JULIUS THOMSON are correct. OSTWALD asked NERNST to check these values of the reaction heat. The latter then could show that the data of THOMSON were incorrect and that the theory of HELMHOLTZ was satisfied also for mercury compounds [Nernst (1888a)].

The great achievement of NERNST during his time in Leipzig was directly connected with the perceptions of VAN'T HOFF and ARRHENIUS. Regarding VAN'T HOFF, it concerned the theory of dilute solutions. Its thermodynamic foundation for the comprehension of chemical equilibria had been established by AUGUST FRIEDRICH HORSTMANN, however, being restricted to gases. A certain similarity between gases and dilute solutions had been recognized primarily by JULIUS THOMSON, without explaining its specific nature. In an ingenious way VAN'T HOFF was able to demonstrate the analogy between ideal gases and dilute ideal solutions, and the common validity of the thermodynamic laws for gases and for dilute solutions, where the key was provided by the osmotic pressure properly studied by WILHELM PFEFFER. ARRHENIUS recalled the words of VAN'T HOFF in his lecture *«Une propriété générale de la matière diluée»* in 1885 at the Swedish Academy of Science in Stockholm, which explained this matter, and which ARRHENIUS called an *"extraordinarily important generalization of Avogadro's Law. ... The pressure of a gas at a given temperature, if a certain number of molecules are contained in a certain volume, is equal to the osmotic pressure developed under the same conditions by the majority of substances, if they are dissolved in an arbitrary liquid, no matter which."* [Arrhenius (1887): 631]. So according to VAN'T HOFF one can formulate the equation $pV = iRT$ between the osmotic or the gas pressure p, the volume V, and the temperature T at constant amount of material (R universal gas constant). He has devoted many studies to the coefficient i, and he could show that for ideal gases it is equal to one and for dilute solutions close to one. ARRHENIUS could indicate its true nature in his dissociation theory, briefly sketched above, which in turn follows from the thermodynamic theory of VAN'T HOFF and the measurements of the ion mobility by WILHELM HITTORF and FRIEDRICH KOHLRAUSCH.

As stated by JAMES RIDDICK PARTINGTON [Partington (1953): 2856], the theory of ARRHENIUS *"gave a convincing picture of mechanism of electrolytic conduction, but a large and important branch of electrochemistry still awaited theoretical explanation.*" It still remained unexplained, how the potential difference is generated in electrochemical processes. However, the experimental experience has shown, that this can be generated by means of electrolytic solutions with different concentrations or of different kinds, without a metallic contact or a chemical action. The far-reaching solution of the problem completing the thermodynamic electrochemistry was found by NERNST during 1888/89 in the laboratory in Leipzig.

Before NERNST published his fundamental theory of the electrode potential he wrote a paper on the theory of diffusion, which appeared in 1888 in the *"Zeitschrift"* [Nernst (1888b)], and a second paper together with MORRIS LOEB [Loeb and Nernst (1888)].

The starting point of this was the general theory of diffusion by ADOLF FICK and the papers by VAN'T HOFF and ARRHENIUS we have just mentioned. In the case of solutions of an electrolyte with different concentrations being in contact with each other he assumed that the driving force for each ion is proportional to the gradient of the osmotic pressure $-\partial p/\partial x$. In this case the moving ion experiences a frictional resistance. This can be understood in terms of the force which is necessary to move an ion at unit velocity. The mobility (or the velocity per unit force) of the cation u_+ or of the anion u_- is inversely proportional to this force. The diffusing amount per unit time is equal to the product of the number of ions, the force acting upon each ion, and its velocity. Hence, for single-valent ions with the concentration c this force per mol of ions is given by $\dfrac{1}{c}\dfrac{\partial p}{\partial x}$. For the amount diffusing through the cross section q during the time dt due to the osmotic pressure this yields $-u_\pm q \dfrac{\partial p}{\partial x} dt$.

Furthermore, NERNST assumed that the ions moving with different velocities for a short time generate an electrostatic force which equalizes the velocities. Denoting the gradient of the potential φ by $-\partial\varphi/\partial x$, we can write the diffusing amount due to the electrostatic force as $\mp u_\pm qc \dfrac{\partial \varphi}{\partial x} dt$.

For the equalized velocity, the combined action of the osmotic pressure and the potential gradient yields the diffusing amount

$$-u_+ q\left(\frac{\partial p}{\partial x} + c\frac{\partial \varphi}{\partial x}\right)dt = -u_- q\left(\frac{\partial p}{\partial x} - c\frac{\partial \varphi}{\partial x}\right)dt \qquad (1)$$

If according to VAN'T HOFF we take for the osmotic pressure p in the case of single-valent ions $p = cRT$, one finds

$$\frac{\partial \varphi}{\partial x} = RT \frac{u_- - u_+}{u_- + u_+} \frac{\partial \ln c}{\partial x} \qquad (2)$$

and after integration

$$\varphi_2 - \varphi_1 = \Delta\varphi = RT \frac{u_- - u_+}{u_- + u_+} \ln \frac{c_2}{c_1} \quad (c_2 > c_1). \qquad (3)$$

For this formula at the same time NERNST provided also a simple thermodynamic derivation.

According to FICK's law the diffusing amount is given by $-Dq\frac{\partial c}{\partial x}dt$. If we insert the expressions for the osmotic pressure p and for the gradient of the potential φ (2) into equation (1) in combination with FICK's law, we obtain

$$D \cdot q \frac{\partial c}{\partial t} dt = u_+ RT\left(1 + \frac{u_- - u_+}{u_- + u_+}\right) \cdot q \frac{\partial c}{\partial t} dt .$$

Then we can derive the relation

$$D = 2RT \frac{u_+ u_-}{u_+ + u_-}$$

for the diffusion coefficient D. By means of measurements NERNST could confirm its validity and thereby also that of equation (3).

NERNST developed his theory of the electrode potentials following from these results for the case of binary electrolytes and under the assumption of ideal solutions. For the additional electric forces which must be added to those arising from the osmotic pressure he derived a principle and provided a "proof", *"which much facilitates the summarizing overview and which saves a lot of calculations"*. He referred to this as the principle of superposition and defined it as follows: "*The electromotoric forces acting in two equally formed systems consisting of a solution of electrolytes, in corresponding locations have the same magnitude and direction, if both systems only differ in the fact, that, if in one system the partial pressure of the positive ions at any point is $p'_1, p'_2 \ldots$, of the negative $p''_1, p''_2 \ldots$, then in the other system the corresponding quantities are $np'_1, np'_2 \ldots$ and $np''_1, np''_2 \ldots$*" [Nernst (1889a): 133–134]. If φ_I and φ_{II} are the electrostatic potentials of the additional forces in the two systems, respectively, according to the principle of superposition φ_I and φ_{II} in corresponding locations of the two systems should be equal to each other or should differ only by a constant.

In the case of solutions of the same electrolyte with different concentrations c_I and c_{II}, for the generated potential difference NERNST could derive the formula

$$\varphi_I - \varphi_{II} = RT \frac{u_- - u_+}{u_- + u_+} \ln \frac{p_I}{p_{II}} \tag{4}$$

which is equivalent to the relation (3) obtained in his paper [Nernst (1888b)] because of $p_{I,II} = c_{I,II}RT$, however, now derived in a different way. In this case he started from the work $(\varphi_{II} - \varphi_I)(\eta_+ + \eta_-)$, which is needed to bring a positive electric charge $+\eta_+$ from the potential φ_I up to the potential

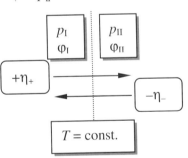

φ_{II} and a negative charge $-\eta_-$ in the opposite direction. If one writes the pressures, which according to VAN'T HOFF are proportional to the concentrations, as p_I and p_{II}, the situation discussed by NERNST can be sketched using the scheme shown in the adjoining figure. Noting that the charge transport is effected by ions, one can find the potential difference $\varphi_{II} - \varphi_I$ under the assumption that the condition $\eta_+ + \eta_- = 1$ enters the calculation of the work. From the results of HITTORF and KOHLRAUSCH one obtains $\eta_+ = u_+/(u_+ + u_-)$ and $\eta_- = u_-/(u_+ + u_-)$, if u_+ denotes the mobility of the cations and u_- that of the anions. If V is the volume occupied by a cation of charge +1 or by an anion of charge −1, according to the sketched conditions for the charge transport the work $\frac{u_+}{u_+ + u_-} \int_{p_1}^{p_{II}} V \, dp$ or $\frac{u_-}{u_+ + u_-} \int_{p_{II}}^{p_1} V \, dp$ is required. Using the BOYLE-MARIOTTE law $pV = RT$, by summation of both contributions to the work and integration NERNST obtained equation (4).

The ideas of NERNST about the electromotoric forces appearing at the interface of reversible electrodes gained special fundamental importance. In this case he defines such an electrode such *"that no work is required for this process, if an amount of electric charge enters from the electrode into the surrounding electrolyte, and if the same amount of charge passes through the system in the opposite direction."* [Nernst (1889a): 147]. This means that equilibrium conditions should be satisfied. It was known for these electrodes, that the transport of electric charge is exclusively due to the ions. Initially, the example raised by NERNST is the silver electrode Ag | Ag$^+$. It is an *"electrode of the first kind"*. However, he also discussed such electrodes, for which the concentration or the pressure of an anion instead of that of a cation is considered. For these *"electrodes of the second kind"* he mentioned the silver/silver chloride-electrode (Ag | AgCl | Cl$^-$) and the calomel electrode (Hg | Hg$_2$Cl$_2$ | Cl$^-$) as examples.

The work needed to raise for such an electrode an amount of electric charge $+\eta$ from the potential φ_I to the potential φ_{II} is equal to $(\varphi_{II} - \varphi_I) \cdot \eta = \varepsilon \eta$. In the case of metal/electrolyte electrodes, ε is the potential difference between the metal and the electrolyte, and $\varepsilon\eta$ is the work needed for transferring the amount of charge η from the solution into the metal. In the case of electrodes of the type of the silver-electrodes because of $\eta = +1$ this work is equal to ε. If p denotes the osmotic pressure of the metal ions in the electrolyte, its change from p_+ to $p_+ + dp_+$ for this type of electrode results in the change of work from ε_1 to $\varepsilon + d\varepsilon_1$. Here $d\varepsilon_1$ is given by $d\varepsilon_1 = -V dp_+$, where V denotes the volume of the amount of cations at the pressure p_+. With the BOYLE-MARIOTTE law one finds $d\varepsilon_1 = -RT \cdot d\ln p_+$ and after integration

$$\varepsilon_1 = RT \ln \frac{P_+}{p_+}, \tag{5}$$

where the integration constant was written as $\ln P_+$. In the case of electrodes of the second type analogously one obtains the relation

$$\varepsilon_2 = RT \ln(p_- / P_-).$$

Based on the analogy of the evaporation and the solution process following from the theory of VAN'T HOFF, NERNST was looking for a physical explanation of the integration constant P_-. For this purpose he introduced the concept of the *"solution tension"* (*Lösungstension*): *"To a substance dissolved due to the contact with a solvent we must also attribute the possibility of expansion, since also in this case the molecules are driven into the space in which they reach a certain pressure; apparently, each substance will become dissolved until the osmotic partial pressure of the molecules during this process becomes equal to the 'solution tension' of this substance."* Then he stated: *"According to this, in water every metal shows a particular 'solution tension', the magnitude of which may be denoted by P."* [Nernst (1889a): 150–151].

NERNST discussed the three cases $P > p_+$, $P < p_+$, and $P = p_+$, in connection with the double layer or potential difference generated at the phase boundary, the force components generated by this, and their relation to the solution tension, as well as the nature of the electromotoric force resulting in the equilibrium state. Then he noted that the mathematical formulation of these ideas again leads to equation (5). Therefore, the integration constant P_+ is apparently identical to the solution tension P.

By means of an analogous consideration also the integration constant P_- obtained from the treatment of the electrodes of the second type should be traced back to an electronegative solution tension.

Hence, NERNST had proposed a theory, which allowed *"to illustrate for oneself the experimentally confirmed fact of the generation of a potential difference between metals and electrolytes."* [Nernst (1889a): 153]. In this kind of treatment of his results typical of him he recognized the actually existing limits of this theory. He clearly stated what at the time was needed but not yet possible for the further development: *"As long as questions such as that of the origin of the osmotic pressure, further, that of the way in which the electricity is attached to the ion, finally, that of the nature of the latter, are so far removed from their solution, I feel that we must avoid a discussion about the nature of the forces which drive the ion out of the metal into the solution, and we must be satisfied with testing only if the formal relations resulting from the introduction of the electrolytic solution tension agree with the facts."* [Nernst (1889a): 153–154].

Furthermore, in the Laboratory in Leipzig NERNST was able to develop a general theory of the concentration chains. The electromotive force of these turned out as the sum of the potential differences at the two electrodes and at the location at which both electrolytic solutions are in contact with each other. NERNST could verify this theory by means of a large number of measurements.

NERNST had emphasized that his theory of the electromotoric force is connected with those of GUSTAV WIEDEMANN and HERMANN VON HELMHOLTZ. The latter and also already before him, however, not noted in Europe at the time, JOSIAH WILLARD GIBBS had found a relation between the electromotoric force, its temperature coefficient, and the reaction heat of the chemical process generating the current. Now NERNST could prove that in the case of dilute solutions his theory agreed perfectly with that of VON HELMHOLTZ, if one took into account the vapor pressure reductions of such solutions calculated by VAN'T HOFF and PLANCK. NERNST admitted that the theory of HELMHOLTZ was more general, since it was not restricted to dilute solutions such as his own. However, contrary to the theory developed by him, it did not allow to separate the electromotoric force into its components. In fact, the theory of NERNST provided still much more by allowing an insight into the mechanism generating these potential differences.

NERNST also investigated the potential difference appearing at the contact between two different electrolytes with equal concentration, and he performed corresponding measurements. However, the exact mathematical treatment of this case became possible only in 1890 by MAX PLANCK.

NERNST applied his superposition principle to the electrolytic thermochains, which he studied experimentally and theoretically. For the case of similar electrodes and the same electrolyte in the two communicating solutions with different temperatures, he could derive and verify experimentally a simple formula. Based on the other concept developed by him in Leipzig, the solution tension, NERNST discussed a general theory of the galvanic elements.

In 1889 WALTHER NERNST documented his impressive results achieved in such a short time in his habilitation thesis *«Die elektromotorische Wirksamkeit der Ionen»* (*"The Electromotoric Action of the Ions"*) (Fig. 3.2) [Nernst (1889b)]. In the same year he published them partly in the Proceedings of the Royal Prussian Academy of Science in Berlin [Nernst (1889c)] and in more detail in the *«Zeitschrift»* [Nernst (1889a)]. His latter article submitted during May 1889 appeared on August 20 of that year.

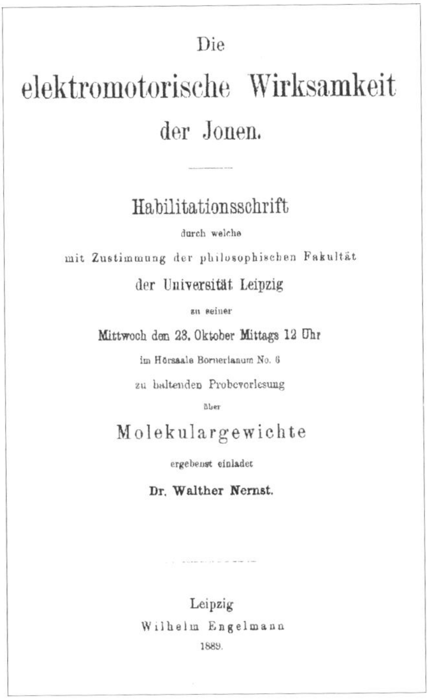
Die
elektromotorische Wirksamkeit
der Jonen.

Habilitationsschrift
durch welche
mit Zustimmung der philosophischen Fakultät
der Universität Leipzig
zu seiner
Mittwoch den 23. Oktober Mittags 12 Uhr
im Hörsaale Bornerianum No. 6
zu haltenden Probevorlesung
über
Molekulargewichte
ergebenst einladet
Dr. Walther Nernst.

Leipzig
Wilhelm Engelmann
1889.

Fig. 3.2 Title page of the habilitation thesis of WALTHER NERNST.

Referring to this publication and to the two previous ones, representing in some sense its preparation, *«Zur Kinetik der in Lösung befindlichen Körper»* (*"Kinetics of the Substances Existing in Solution"*) [Nernst (1888b), Loeb and Nernst (1888)], MAX BODENSTEIN summarized: *"With these three fundamental papers Nernst placed himself next to the three stars of the slightly older scientists, who at the time have created the foundation of our physical chemistry: van't Hoff, who found the theory of the solutions in addition to many other important things, Arrhenius, who related the observed deviations to the electrolytic dissociation, and Ostwald, who performed the collection and organization of the research in addition to excellent works. Subsequently, Nernst has become by far the most active member of this union."* [Bodenstein (1942a): 83].

50 years after the publication of the great scientific results from Leipzig of his teacher, FRIEDRICH KRÜGER, a student of NERNST in Göttingen, stated: *"His youth article already indicates those aspects of its author, which distinguished all of his subsequent works: the clear logical-mathematical treatment of the problems, the intuitive understanding of physical connections with great phantasy, the experimental inventiveness and skill, always supported by an indefatigable working power."* [Krüger (1939): 554].

Although further experimental and theoretical work still had to be done, as NERNST himself emphasized in 1889, one can say, that the theory of the electrode potential created by him and the equation derived in this connection represent a certain culmination and completion of the thermodynamic electro-chemistry, which assumes the state of equilibrium and the absence of an electric current. In some sense equation (5) represents the original version of the equation subsequently developed further, which occupies the center of this theory, and which today is known as the NERNST equation.

In the case of a metal electrode Me | Me^{z+} with a potential generating process Me $\rightleftharpoons$ Me^{z+} + ze^- (Me: metal, e^-: electron) the NERNST equation in the form used presently can be obtained from (5) in the following way: since the solution tension $P = P_+$ represents a quantity specific for each substance, depending only upon the temperature but not upon the concentration, one can introduce the function $-\kappa(T) = RT \ln P$. The work per mole of the motion of the electric charge against the potential difference ε of the generated electric double layer is given by $z\varepsilon F$, where F is the

FARADAY constant indicating the amount of charge of one mole of elementary charges. If we replace the osmotic pressure by the concentration c_{Me}, from the equilibrium condition

$$z\varepsilon F - \kappa - RT \ln c_{Me}^{2+} = 0$$

we obtain the NERNST equation in the form

$$\varepsilon = \varepsilon_0(T) + \frac{RT}{zF} \ln c_{Me}^{2+}.$$

Here $\varepsilon_0(T) = \kappa(T)/zF$ is the normal potential in the case $c_{Me}^{2+} = 1$.

We do not discuss the modern thermodynamic derivation of the NERNST equation and its application to electrodes of the second type, redox electrodes, and gas electrodes, etc., since this would be beyond the scope of this Chapter.

3.4 The "Ionists" versus the "Anti-Ionists"

MAX BODENSTEIN commented the acceptance by WALTHER NERNST of the position as assistant professor with WILHELM OSTWALD in Leipzig with the words: *"In this way Nernst joined the community of the 'Ionists', their doctrine of the electrolytic dissociation at the time finding only little understanding and meeting strong opposition in the scientific community, and he helped to clarify and consolidate this subject by means of works, which were equally outstanding in their theoretical concept and in their experimental realization, and at the same time appeared in an abundance and with such a speed, as it can be accomplished only by a specially gifted and hard-working researcher."* [Bodenstein (1942a): 80].

So NERNST had joined the *"wild army of the ionists"*, using an expression by AUGUST FRIEDRICH HORSTMANN, its stronghold being the 2nd Physical Chemical Laboratory of OSTWALD in Leipzig, at which not only in this direction but in a very general way the research was concentrated *"with much energy on the further development of physical chemistry"* [Riesenfeld (1930): 14]. In addition to OSTWALD and his three assistant professors WALTHER NERNST, ERNST BECKMANN, and JULIUS

WAGNER, in one way or another international visitors participated in this effort, whose names became well known in science: ARTHUR AMOS NOYES (USA), JAMES WALKER (Scotland), GUSTAV TAMMANN (from Dorpat, then Russia), as well as WILHELM MEYERHOFFER and GEORG BREDIG. Also SVANTE ARRHENIUS, coming from Amsterdam, worked with OSTWALD in Leipzig during the summer term of 1888. In this way three of the most important "ionists" were united in one laboratory for a short, but very important and fruitful period.

The concept of the ion had been introduced into science by MICHAEL FARADAY in 1834. It originates from the Greek verb εῖμι (*eími*) to be translated with 'walk' (German 'wandern') in this connection. Its participle forms ἰών and ἰόν (*ión*) mean 'someone, who is walking' and 'something walking', respectively. FARADAY did not provide more special information about the structure of the ions. The fact that the ions are electrically charged could be demonstrated in particular by the experiments performed by WILHELM HITTORF since 1853. Here, anions (Greek ἀνα-/ἀν- = 'upwards') are negatively and cations (Greek κατα-/κατ-/καϑ- = 'downwards') positively charged.

Subsequently, RUDOLF CLAUSIUS introduced the assumption that a fraction of the molecules in an electrolytic solution decayed into ions, where the emphasis is on the fraction which has decayed. This theory had been accepted by the scientists almost without exception. However, after SVANTE ARRHENIUS had found out, how the ratio of the amount of ions and that of the molecules, from which they originated, can be determined, and after several corresponding methods gave the result, that most of the salts showing a neutral solution and also the strong acids and bases are practically completely disintegrated into ions, these ideas were accepted only by a relatively small number of scientists called the "ionists".

It must be emphasized, that there were essentially two papers representing in some sense the starting point for the formation of the smaller group of the "ionists" and the larger group of their opponents, the "anti-ionists". Both papers appeared in the first volume of the *Zeitschrift*. One author was MAX PLANCK [Planck (1887)], the great merits of whom regarding the theory of the electrolytes being nearly ignored, and whose results obtained independently were published later several times only shortly after the results obtained by NERNST. The other author was SVANTE ARRHENIUS [Arrhenius (1887)], the true founder of the dissociation theory, for the validity of which the *"wild army of the ionists"* had been fighting hard.

WILHELM OSTWALD certainly belongs to the most prominent defenders of the dissociation theory. He remembered: *"I was warned several times not to defend it so absolutely. ... We obtained unexpected assistance from the independent investigations of the excellent mathematical physicist Max Planck."* [Ostwald (1927): 30]. His student NERNST and, of course, ARRHENIUS perhaps were his most outstanding co-defenders.

One of the main reasons why many opponents of the dissociation theory rejected it, may have been their difficulty to understand, that exactly the substances held together because of the strongest "relationships" such as the salts, acids, and bases in solution mentioned above, can be completely dissociated. OSTWALD pointed out the mistake in this case, which comes from the fact, that the relationship holding together a chemical compound and that of the compound versus another one are opposite in magnitude. Furthermore, many "anti-ionists" did not understand that the products of the split appear during the dissociation not in form of free atoms which should react strongly with the solvent, but instead in form of electrically charged particles with completely different properties.

Much criticism was likely expressed mostly by chemists, who at the time could not follow the mathematical treatment of the problem, as it was carried out by ARRHENIUS and PLANCK, and subsequently also by NERNST and other physicists. However, physicists were also among the opponents, although *"because of the mathematical formulation, the theory"* was *"rather acceptable."* [Riesenfeld (1930): 15].

OSTWALD had to complain that his request to clearly formulate the expressed criticism was ignored in most cases. An exception was the son of his predecessor GUSTAV WIEDEMANN. This EILHARD WIEDEMANN was Professor of Physics since 1878 in Leipzig and since 1886 in Darmstadt. Actually, he submitted an article specifying his objections to the *Zeitschrift*, which OSTWALD published quickly together with his comment easily showing the invalidity of the objections [Ostwald (1888)]. EILHARD WIEDEMANN refrained from an answer, and in particular from the announced discussion *"about the reliability of some physical-chemical conclusions at all."* [Ostwald (1927): 30]. PLANCK also entered this controversy and showed, that the polymerization of water assumed by WIEDEMANN as an argument against an increased particle number in electrolytic solutions is irrelevant.

Referring to results by ARRHENIUS, ERNST HERMANN RIESENFELD noted at the time, that the attacks by the "anti-ionists" increased, *"although one should think, that because of these investigations and the many other contributions, simultaneously coming from the Institute of Ostwald and based on the dissociation theory, the foundations of the latter should be proved sufficiently."* [Riesenfeld (1930): 15].

Also WALTHER NERNST had been included by OSTWALD into one attempt to overcome this ignorance. In his paper for justifying the dissociation theory [Ostwald (1888)] OSTWALD could show theoretically by means of a «*Gedankenversuch*», that the existence of free ions in an electrostatically charged electrolyte is a necessary consequence of the valid laws of electrodynamics. *"However, the possibility to realize experimentally the scheme indicated there has been questioned in a conversation by such an authority, that we felt it would be our duty to remove any possible doubt also experimentally."* [Ostwald and Nernst (1889): 120]. Here *"we"* indicates OSTWALD and NERNST, and *"authority"* perhaps the otherwise highly honored physicist AUGUST KUNDT, who in 1888 had become the successor of HERMANN VON HELMHOLTZ as the Chair of Experimental Physics in Berlin.

The set-up of the experiment performed in Leipzig consisted of a flask filled with dilute sulfuric acid, surrounded on the outside by a tinfoil and electrically connected with a "capillary electrode" by a wet thread. About half of this electrode was also filled with the dilute sulfuric acid and the other half with mercury. The content of the flask was connected with the acid by means of the wet thread, the mercury was grounded by a platinum wire, and the tinfoil was connected with the positive pole of an electrical machine. Because of the positive charging of the tinfoil, the SO_4^{2-}-ions of the acid are attracted by it, whereas the H_3O^+-ions are repelled. The latter ions reach the capillary electrode via the thread and, hence, they get to the mercury, where they are discharged generating gaseous hydrogen. Their charges then disappear into ground. The generated gas could be observed by the shift of the meniscus of the mercury. The transport of the electric charges detected by means of this generation of gas did not happen within a closed circuit, but instead because of electrostatic induction (*Influenz*). Based on the fact generally accepted since FARADAY, that in an electrolyte the transport of charges is connected exclusively with the motion of ions, OSTWALD and NERNST concluded: *"If during the 'influenz action' metallic conduction through the electrolyte would be involved, as it was argued critically from a well-known corner as a possibility, there would be no reason for the appearance of hydrogen at the electrode; on the other hand, the latter proves that electrolytic conduction happened, i.e., that free ions did exist and were moving."*

Furthermore, in their article *"About free Ions"* («*Über freie Ionen*») [Ostwald and Nernst (1889)] OSTWALD and NERNST experimentally and theoretically invalidated the argument, that the ions only appear at the time of the electric charging, and they *"have presented conceptions which one needs of the phenomenon of electrolysis based on the assumption of the preexistence of free ions."* [Ostwald and Nernst (1889): 128].

The established fact in modern science, that the dissociation theory as presented and defended by the "ionists" is believed to be not only perfectly confirmed but even represents one the foundations of electrochemistry, last not least must be credited to the works of WALTHER NERNST, which he began in Leipzig and continued in Göttingen.

Chapter 4

The Göttingen Period: The Rise to World Fame (1890 – 1905)

Immediately following his habilitation, for the summer term in 1889 NERNST went to the University of Heidelberg as an assistant of JULIUS BRÜHL.

In this way for a short time he was employed at the oldest University of Germany. It had been founded in 1386 by the Palatine Elector RUPRECHT I in his town of residence, following the example and in some sense also as a replacement of the Sorbonne in Paris, which had been lost to the Roman Pope because of the schism of 1378. At the same time it was only the third University within the Holy Roman Empire of German Nation after Prague (1348) and Vienna (1365). As expected, its first Rector, MARSILIUS VON INGHEN, run it as a roman-catholic Institution. However, in 1558/59 it was reformed by PHILIPP MELANCHTHON, and it became protestant in character. After the Thirty Years' War in 1652 it became newly organized as a secular institution. In the 18th century it lost its former importance, which it recovered again only after 1803, when Heidelberg became part of the territory of Baden.

The fact that since 1852 a germ-cell of Physical Chemistry had developed at this University because of the appointment of ROBERT WILHELM BUNSEN, may have played a role in the decision of NERNST, to accept a position at Heidelberg. Together with GUSTAV ROBERT KIRCHHOFF, who worked and taught in Heidelberg from 1854 until 1874, in 1859 he had developed the method of spectrum analysis. BUNSEN retired at the end of the winter term 1888/89. At the beginning of the following winter term VICTOR MEYER, another pioneer of Physical Chemistry, became his successor. During the interregnum, in the summer term of 1889, the most senior Professor of Chemistry at the time in Heidelberg, JULIUS BRÜHL, had given the main chemistry course with NERNST as his assistant.

4.1 The *Georgia Augusta* University in Göttingen

For NERNST Heidelberg was only a brief transition period. EDUARD RIECKE, the Full Professor of Physics at the University of Göttingen, had invited him to work at his Institute as an instructor, and he promised him the position as a Professor of Physical Chemistry. NERNST took up this position in 1890.

Compared to Heidelberg and Leipzig, the University of Göttingen was relatively young. It was founded in 1737 by the Elector GEORG II AUGUST of Hannover, who was King GEORGE II of England at the same time. Therefore, as *Georgia Augusta* it carries his name. In 1866 it became a Prussian Institution.

Initially being among the highest ranks within the German universities, since the 18th century it gained European stature, which last but not least was increased further because of the activities of NERNST. The «*Göttinger Gelehrten Anzeigen*» became a leading scientific journal in Europe.

GEORG CHRISTOPH LICHTENBERG had studied at this university, which was strongly influenced by the ideals of the Age of Enlightenment in Germany, and from 1770 until the end of his life he had been active as Professor of Applied Mathematics. In addition as a writer, he is recognized as a multi-talented scientist and leading experimental physicist of his time. In Göttingen he had given the first course in Experimental Physics in Germany. In physics LICHTENBERG in particular gained international recognition because of his contributions to the field of electricity. In 1777 he had reported the electric patterns named after him, which allowed to visualize the action of electricity.

Also CARL FRIEDRICH GAUSS, the *Mathematicorum princeps*, worked as Professor of Astronomy and Director of the Observatory in Göttingen from 1807 until the end of his life. To him we owe important contributions to the mathematics of his time, but also to astronomy, geodesy, and physics. In the context with the activities of NERNST we have to mention his studies of electromagnetism performed together with WILHELM WEBER. Also we must refer to the invention of the first electric telegraph achieved by the two scientists in Göttingen in 1833. WILHELM WEBER occupied a position as Professor in Göttingen from 1831 until 1837 and again since 1849. From 1843 until 1849 he taught and worked in Leipzig.

Although they are not directly connected with NERNST, we should mention also JACOB and WILHELM GRIMM, who are the founders of *Germanistik* as the science of German language and literature and who occupied Chairs at the *Georgia Augusta* in the beginning of the 1830s. They increased the fame of this University, although in 1837 they belonged to the so called "Göttingen Seven", who protested against the repeal of the fundamental state law of 1833 by the new sovereign ERNST AUGUST II and were removed from their office. The other members of the "Göttingen Seven" were GEORG GOTTFRIED GERVINUS, historian of literature, the historian CHRISTOPH DAHLMANN, the jurist WILHELM EDUARD ALBRECHT, the orientalist, philologist, and theologian HEINRICH VON EWALD, and WILHELM WEBER. Subsequently, since 1841 they could continue their important work in Berlin as members of the Prussian Academy.

4.2 Eduard Riecke, Felix Klein, and Mathematics in Göttingen

EDUARD RIECKE, who had brought NERNST to Göttingen, had studied mathematics and physics in Tubingen with CARL NEUMANN. Subsequently, he continued his studies of physics with WILHELM WEBER in Göttingen, where he obtained his PhD and then his habilitation. Since 1881 he was Full Professor of Physics at the University of Göttingen. The research of RIECKE dealt with the physics of electrons. Already in 1881 he proposed a formula for the deflection of the cathode rays, which he treated as rapidly propagating electric particles, in a magnetic field. He developed a theory for the electric conduction in metals, he investigated the motion of electric particles in gases, and he let himself become inspired by NERNST for studies of the thermodynamic potential. RIECKE's teacher WEBER had described the electric current in metals in terms of the transfer of electric particles from the sphere of interaction of one metal atom to that of another, where the direction is given by the applied voltage. On the other hand, RIECKE assumed the existence of free electrons behaving like a gas in the metallic lattice. Also we should mention the two volumes of the *"Textbook of Physics for Independent Study and for Use together with Courses"* («*Lehrbuch der Physik zu eigenem Studium und zum Gebrauch bei Vorlesungen*») of RIECKE [Riecke (1895)], which had reached several editions.

In Göttingen since the activities of CARL FRIEDRICH GAUSS, mathematics has remained a subject of high rank and esteem. Also for NERNST it occupied a high level of importance, and especially at his new place of employment it gained special significance in several ways. Subsequent to his highly successful period in Berlin, PETER GUSTAV LEJEUNE DIRICHLET, the founder of the analytic number theory and a highly respected promoter of mathematical physics, had become the successor of GAUSS in Göttingen.

Also BERNHARD RIEMANN, one of the most important mathematicians of the 19th century, was most closely connected with the University of Göttingen. Here he had studied, then became assistant of WILHELM WEBER, in 1854 was promoted to instructor, and in 1855 to professor when DIRICHLET became Full Professor. After the death of DIRICHLET he became his successor. The multiple and fundamental contributions of RIEMANN dealt among others with the theory of numbers, the modern theory of functions, topology, and mathematical physics, and still have strongly influenced the mathematics and theoretical physics of the 20th century.

From 1868 until his early death in 1872 ALFRED CLEBSCH, the founder of algebraic geometry, taught in Göttingen. He had studied in Konigsberg (now Kaliningrad/Russia) with FRANZ NEUMANN, the father of CARL NEUMANN, the teacher of RIECKE. FRANZ NEUMANN belongs to the cofounders of mathematical physics in Germany. He gained some importance for NERNST in connection with the Third Law of Thermodynamics because of the rule established by him and HERMANN KOPP for calculating the values of the molar heat from the atomic heat. In 1868, together with CARL NEUMANN, CLEBSCH founded the «*Mathematischen Annalen*», an important journal existing still today.

Although he had studied in Bonn with JULIUS PLÜCKER, the famous expert in analytic geometry and an experimental physicist, and also had obtained there his PhD, the multi-talented mathematician FELIX KLEIN (Fig. 4.1) can also be considered a student of CLEBSCH. Already at the early age of 23 KLEIN became Full Professor in Erlangen. Subsequent to appointments as Professor at the Technische Hochschule in Munich and at the University of Leipzig, starting on April 1, 1886 he was appointed at the University of Göttingen, where he had acted as an instructor already in 1871/72. KLEIN's primary interest was group theory. Because of his scientific research, but also as an organizer of science and a reformer of education he had a lasting effect on the developments of mathematics, its applications, and on the instructions within this discipline. In Göttingen he was able to create an international center of mathematics.

Fig. 4.1 FELIX KLEIN.

Since 1875 HERMANN AMANDUS SCHWARZ worked in Göttingen. The inequality of thermodynamics named after him plays an important role. He turned out to become an opponent of the goals of KLEIN regarding the reforms and the organization of science. KLEIN could succeed with these goals only after SCHWARZ accepted an offer from Berlin in 1892. So still in the same year together with HEINRICH WEBER, the successor of SCHWARZ, he founded the Mathematical Society of Göttingen. WEBER generated important contributions to mathematical physics, the theory of numbers, and algebra.

Five years after NERNST, in 1895 DAVID HILBERT came to Göttingen. This appointment had been arranged again by FELIX KLEIN, who had recognized HILBERT as the founder of a new field of research within the theory of numbers. Because of HILBERT, who remained faithful to Göttingen until his death in spite of honorable offers from other places, this university became a world center of mathematics. The mathematics and theoretical physics – here we mention only the fundamental contributions to quantum theory (HILBERT-space) and General Relativity – of the 20th century were strongly influenced by him and his school. It is correct to refer to DAVID HILBERT as one of the or even the single most important mathematician of this period. In 1902 he could bring HERMANN MINKOWSKI, his fellow student from Konigsberg, to a newly created position in Göttingen. MINKOWSKI's name is closely connected with Special Relativity.

Beyond the common professional interest, during his whole life NERNST kept a friendly relation with his colleague HILBERT having nearly the same age. This friendship also extended to the two families.

4.3 Early Studies in Göttingen: The Nernst Distribution Law

At Easter 1890 NERNST started as an assistant and instructor (*Privatdozent*) at the Physical Institute of the University of Göttingen directed by EDUARD RIECKE. In this Institute great advances in the study and knowledge of electricity and magnetism had been and still were achieved, and here NERNST could optimally utilize his capabilities of which he was well aware. He not only continued in an excellent way the tradition of the *Augusta Georgia* in electromagnetism. During his 15 years in Göttingen he pursued further his research started in Leipzig in the field of electrochemistry and physical chemistry, leading to an extremely high scientific and internationally recognized level and preparing the final step of reaching the summit in Berlin.

On March 10, 1890 the Faculty of Philosophy in Göttingen informed the curator of the university, that based on his eleven papers published so far and on his public lecture presented on the previous day *"About the Participation of Electric Forces in Chemical Reactions, Dr. phil. Walther Nernst has been given the venia legendi for the subject of physics"* [UAHUB: I, 1]. This represented the starting signal for a brilliant career as a university teacher, the further details of which will be discussed in a separate section. At this point we mention only that already in September 1891 NERNST obtained the position as Professor of Physical Chemistry promised by RIECKE in his Institute, and that about three years later he became even Full Professor of this discipline and Director of a corresponding Institute.

Therefore, it seems reasonable to divide the period of NERNST in Göttingen into the early one from 1890 until 1894 and the subsequent decade until 1905.

A new method for determining the molar mass represented the focal point of the work during the first year 1890 in Göttingen [Nernst (1890a)]. NERNST was able to show, that the molar mass of a substance can be found by connecting its value with the lowering of the partial miscibility of ether and water. He investigated the latter by measuring the freezing point of the mixture of both solvents.

In 1890 and 1891 the papers were also prepared as mentioned in Section 2.4.2, which NERNST wrote together with PAUL DRUDE, continuing his studies of the NERNST-ETTINGSHAUSEN effects. DRUDE had studied in Göttingen and obtained his PhD with a theoretical thesis on the optical behavior of opaque crystals. His advisor was WOLDEMAR VOIGT, a student of FRANZ NEUMANN, similar to ALFRED CLEBSCH. In 1887 he obtained his habilitation and became an assistant of VOIGT who represented the field of crystal physics. Since 1883 VOIGT taught theoretical physics at the *Georgia Augusta*. For example, by means of symmetry principles and by inventing the tensor concept he wanted to renew the relation between mathematics and physics. However, he disapproved atomic models and quantum theory.

On November 27, 1891 NERNST announced to WILHELM OSTWALD (pointing out in his introduction *"By the way, recently I have been more of a physicist"*) the completion of another paper written together with

DRUDE dealing with the *"fluorescence of extremely thin films"* [Zott (1996): 32; Drude and Nernst (1891)]. Both authors could show that the fluorescence of standing waves is due to the wave crests by directly visualizing the light wave in terms of the fluorescence of thin films. This technique had been invented by OTTO WIENER.

On August 9, 1890 NERNST wrote in a letter to WILHELM OSTWALD: *"In the meantime I have submitted a paper 'on the distribution of a substance between two solvents' to our Provincial Academy with the friendly support of my superior* [RIECKE]*."* [Zott (1996): 14]. This publication and a subsequent one from 1891 [Nernst (1890b)] contain a law, which is referred to today as the NERNST Distribution Law.

MARCELIN BERTHELOT had recognized already the underlying phenomenon and had published his ideas in 1872. He had noted that the ratio of the concentrations, which result if a substance is distributed between two immiscible solvents, remains constant based on the same temperature. If c_A and c_B denote the concentrations in the solvents A and B, respectively, we can write

$$\kappa(T) = \frac{c_A}{c_B}.$$

Here we must note that this expression cannot be explained in terms of the equilibrium of a chemical reaction, although mathematically it corresponds to the mass action law. Instead, the distribution is based on an equilibrium between phases, from which this formula can be derived thermodynamically.

NERNST could interpret this law more deeply than BERTHELOT and he was able to discuss exceptional cases. So he treated the possibility that one substance does not have the same molecular size or molar mass in the two phases. In the case of the distribution of benzoic acid between water and benzene the obtained experimental data only agreed with the law, if one assumed that in water the acid exists as a monomer and in benzene as a dimer. NERNST could take into account the location of the equilibrium of the depolymerization by means of the mass action law. Already in 1890 regarding his papers dealing with the Distribution Law NERNST had stated: *"The results which I obtained were really quite sur-*

prisingly smooth, as long as an experimental verification of the solution laws still can surprise us at all." [Zott (1996): 14].

Since his early years in Göttingen the research of NERNST essentially dealt with nearly all directions of physical chemistry at the time. To the subjects of thermodynamics we have mentioned we must add, for example, the solubility of mixed crystals, the osmotic pressure, and the theory of the boiling and melting point (together with ALBERT HESSE). Together with GUSTAV TAMMANN, who was at the time still in Dorpat and became Professor in Göttingen only in 1903, he worked on «*Maximaltension*» [Tammann and Nernst (1891)]. During February of 1892 NERNST wrote to OSTWALD regarding his investigations of concentrated solutions: *"Against all expectations, the matter is very simple, such that so to speak I can develop a theory of 'ideal conc. sol.', which in its simplicity does not fall much behind the others, but which is much different. Only the state remains complex, in which a soln. passes from the 'ideal dilut.' into the 'ideal conc.' state."* [Zott (1996): 40]. The corresponding publications appeared in 1892 and 1894 [Nernst (1892a)].

Papers dealing with chemical kinetics were prepared also, such as the one by NERNST and C. HOHMANN on the formation of amyl esters from carbonic acids and amyls, and the study of NERNST on the role of the solvent in the chemical reaction, both of which appeared in print in 1893.

Of course, electrochemistry was also taken into consideration. In addition to concentration chains, the dielectric constant occupied the center of attention [Nernst (1893a)]. For determining the latter NERNST developed a measuring technique, which subsequently remained in use for a long time, and which represents an early example of his distinct capability to develop instruments and technical equipment based on new principles also conceived by him. In this case he started from the WHEATSTONE bridge. Simultaneously, the dielectric and the electrolytic conductance were compensated by a variable capacitor and a variable resistance, respectively. In this way, even the conductance of highly diluted solutions could be determined. Characteristically, NERNST felt, that the development of his measuring technique, of the high value of which he was quite convinced, was achieved by him as a physicist. So on April 30,

1893 he reported to OSTWALD: *"At the moment I work out a method for determining the dielectric constant (also of conducting materials) as simply as the conductance; I determine reliably resistances of liquids of 20 000 Ohm, perhaps I replace even the method of Kohlrausch. To revel in pure physics from time to time is a real treat, and I think before long I will deal also with the oscillations of Hertz."* [Zott (1996): 57].

At the 65th Congress of the *Gesellschaft Deutscher Naturforscher und Ärzte* in Nuremberg in 1893 NERNST presented his papers on the dielectric constants. Their reception was described by RICHARD WILLSTÄTTER, the later Nobel Prize winner, with the words: *"In a combined session of physics and chemistry Nernst presented a learned and important paper; his star was on the rise. The applause was very thin."* [Willstätter (1958): 86]. However, it met general interest. The investigations turned out to be so promising, that they were continued still further. On February 5, 1894 NERNST told OSTWALD regarding this matter: *"I am still stuck in the 'dielectricity', and it appears as if a lot may come out for the solutions of salts."* [Zott (1996): 68].

The example of this research perhaps emphasizes for the first time a typical feature of NERNST: His ingenious developments of measuring instruments always were subjected to the higher goal of the improved theoretical insight into the physical laws, without impairing its importance in his own assessment. Based on these studies NERNST was able to demonstrate for the first time that solvents with a large value of the dielectric constant always display a strong electrolytic dissociation. However, he was only a little ahead of others, since the publication of his results in the *Göttinger Nachrichten* in July 1893 [Nernst (1893a)] was followed already in October by a paper by JOSEPH JOHN THOMSON, in which the latter reported similar results obtained independently of NERNST. However, the young scientist in Göttingen could also show further, that solvents strongly tending toward ionization also have a tendency to association, and he could point out additional effects which influence the ionization. In particular, here we must mention the association of ions and of solvent molecules.

We still have to mention another paper, which NERNST wrote together with PAUL DRUDE in 1894 [Drude and Nernst (1894)]. Already on July 17, 1893 NERNST reported to WILHELM OSTWALD in Leipzig: *"I am after certain electric properties of the ions, from this perhaps a lot may develop. For example, the considerable contraction associated with the dissociation of the ions, according to all previous experience (see, for example the new measurements of Kohlrausch) is likely to be understood in terms of electrostriction (resulting from the electric field of the ions)."* [Zott (1996): 64]. In fact, DRUDE and NERNST then were able to show that an ionized substance could cause a contraction of the solvent by means of electrostriction because of the existence of charged particles, if the dielectric constant of the solvent increases with increasing pressure.

From the example of this collaboration we can recognize a typical feature of NERNST consisting of the fact that always he emphasized the share of other people in a joint publication, and that, when needed, he acted on behalf of the professional advancement of a colleague clearly deserving it. So on July 10, 1894 NERNST announced to OSTWALD his paper written together with DRUDE in the following way: *"I very much hope, that before long the electrostatic field will be placed next to the electromotoric effects of the ions. Here a large field appears to be ready for harvesting! These ideas were generated in a discussion with Drude; I will take this opportunity to recommend Drude to you warmly. Ebert has left and likely you must find a replacement. In D. you would get a physicist highly stimulating in the personal contacts; furthermore, it would be an act of kindness if you could provide an existence for D."* [Zott (1996): 76]. HERMANN EBERT had left Leipzig in 1894, in order to accept a position as Professor in Kiel. Actually, still in the same year PAUL DRUDE became Professor of Technical Physics in Leipzig, although ARTHUR VON OETTINGEN, the former teacher of OSTWALD in Dorpat, was appointed as the direct successor of EBERT.

Finally, we mention still two electrochemical papers from the year 1892. In one paper NERNST dealt with the potential difference in dilute solutions. The other paper, published together with his student ROBERT PAULI, was entitled *"More on the Electromotive Activity of the Ions"*

[Nernst and Pauli (1892)]. PAULI being only two years younger than NERNST, was sent by his teacher from Göttingen to OSTWALD in Leipzig, with the announcement from October 16, 1892: *"Perhaps you give him a project which requires a bit of calculations, since he is educated more mathematically. However, so far he was occupied more by student affairs instead of pure science, and in order that at last he gets rid of the 'Allotriis'* [skylarking], *I advised him to take this change of air."* [Zott (1996): 48]. OSTWALD answered one week later: *"Mr. Pauli showed up, but he disappeared again immediately to attend a wedding."* [Zott (1996): 50]. Nevertheless, due to the guidance of RIECKE and NERNST it could be achieved, that PAULI got his PhD in Göttingen in 1893 [Pauli (1893)].

At any rate, PAULI can be counted as the first PhD student of NERNST, to whom during the early Göttingen period of the latter only MAX ROLOFF can be added. ROLOFF got his PhD in 1894 with a thesis dealing with the *"Photochemical Action in Solutions"*. NERNST commented the corresponding publication [Roloff (1894)] with the words: *"Some details were obtained which are not bad. At any rate, it is a very industrious publication."* (Letter to OSTWALD from October 6, 1893 [Zott (1996): 65]).

4.4 Marriage with Emma Lohmeyer, and the Walther Nernst Family

Within the same context of remarks by NERNST about his investigations of ion mobility, diffusion, concentrated solutions, and the potential difference in dilute solutions, in a letter from March 6, 1892 to his former student, WILHELM OSTWALD wrote at the beginning: *"At first, please accept congratulations from my wife* [HELENE] *and myself to your engagement. I hope that this event reconciles you perfectly, if it still should not have happened, with your move to Göttingen and my role in this."* [Zott (1996): 41].

About half a year after the engagement with EMMA LOHMEYER, seven years younger than NERNST, on September 1, 1892 the wedding took place (Fig. 4.2). There followed nearly five decades of a harmonious marriage.

A few months after his arrival in Göttingen WALTHER NERNST had met EMMA LOHMEYER at a ball. After this first encounter, prior to the engagement there were frequent joint walks in the park of the University, further attended balls, and a sleigh-ride. The honeymoon led to Italy via Kassel and a visit of ALBERT VON ETTINGSHAUSEN in Graz.

Fig. 4.2   Wedding picture of EMMA and WALTHER NERNST.

In some sense EMMA's father was a colleague of NERNST. Like his son-in-law, the *Geheime Medizinalrat* KARL FERDINAND LOHMEYER was an *Extraordinarius* at the University of Göttingen, however, for surgery. He operated a private clinic established in his house. He had studied in Göttingen. In 1852 he got his PhD with a thesis *"About Poisoning by Copper"*, and he habilitated also in Göttingen. He gave courses on surgery, but also on eye diseases, medical jurisprudence, and public pharmaceutics. He also edited several books such as the *"Textbook of General Surgery"* (1858), *"The Gunshot Wounds and Their Treatment"* (1859), and many others. LOHMEYER was also a gifted pianist and cellist, who often performed music together with his friend JOSEPH JOACHIM, the famous violinist. In addition, with a large financial effort he had assembled a large collection of carpets and paintings.

With respect to surgery and ophthalmology, FERDINAND LOHMEYER followed a great tradition of the *Georgia Augusta*. Here CONRAD MARTIN JOHANN LANGENBECK had worked and taught, as a highly important surgeon and anatomist, who had succeeded to improve many operation techniques by careful anatomic studies and simplifications of the instruments. In 1804 he became *Extraordinarius*, and in 1807 he opened his own clinic for surgery and ophthalmology. In 1808 together with ADOLPH FRIEDRICH HEMPEL he was appointed to the Chair of Anatomy, and in 1814 to the position of *Ordinarius* for anatomy and surgery. To him ophthalmology owes a method for the cataract operation and a technique for the fabrication of an artificial pupil developed together with CARL FERDINAND VON GRAEFE, the father of the famous ophthalmologist ALBRECHT VON GRAEFE in Berlin. LANGENBECK had written popular anatomic textbooks, and he emphasized in particular the fact, that practical anatomic knowledge is a necessary requirement for the activity of a medical doctor or surgeon.

Also his nephew, BERNHARD VON LANGENBECK, one of the most eminent surgeons of the 19th century, had studied medicine in Göttingen, and in 1835 got his PhD with a prize-winning thesis about tumors of the eye. Later on he worked in Kiel and Berlin.

KURT MENDELSSOHN, a later student of NERNST, characterized his young wife with the following words: *"Emma Lohmeyer was one of the prettiest girls in Goettingen. In addition to the bloom of youth her face showed intelligence, a happy disposition and a keen sense of humour."* [Mendelssohn (1973): 42; 60]. Because of the early death of her mother MINNA LOHMEYER, already at the age of 16 EMMA had to manage the household of her not so affluent father after her oldest sister META had married. Therefore, she had learned to work disciplined and assiduously. She kept this habit during her whole life.

All five children of the NERNST couple were born in Göttingen. The birth of the first three happened during the years, which also professionally were highly important for WALTHER NERNST. In 1893, when his textbook *"Theoretical Chemistry"*, soon to become world famous, appeared for the first time, his first son was born, who was named RUDOLF after the uncle NERGER. In the following year, in which NERNST was appointed to the first *Ordinariat* of Physical Chemistry in Göttingen, the first much desired daughter HILDE was born on October 27, 1894. Five days before, her father had received the letter from Munich inviting him to become the successor of LUDWIG BOLTZMANN. The position of *Ordi-*

narius in Göttingen was connected with a separate Institute, which opened in 1896. During the same year the second son GUSTAV was born, named after the father of NERNST.

There followed two more daughters: EDITH in 1900 and ANGELA in 1903. So the youngest child similar to the oldest had been given a name by NERNST, which recalled the happy times at the demesne Engelsburg of the uncle RUDOLF NERGER. *Angela* is the feminine form of the Latin word *angelus*, meaning angel.

Both sons were killed during the First World War before they could finish their education. Their father suffered much because of this heavy loss. In 1917 NERNST sold the estate in Rietz near Treuenbrietzen in the province of Brandenburg, which he had purchased in 1907 (Fig. 4.4). He just could not stand any more the memory of the many happy days he had spent there together with his sons. The War Memorial of the church of the inconspicuous village carries the names of the sons of the great scientist.

Fig. 4.3 EMMA NERNST in 1899 with the children RUDOLF, HILDE, and GUSTAV.

In 1920 NERNST's oldest daughter HILDE married the chemist HEINZ CAHN, who had been a student of EMIL FISCHER and had contacts to the international banking community.

Also the husband of ANGELA, ALBERT HAHN, was of Jewish origin. He was a judge and owned parts of the assets from his family business, until it was expropriated by the Nazi Regime. Since HAHN was baptized and had distinguished himself during the First World War, his dismissal in 1933 was cancelled in 1935, until the anti-Semitic Nuremberg Laws became effective. For the people of Jewish origin, the latter eliminated the right for employment within the public service. Only in 1938 the family HAHN emigrated to England and one year later to Brazil. Shortly before the departure to South America in July 1939, ANGELA HAHN visited Germany once again for a short time. During this visit WALTHER NERNST saw his youngest daughter for the last time.

Fig. 4.4 Estate and church in Rietz during spring 2005.

Soon after the assumption of power by the NS Regime HILDE and HEINZ CAHN left their native country together with their children and emigrated to England. As a chemist and because of his contacts to banks, the professional integration as a foreigner was much easier for him than for ALBERT HAHN, being a German lawyer practically without any means.

As reported by KURT MENDELSSOHN [Mendelssohn (1973): 44; 64], February 25, 1900, the day on which the third daughter of NERNST was born, must have been a mild day of early spring, when her father noticed already the first flowers of the year, such that he let her be baptized by the name EDITH PRIMULA. If this is really true, NERNST had in mind probably the primrose *primula elatior*.

EDITH left her parents home in 1922, when she married the lawyer RUDOLF ERNST VON ZANTHIER in Greifswald. In 1912 he became *Dr. jur.* and in 1925 the *Dr. rer. pol.* was added. In 1920 VON ZANTHIER entered the agricultural administration of Prussia, and in 1926 he became instructor at the University of Greifswald. From 1929 until 1931 he was consultant for agriculture of the Central Government of China in Nanking. Their daughter born in 1923 was given the name ANGELA like her aunt who was 20 years older.

In 1942 in his necrology for the great colleague, in a highly courageous way despite the political situation of the time, MAX BODENSTEIN spoke about his marriage and the heavy blows NERNST had received regarding his children: *"In 1892 in Göttingen with Emma Lohmeyer ... he found the partner for life, standing by his side understandingly and faithfully, during times of glory, but also during the days of deep sorrow, from which he could not escape. Five children originated from the marriage, two sons, who at their most promising age became victims of the First World War, and three daughters, two of whom with their families live far away, impossible to reach because of the difficult times, such that during the last years of the deceased only one of the children could be near the parents, a heavy burden, which he also felt as such, although he never complained about it."* [Bodenstein (1942a): 81].

4.5 The Textbook "Theoretical Chemistry from the Standpoint of Avogadro's Rule and Thermodynamics"

The position of *Extraordinarius* allowed NERNST, who was already well known in the scientific community because of his achievements, to found an academic school. In addition to his direct corresponding activities in Göttingen, to a high degree also his textbook *"Theoretical Chemistry from the Standpoint of Avogadro's Rule and Thermodynamics"* contributed to its realization in only a few years. His textbook appeared for the first time in 1893 and spread worldwide NERNST's fame as a scientist and academic teacher.

The prehistory of this textbook is typical for the status which had been reached by the physical chemistry within the field of chemistry and by the corresponding contributions of NERNST during the last decade of the 19th century. It was the goal of the young physicist in Göttingen, to make this modern discipline better known. Already on June 6, 1890 in a letter to WILHELM OSTWALD, NERNST could add a *post scriptum*: *"I have been asked by Dammer to write a physical chemistry of 20 printed sheets for the 'Dictionary of anorg. Chemistry' (an undertaking à la Beilstein). The fee is good, and I am nearly willing – what would you say? Also I am supposed to write for the dictionary of Hell."* [Zott (1996): 13–14].

Apparently, the contribution to the *"New Dictionary of Chemistry"*, which was edited by CARL HELL and CARL HAEUSSERMANN after the death of HERMANN VON FEHLING, and which appeared in Stuttgart in ten volumes from 1874 until 1930, did not materialize. On the other hand, in 1892 the *"General Part"* (*«Allgemeiner Teil»*) from NERNST in fact became the introduction to OTTO DAMMER's inorganic-chemical version of FRIEDRICH KONRAD BEILSTEIN's handbook of organic chemistry [Nernst (1892b)]. It then became the germ-cell of the famous textbook *"Theoretical Chemistry"*. In the same year there appeared also the first essay *"Physical Chemistry"* written by NERNST for the *"Yearbook of Chemistry"* edited by RICHARD MEYER, in order to popularize his scientific discipline [Nernst (1891)].

In critical reviews WILHELM OSTWALD praised both contributions [Ostwald (1892)]. In his first book, the first part of DAMMER's handbook, according to his own words NERNST had tried to treat the actual status of physical chemistry and its most important goals, based on the leading aspects as judged by him. This was accomplished in such an excellent way, that OSTWALD proposed to his former assistant to expand this chapter into a separate textbook.

In fact, this textbook appeared already in 1893 in its first edition with the title given above [Nernst (1893b)] (Fig. 4.5). According to the understanding at the time, the title 'Theoretical Chemistry' indicates General and in particular Physical Chemistry [Bartel (1988)]. By mentioning the rule of AMADEO AVOGADRO and by including thermodynamics NERNST touches exactly upon the leading aspects of these directions of chemistry. The former, according to which for the same values of pressure and temperature the same volumes of an (ideal) gas always contain the same

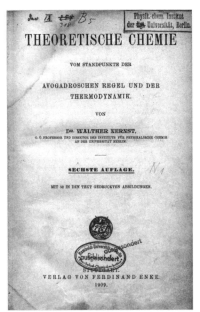

number of particles, represents *"a nearly inexhaustible horn of plenty donated by the molecular theory"* [Nernst (1893b): IX] for the atomic or molecular concept of matter. At the time the latter was criticized and opposed even by some of the leading scientists such as ERNST MACH and WILHELM OSTWALD. Hence, NERNST had taken clearly the side of his teacher LUDWIG BOLTZMANN, who was much attacked regarding this matter.

Fig. 4.5 Title page of the 6th edition of the «*Theoretische Chemie*» by WALTHER NERNST.

Fifty years after the first publication of the book JOHN EGGERT, a student of NERNST in Berlin, well characterized this exposition of thermodynamics and of the atomic structure of matter with the words: *"Today the emphasis of those two mighty pillars, upon which Nernst constructed his theoretical frame-work, may sound nearly trivial – at the time these concepts were bold and guaranteed for the book 'an independent and unique position within the literature', as it was expressed by Ostwald in his review."* [Eggert (1943a): 413]. This remark by OSTWALD must be understood from the background, that still in 1899, instead of "molecular" or "atomic weight", he spoke of the "weight of the compound" or at best of the "hypothetical atomic weight". However, contrary to the hypothesis of the atoms, since the works of HORSTMANN and VAN'T HOFF the importance of thermodynamics for physical chemistry was not in doubt any more.

The contrast to his *"boss"* in Leipzig, OSTWALD, who according to NERNST had introduced him to physical chemistry, and the general agreement reached in Graz regarding the hypothesis of the atoms, may have been the main reason, that in the book it is stated: *"The author dedicates this book to Prof. Dr. Albert von Ettingshausen in Graz in faithful memory of his learning and wandering years"*.

In the preface of the textbook NERNST emphasized: *"At any rate I believe that at present an epoch of quiet but successful work has arrived for the scientists in the field of physical chemistry; the ideas are not only available, but are also ripe up to a certain completion. Lucky new ideas always lead to fruitful results, since they are followed by a time of increased creativity, and so presently one sees with rare unanimity the research of the different developed nations being eagerly and successfully occupied with the construction of the frame-work of theoretical chemistry. At such times the need for a presentation of the leading ideas, teaching the student and advising the scientist, becomes particularly vivid."* [Nernst (1893b): VII].

In fact, during the subsequent three decades NERNST had worked on and also with his book, which since its first appearance always showed a balanced ratio between experiment and theory. The excellent selection of the material and its arrangement, as well as the clear, impressive, and fresh presentation always remained, if new and current results of the research appeared. The author himself contributed to the latter to an appreciable degree and did not hesitate to express his explicit personal criticism. The book contributed significantly to the fact that chemistry became recognized as an exact science. For a long time not only physical chemists and physicists, either as students or as researchers, have benefited greatly from the *«Theoretische Chemie»*, but also other people such as engineers, pharmacists, and medical people working in other directions of chemistry. The textbook provided them with the capability, to extract the general underlying laws from the experimental data and, in turn, to derive predictions from them.

The success of the book is demonstrated by its many editions, which continuously accomplished to integrate the increasing amount of the research results. Therefore, the editions always increased in size, as can be seen from Fig. 4.6.

The book enjoyed a high reputation also in foreign countries. Already prior to its second edition, in 1895 there appeared an English translation done by CHARLES SKEELE PALMER [Nernst (1895)]. It was followed by four additional translations into this language by other people, and since 1911 by two French editions. It is interesting, that at the time in England in contrast to the USA there were quite a few older scientists, who rejected the recent developments in chemistry treated by NERNST, and polemized against them. Nevertheless, the modern German books covering this field found acceptance and sold well.

In 1912 *"an authorized piano-score of the Nernst"* [Eggert (1943a): 413] was produced, as it was jokingly referred to. Here we mean the *"Guide-book of Theoretical Chemistry"* («*Leitfaden der theoretischen Chemie*») written be the physical chemist WALTER HERZ from Breslau [Herz (1912)] and following closely the textbook of NERNST. Parallel to its original model it grew from 271 pages in the beginning to 309 pages of its last edition prepared together with LOTHAR LORENZ, a student of HERZ. It is typical of the *"guide-book"*, that its third edition is identical to the second edition of HERZ' book *"Physical Chemistry as the Basis of the Analytical Chemistry"*.

Edition(s)	Year	Pages
Dammer	1892	359
1st	1893	589
2nd	1898	703
3rd	1900	710
4th	1903	749
5th	1906/07	784
6th	1909	794
7th	1913	838
8th–10th	1921	896
11th–15th	1926	927

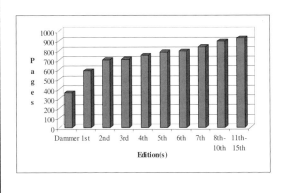

Fig. 4.6 Statistics of the editions of the «*Theoretische Chemie*» by WALTHER NERNST ('Dammer' = [Nernst (1892b)]).

Sometimes the scientific community looked critically at the considerable subjectivity of NERNST in the selection and evaluation of the material included in his textbook. The physical chemist RUDOLF WEGSCHNEIDER from Vienna, who earlier had also worked with OSTWALD, is an example. His remark of 1913 is characteristic for the scientific personality of NERNST: *"In the case of the books of most other authors ... one would not feel such a strong emphasis of the personal opinion to be an advantage. It is partly different, if the author is the man who presently without any doubt represents the leadership in the physical-chemical research. In this case any insight into the world of his ideas is interesting, and even also for those, who do not let their judgments be determined only by his authority."* (quoted after [Eggert (1943a): 413]).

Colleagues in the field, representing the pure chemistry, like Professor HUGO ERDMANN, working in Halle at the river Saale and in Charlottenburg (near Berlin at the time) in the fields of inorganic, organic, and physical chemistry, were missing the treatment and clarification of currently discovered phenomena in chemistry. From a «*Theoretische Chemie*» one expected instructions and hints for these, in particular in connection with the chemical-synthetic work and with the development or improvement of the technical processes. In the eyes of these people NERNST's book offered a physical chemistry without any use for the practical chemical problems, being only the result of some hobby with an academic character. In particular, the level-headed physical concept of NERNST of atoms, ions, and electrons did not correspond to the ideas frequently expressed by chemists regarding this matter.

NERNST could not and did not want to include the many theoretically unexplained problems and individual facts in his textbook, in particular since it increased in volume continuously. On the other hand, he included in his book such developments in their early phase, which should become highly important for the frame-work of physical chemistry. To these belong quantum theory, radioactivity, electron theory, the electrostatic theory of the solution of strong electrolytes, and others. *"Of course, in a treatment of theoretical chemistry different chapters from physics and chemistry must find room; in the final analysis the former just represents the essence of what the physicist necessarily must know about chemistry and the chemist about physics, except for the case in which the physicist wants to perform physics and the chemist chemistry explicitly as a specialist. Therefore, the development of physical chemistry into a separate branch of science means (and I want to emphasize this) not only the creation of a new science, but instead much more the combination of two sciences which so far were rather separate."* [Nernst (1893b): VIII]. With these words, already in 1893 the physicist WALTHER NERNST described his opinion, his interest, and his program, which he had adopted not only in connection with his textbook, and which could be summarized by the motto of his lecture course *"The physical chemistry pursues the aims of chemistry with the weapons of physics."* [Eggert (1943a): 414].

At this point one should also mention the merits earned by the publishers founded by FERDINAND ENKE in Stuttgart in producing such books as NERNST's «*Theoretische Chemie*» or HERZ' *"Guidebook"* etc. in connection with the development of physical chemistry.

During the third decade of the 20th century the interest of NERNST turned to other fields away from physical chemistry, and the editions of his textbook were terminated. Now, characteristic of his important role within this discipline, there were the corresponding works of two of his former students in Berlin, which continued his «*Theoretische Chemie*», in some sense exemplifying the statement of ARNOLD SOMMERFELD, that an academic teacher is as good as he can make himself replaceable. Here we have in mind the *"Compendium of Physical Chemistry"* («*Grundriss der physikalischen Chemie*») by ARNOLD EUCKEN [Eucken (1922)], which saw many editions and later was republished by EWALD WICKE, and the *"Textbook of Physical Chemistry"* by JOHN EGGERT, which from the beginning was produced together with LOTHAR HOCK, EGGERT's colleague in Giessen [Eggert (1926)].

Contrary to the different opinion expressed by JOHN EGGERT, last not least it was the point of view documented in the textbook of NERNST about the relation between physics, mathematics, and chemistry, which prepared the fact, that in the sense of a paradigm change just his student

ARNOLD EUCKEN conceptually developed chemical physics from physical chemistry in form of a new textbook [Eucken (1930)]. Today 'chemical physics' still has the same meaning as 'theoretical chemistry', whereas in the sense of the notion of 'theoretical physics' the latter still represents a promising research subject in spite of some progress [Bartel (1988)].

4.6 The First Professorship and the Establishment of a Chair of Physical Chemistry

When in 1893 the textbook of NERNST appeared in print for the first time, the university career of its author had started already three years earlier with the *venia legendi* and was very soon to reach the highest academic level. On March 10, 1890 the Faculty of Philosophy of the University of Göttingen informed the Chancellor of the University, that based on the eleven papers which so far had appeared in print, including the habilitation thesis, and on *"the public lecture About the Role of Electric Forces in Chemical Reactions"* given on the previous day, *"Dr. phil. Walther Nernst ... received the venia legendi for the field of physics"* [UAHUB: I, 1]. In this way NERNST had obtained the right to teach at the Physical Institute as a lecturer. Its *Ordinarius*, EDUARD RIECKE, had invited him to Göttingen with the promise to get the position of *Extraordinarius*.

Before RIECKE could realize his promise, NERNST had the opportunity to become *Extraordinarius* in Giessen. Regarding this matter, on March 1, 1891 he mentioned to WILHELM OSTWALD: *"Regarding your inquiry about Giessen, today I want to let you know very briefly what happened so far. 8 days ago Prof. Himstedt asked me, if I would perhaps accept a professorship (Extraordinariat) in G.; there will be proposed: Nernst, Arrhenius, Beckmann; at least H. and Naumann would agree. Of course, I answered positively, that I would like to become Professor of Physical Chemistry, however, that I could not say, if I would leave from here, and I informed Prof. Riecke about the letter."* [Zott (1996): 25].

In the 19th century the University, founded in 1607 as the Lutheran *Ludoviciana* in Giessen by the Landgrave LUDWIG V from Hesse-Darmstadt, had acquired a high international reputation. Because of JUSTUS VON LIEBIG, who worked from 1824 until 1852 at this university named after him today, Giessen had become the world center of Chemistry. Also WILHELM CONRAD RÖNTGEN worked here from 1879 until 1890.

Because of this background it could have appeared attractive for NERNST, to represent Physical Chemistry as a researcher and instructor of this interdisciplinary field between physics and chemistry just at this University of the Province of Hesse-Darmstadt. However, he preferred to remain in Göttingen, if the position of *Extraordinarius* promised by RIECKE could be realized. In fact, already on March 3, 1891 RIECKE had contacted directly the Department of Culture in Berlin regarding this matter, and he emphasized, that with the establishment of the position of *Extraordinarius* for NERNST, it would be guaranteed, that he would remain at the University of Göttingen continuing his activities in the field of Physical Chemistry. If the offer from Giessen would become official, NERNST himself also wanted to ask the Department of Culture in Berlin, if he may decline an offer from outside Prussia. In the case of a positive answer which he had hoped for, he could expect an early advancement in Göttingen.

On August 14, 1890 NERNST received from Darmstadt the question if he intended to accept the offer from the University of Giessen, and he was asked to reply shortly. Twelve days later NERNST signed two documents in Berlin. In one document he committed himself that after an appointment as *Extraordinarius*, at least for six semesters he would not leave the University nor would he change his position at all without the approval of the Secretary. The other document covered the details of the appointment, which NERNST approved. Here it was agreed that he would take the position as assistant at the Physical Institute as long as RIECKE was its director.

The document from the Prussian *Ministry of the Spiritual, Teaching, and Medical Matters*, which confirmed the appointment as *Extraordi-*

narius, is dated from September 7, 1891. This position within the Philosophical Faculty was transferred to NERNST *"with the commitment to represent Physical Chemistry in courses and laboratory sessions"*, [UAHUB: I, 12]. In order to be able to pay NERNST's annual salary of 2500 *Reichsmark* effective from September 1, 1891, the position of *Extraordinarius* had been transferred from the Medical to the Philosophical Faculty. In addition, NERNST received annually 540 *Reichsmark* as a regular allowance for housing expenses. The young Professor was sworn-in on October 12, 1891.

On September 15, 1891 from the «*Dom. Engelsburg b. Graudenz*», where he spent his vacation, NERNST reported to OSTWALD: *"I just have obtained my appointment to Extraordinarius in Göttingen, with the same salary, which I had been offered from Giessen, and in addition with a 600.– RMk budget for physical-chemical exercises etc. One cannot ask for more!"* [Zott (1996): 30]. On October 29, 1891 NERNST wrote to his friend ARRHENIUS, who in the meantime had obtained the position of lecturer at the *Hoegskola* of Stockholm: *"... however, one must leave it to the title of Professor, that one feels very well with it, in particular, here in Gött., where the step from lecturer to Prof. is quite large, because according to an old local tradition the caste-spirit has opened a certain gap in-between."* [Zott (1996): 30–31].

The *Extraordinarius* position in Giessen was obtained by ERNST BECKMANN, who in the previous year had been appointed already in Leipzig as *Extraordinarius* of Physical Chemistry. In this way the strategic plan of NERNST of further establishment and expansion of Physical Chemistry by the appointment of suitable young scientists was realized to a large extent: ARRHENIUS in Stockholm, BECKMANN in Giessen, and NERNST himself in Göttingen. The possibility of an appointment in Giessen, certainly also influenced by NERNST, probably caused the Swedish authorities to offer to SVANTE ARRHENIUS still in 1891 the position as lecturer of Physics in Stockholm. Because of skeptical and doubtful feelings of his Swedish colleagues and in particular also of Lord KELVIN regarding his qualifications, he was appointed as Full Professor at this University only in 1895. Already in 1892 ERNST BECKMANN left Giessen in order to accept the offer as Full Professor of Pharmaceutical Chemistry in Erlangen. In 1897 he obtained the Chair of Applied Chemistry especially established for him in Leipzig, and in 1912 the position as Full Professor at the University of Berlin.

Already three years after his appointment as *Extraordinarius* NERNST reached the highest level in an academic career: Full Professor. He was only thirty years old. This high recognition was due to LUDWIG BOLTZMANN, the teacher of NERNST in Graz, who since 1890 occupied the position of *Ordinarius* of Mathematical Physics in Munich, but in 1894 wanted to change again to Vienna. Apparently, for some time BOLTZMANN had lost sight of the career of NERNST, since on July 22, 1891 he wrote to WILHELM OSTWALD: *"I would like to ask you to let me know, where Mr. Nernst is at the moment and in which position."* [Körber and Ostwald (1963): 23]. One can only presume, that in 1894 he had proposed him to the Faculty in Munich as his successor. At any rate, on October 23, 1894 NERNST could report to OSTWALD: *"But yesterday I was in the lucky possession of an offer from Munich to Boltzmann's position; the offer came as quite a surprise, and the delight of my wife and myself was very large."* [Zott (1996): 77].

However, already on May 9, 1893 OSTWALD had hinted to NERNST about an offer from Munich: *"P.S. I have completely forgotten to tell you, that Baeyer asked me in Munich, if and for what you could be won over for Munich (as Extraordinarius). I felt, that the chance for you to be together with Boltzmann would make the winning of your person for Munich much easier."* [Zott (1996): 61].

WALTHER NERNST was enthusiastic about his opportunities for advancement in Munich, in particular, since as *Ordinarius* he would obtain working space in the newly created Physical Institute. Only the pregnancy of his wife kept him for the moment from traveling to the Capital of Bavaria, *"in order to conclude the matter and to present myself to the Secretary."* [Zott (1996): 77]. On the other hand, the University of Munich wanted NERNST to start his position as Professor already on November 1, 1894.

As in 1891 in the case of the position as *Extraordinarius*, the Philosophical Faculty of the Georg-Augusts-University and in particular the Dean FELIX KLEIN acted immediately to try to not lose NERNST to Göttingen or at least not to Prussia. On October 25, 1894 it composed two letters regarding this matter. The shorter letter is addressed to the University Curator ERNST HÖPFNER in Göttingen with the following text: *"Re[:] Promotion of the Extraordinarius Professor Dr. Nernst to Ordinarius Professor. I am honored to present to Your Honor in the attachment an idea, which the Philosophical Faculty wants to propose to Mr. Secretary regarding the offer to Professor Nernst from Munich. We hope that based on the steps here outlined it may be possible to turn the threatening loss into a large gain for our University. Asking you to strongly support our arguments at the Mr. Secretary, respectfully the Dean of the Philosophical Faculty Prof. Dr. F. Klein."* [UAHUB: I, 14]. The long document, the attachment mentioned, was presented to the Secretary *"of Spiritual, Teaching, and Medical Matters"*, ROBERT BOSSE in Berlin.

Its content demonstrates the great reputation NERNST had already enjoyed at the time such that this document shall be reproduced here nearly completely [UAHUB: I, 25–28]:

"The Philosophical Faculty permits itself to present to Your Excellency respectfully the request for promotion of the Extraordinarius Professor Dr. Nernst to Ordinarius in the Philosophical Faculty of our University. Your Excellency knows that Professor Nernst has received an offer of the Position of Ordinarius of Theoretical Physics in Munich to the Chair being vacant because of the departure of Boltzmann. The imminent danger, that he accepts such an honorable offer, can hardly be eliminated by only offering him in Göttingen an equivalent; therefore, if it would be possible to generate here a position, in which he could pursue higher goals in the field of his own talent and interest and could take up new and larger tasks, we believe that it would be possible to keep him at our University.

However, his loss would mean a severe damage of our scientific life. He is one of the most successful and most lucky scientists, one of the leading minds in a field, which during the recent years, not in the least due to his contributions, has developed in an unexpected way, which continuously supplies the sister sciences of physics and chemistry with plenty of ideas, and which also in a practical sense gains an ever increasing importance. In an excellent way Professor Nernst was able to introduce Physical Chemistry into our University by means of courses and laboratory exercises; because of the clear and well-structured presentation in his book 'Theoretical Chemistry from the Standpoint of Avogadro's Rule and Thermodynamics' he has become a teacher of his science, acting beyond the borders of our country, and students arrive already from far away in order to devote themselves to the new science under his personal guidance. There exists the germ for a rich and fruitful development, and only with sorrow we would see the one leave, who planted it and who is mostly destined to harvest the fruits. ... However, the loss of Nernst would hit us even harder, since among the available younger scientists there is nobody, whom we could consider immediately as successor.

For the direction and the success of Nernst's scientific activity it was extremely important, that before he turned to Physical Chemistry, he occupied himself with investigations of complex phenomena of electricity. In this way he gained a most perfect experience with the methods of electrical measurements, the most exact knowledge of the theoretical concepts in the field of electricity. These bore their most beautiful fruits in his papers on the electromotoric effects and on the dielectric constants of the salt solutions, papers which were important because of the achieved results as well as highly promising in their further development. Within the scientific community there will be no doubt about the fact that exactly in the field of electrochemistry rich treasures still will be found. None of our younger scientists shows the capabilities necessary for success in the same way such as Nernst, nobody would be suitable like him to become the head of an Institute of Electrochemistry.

The establishment of such, the detailed study of electrochemistry in a high and comprehensive spirit, this is the task, through which we hope to bind Professor Nernst to our University; he himself has said that he would stay here, if such activities would open up. On the other hand, the establishment of an Institute of Electrochemistry is a requirement of the time; its achievements will expand the scientific knowledge, they will be beneficial both for the technology and the well-being of our country, they will guarantee the leadership of Ger-

many also in this field, which it has attained for a long time in the field of general Chemistry. This coincidence is so excellent and is so directly connected with the most important interests of our University, that we cannot refrain from indicating this most strongly, in order to beg Your Excellency urgently to look favorably at our desires.
The Philosophical Faculty. The Dean. (signed) Prof. Dr. F. Klein."

Then NERNST traveled via Berlin to Munich. There he obtained time for reflection until November 20 or even until December 1, 1894, if it was necessary. When he returned to Göttingen on October 28, one day after the birth of his daughter HILDE, he brought along surprising news from Berlin: *"Furthermore, presently I am being considered (difficult to believe! Please do not talk about this) for Berlin. Now we ionists are really on top!"* [Zott (1996): 79].

Already on the following day WILHELM OSTWALD answered NERNST's letter from October 31, 1894 to him: *"Recently, Fischer told me in Berlin with an important look: 'We think very highly about Nernst'."* [Zott (1996): 79–80]. Also EMIL FISCHER had indicated to OSTWALD, that the Faculty in Berlin wished to see NERNST as the Chair of Experimental Physics, succeeding AUGUST KUNDT who had just died. Because KUNDT belonged to the influential "anti-ionists", the remark by NERNST we have just quoted emphasizes, that, indeed, in Berlin one wanted to see Physics being represented by leading scientists defending these modern concepts, which gained more and more support due to their evidence of truth. Therefore, also the "ionist" VAN'T HOFF was discussed as successor of KUNDT.

At any rate, at the end of the year 1894 in Berlin, there was much interest to win NERNST for the Prussian capital. Director HENRY THEODORE VON BOETTINGER and FRIEDRICH ALTHOFF, who had been extremely beneficial for the development of the sciences and mathematics at the Prussian Universities, were highly active in this direction. They asked WILHELM OSTWALD to join this effort. The latter emphasized repeatedly, that NERNST would like to go to Munich, and because of the highly attractive conditions for him over there, a substantial offer would be needed, if he should stay in Berlin or Göttingen.

Also the proposal of the Faculty in Göttingen, to establish an Institute of Electrochemistry for NERNST, was pursued in particular by ALTHOFF with support from OSTWALD. So on November 16, 1894 NERNST asked OSTWALD on behalf of ALTHOFF, by all means to come to Göttingen because of a conference in justifying an Institute of Electrochemistry. Since in the meantime he knew, that he was considered perhaps for an offer from Berlin, he added to his request: *"Of course, I feel excellent, I pass my restless time with a lot of 'Allotria'. After all, it is a peculiar situation, that I do not know, if next Semester I will give a course on Physical Chemistry, Mathematical Physics, or finally Experimental Physics, and I cannot decide between three highly different Institutes."* [Zott (1996): 83].

After the meeting in Göttingen with OSTWALD, ALTHOFF asked the latter to explain to him in a document, which can be understood by a lay person, the scientific and economic importance of Physical Chemistry and Electrochemistry, the need for these fields to be represented also at Universities, and the irresponsibility, not to keep NERNST available for these sciences. OSTWALD answered in terms of a longer essay *"The German Chemistry, the Physical Chemistry, Electrochemistry, and Nernst in Göttingen"*, in which he roughly explained what had been expressed already by the Philosophical Faculty and by FELIX KLEIN. Although NERNST could represent Physical Chemistry, Theoretical Physics, and Experimental Physics in an excellent way, his particular talents concerned the first field and electrochemistry. To remove the activity and the unique power of NERNST from these rapidly developing fields, which are extremely important for the German industry, would interrupt their development because of this great loss. *"Only in order to avoid this, I intended to refrain from the offer to Nernst from Berlin, which otherwise I would support in all respects, and I have no doubt that also the local Philosophical Faculty would appreciate the non-consideration of its wish, as long as Nernst remains available for Physical Chemistry."* [Zott (1996): 87].

Dated December 20, 1894 the Secretary of *the Spiritual, Teaching, and Medical Matters*, JULIUS ROBERT BOSSE, sent the generally desired note *"to the Royal Extraordinarius Professor Dr. Walther Nernst Esq. in Göttingen ... that His Majesty the Emperor and King most gracefully has been pleased to appoint you to Ordinarius Professor in the Philosophical Faculty of the University of Göttingen. By sending you attached the certificate of appointment effected at the highest level on December 11 of this year, I commit you to represent the Physical Chemistry and in particular also the Electrochemistry in courses and laboratory sessions and to take up the directorship of the Institute to be founded in these fields."* [UAHUB: I, 38].

On January 1, 1895 even the *Vossische Zeitung* (VOSSian Newspaper) reported on this event in Göttingen: *"At the University of Göttingen a University Institute of Physical Chemistry will enter life in the near future. Its head will be Prof. Walther Nernst. So far Dr. Nernst performed his work in the University Institute of Physics. However, he has tied his stay in Göttingen to the condition, that for his discipline a separate Institute will be established."* After the Institute of NERNST to be founded has been discussed next to the other Institutes existing already in Germany, namely that of HANS LANDOLT in Berlin, of WILHELM OSTWALD in Leipzig, and in some sense also those of VICTOR MEYER and LOTHAR MEYER in Heidelberg and Tubingen, respectively, a little later the newspaper observed: *"Presently Nernst is one of the most outstanding experts of Physical Chemistry."* The newspaper clips were added *"To the personal documents of Prof. Nernst. Göttingen, 1./11. 95."* [UAHUB: I, 42].

So in 1894, in its so-called *"black year of physics"*, Germany had to endure the death of three of its great physicists, HEINRICH HERTZ, HERMANN VON HELMHOLTZ, and AUGUST KUNDT. However, at its end it was allowed to see the bright rise of WALTHER NERNST.

4.7 The New Institute of Physical Chemistry and Electrochemistry

The establishment of a separate Institute together with its directorship and the position of *Ordinarius* of the corresponding field meant for NERNST reaching a summit in an academic career, since now there was the opportunity of an almost unhindered unfolding of his own scientific ideas. Hence, in the sense of ARISTOTLE, the potentially possible (δυνάμει ὄν) did exist, which in itself carries the seed for the actual reality (ἐνεργείᾳ ὄν), which in Göttingen and later in Berlin because of his genius NERNST could form to reality (ἐντελέχεια).

In fact, in Göttingen the working conditions for NERNST were good from the beginning, since already with the appointment to *Extraordinarius* in October of 1891, in the Institute of RIECKE a small section of Physical Chemistry was established for him.

In 1894 together with the position of *Ordinarius* NERNST had been promised a separate Institute. Already on Easter of the following year in a few provisorily prepared rooms, NERNST could start with the operation of the Institute and, hence, could have a preliminary dedication. Such a short term was owed to the engagement of the Prussian Secretaries of teaching and of finances, JULIUS ROBERT BOSSE and JOHANNES VON MIQUEL, respectively, and also to that of FRIEDRICH ALTHOFF, the University Curator ERNST HÖPFNER, as well as to Professors in Göttingen, namely the mathematician FELIX KLEIN, the physicist EDUARD RIECKE, and the chemist OTTO WALLACH, to the representative in the German Diet (*Landtag*) HENRY THEODORE VON BOETTINGER, and last but not least, to WILHELM OSTWALD.

Fig. 4.7 NERNST's Institute in Göttingen before 1906 (top) and in 1991 (bottom).

For this purpose was bought a lot of 3300 m² area of the former University Chancellor ADOLF VON WARNSTEDT including the existing villa (Fig. 4.7) and a small additional building in *Bürgerstraße 50*, located close to the Chemical Institute. During the year 1895 the villa was severely modified, and a larger and a smaller annex were constructed. Already at the end of the year the new space could be put to use at least partly. The total costs were 165 000 *Mark* consisting of 63 000 *Mark* for the purchase, 42 000 *Mark* for the additions, and 60 000 *Mark* for the interior equipment. On the first floor of the Institute the residence of the Director was furnished, into which NERNST moved during August of 1895. The layout of the space in the new Institute was well considered by NERNST (Fig. 4.8).

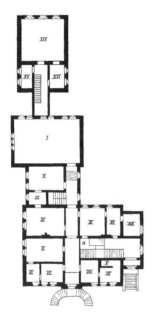

Fig. 4.8 Plan of the ground floor of the Institute of NERNST in Göttingen (from [Nernst (1896a)]):
I: Hall for physical-chemical laboratory sessions, sometimes auditorium; II: room for chemical work; III: preparation room; IV: auditorium; V: library and scale room; VI: room for special studies; VII: workshop; VIII, IX: same as VI; X: room which can be darkened by a curtain for spectrum analytic and photographic work; XI, XII, XIII: office of the Director; a: partition for keeping expensive glass instruments.

Important equipment was also placed in the basement. There existed a machine room, which contained a gas motor, a dc-, ac-, and a three-phase generator from SIEMENS, and an air liquefier from LINDE, which allowed experiments down to temperatures of $-190°C$. Furthermore, there was a room for storage cells, which supplied voltages between 2 and 72 volts to the laboratories, a blower room, a stockroom for chemicals, a room for storing coal, and a packing room. Two smaller and four larger rooms with extra height could be used for special physical-chemical, electrochemical, and physical experiments, and two additional rooms for chemical work. One laboratory, completely surrounded by the other rooms in the basement, was placed lower than the others and carried a double ceiling, such that its temperature could be kept well constant. It was suitable for experiments with instruments which could not be placed in a standard thermostat. Hence, at the time the Institute had the most modern equipment.

The duty regarding his care for the newly established Institute, accepted by NERNST at the appointment as *Ordinarius*, was taken by him very seriously. According to its importance and special direction, which is expressed by its name *"Institute of Physical Chemistry"* with the addition *"and in particular Electrochemistry"* [Nernst (1896a)], on November 25, 1895 NERNST registered it for admission to the German Electrochemical Society, founded just in the previous year, with its first Chairman WILHELM OSTWALD. Prior to this, he had completed a visiting trip for information to the existing similar facilities in Darmstadt, Heidelberg, Munich, Erlangen, and Wurzburg.

The effort for improving the Institute was continued in 1896. So in the middle of August of that year he traveled to the *"Mechanics Day"* in Berlin, in order to collect the organized exhibition material, which could be beneficial for his Institute. Even only after the solemn dedication ceremony on May 15, 1897 NERNST stated in a letter to OSTWALD: *"I myself hope, after I have been completely occupied for nearly two years by the start of the operation of the Institute, that now I can take up again a few things."*, where, however, he had to admit already: *"The Institute is too crowded, soon I will have to add some room or I must leave the official residence, if your new Institute does not provide some relief."* [Zott (1996): 120–121].

In fact, soon the Institute became too small, since the modern field of Physical Chemistry in general and especially its in the meantime famous representative in Göttingen attracted students from all over the world to this location of learning and research. The remark on this matter just quoted indicates that NERNST's ambition was strongly directed to the goal to serve the development of his discipline. Also it is typical of him that he places the practical material support next to the ideal one. So he not only suggested a necessary expansion of his Institute, but in 1898 he financed an annex with three rooms each on the ground floor and the basement. He took the necessary sum of 40 000 *Mark* from the profit he made with his lamp. This contribution was so large, that the approval of the Emperor was needed for being accepted: *"Following the report of the 8th of the month I want to present to the University of Göttingen my sovereign approval for the acceptance of the forty thousand Mark donated to it by Professor Dr. Nernst in the interest of the improvement of its Institute of Physical Chemistry."* [UAHUB: I, 84].

As we can see from the note of the *Vossische Zeitung* quoted above, the new Institute in Göttingen represented a distinguished special facility, which called for a suitable dedication ceremony at the right time. During March of 1896, NERNST was considering the end of the following month for this purpose. However, on April 12, 1896 when he had just returned from a trip to the Riviera, he talked about a shift of this date. In fact, the big event took place only on June 2, 1896. On this day from Berlin there came the Secretary of Teaching, JULIUS ROBERT BOSSE, and the councilor in the Ministry of Culture, FRIEDRICH ALTHOFF. The University of Göttingen was represented by its Curator ERNST HÖPFNER and its Prorector CARL LUDWIG VON BAR. From the colleagues and friends the following people attended as guests: SVANTE ARRHENIUS from Stockholm, ERNST BECKMANN from Erlangen, JACOBUS HENRICUS VAN'T HOFF from Berlin, MAX LE BLANC from Leipzig, WILHELM BORCHERS from Duisburg, FRIEDRICH WILHELM KÜSTER from Marburg, CARL HEIM from Hannover, and GEORG KAHLBAUM from Basel. Unfortunately, the directors of the two important German Institutes of Physical Chemistry in Berlin and Leipzig, HANS LANDOLT and WILHELM OSTWALD, respectively, were unable to come. Corresponding to the dignity of the occasion, WALTHER NERNST focused his speech on the goals of Physical Chemistry, discussing many interesting and remarkable aspects. In the beginning he said: *"This day deserves especially the notion of a day of joy, because it represents an important new proof of the fact, that an inner reunification has happened between two fields of science which were rather separated so far. I say 'reunification', since the separation of these fields of science does not date back to the Old Ages."* [Nernst (1896b): 2]. By discussing the relation between physics and chemistry and its historic development, he raised the question, where these sister sciences would differ from each other. He rejected the classification based on the hypothesis of atoms, according to which chemistry deals with the formation of molecules from atoms, and physics with the molecules generated in this way. He also rejected the opinion, which associates with chemistry the subject of the transformation between materials, and with physics the subject of the changes during which the material properties of the system in question remain the same. However, NERNST could not find a satisfactory answer, and he summarized: *"A definite distinction cannot be found easily."* [Nernst (1896b): 5].

On the other hand, a common feature can be pointed out easily: *"In contrast to all other branches of the sciences,"* chemistry and physics are *"... constructive sciences"*, different from the *"descriptive ones"*, to which belong, for example, astronomy, physiology, and zoology, because: *"Only physicists and chemists create their own systems, and depending on the theoretical concepts, which they test and between which they want to come to a decision, from the unformed raw material of the outside world they build for themselves the particular system, which from their viewpoint looks to them just highly worthy of careful research."* [Nernst (1896b): 5]. If a scientist wants to *"set himself to the task to know the general laws, which dominate the existence and the developments of the outside world, ... hence, he will become either a physicist or a chemist, or from the standpoint of our Institute we might add a physical chemist."* [Nernst (1896b): 5–6].

Furthermore, NERNST emphasized, that a separating feature of the occupation with physics and with the rich chemistry lies in the different method of the two sciences. *"If the physicist and the chemist, each in his own field and with its methods, is working, a large area between the two remains unexplored, namely all which can only be treated by the simultaneous application of both methods of working. And that here physical chemistry finds an immensely large and worth-while field for its activity, is clearly demonstrated by the scientific achievements of the last dec-*

ade." [Nernst (1896b): 9]. The young age of physical chemistry results from the fact, that only within the last fifty years large and general laws have been found, *"if we think of the laws of thermodynamics, of the principles of the kinetic theory of gases, of the Maxwell-Hertz equations of electrodynamics, of the theory of the electrolytic phenomena, or of the system of absolute units, which forms the solid base of the measuring doctrine of nature."* [Nernst (1896b): 9]. In this way NERNST listed nearly all of those subject areas, to which he himself contributed significantly by his actions.

From the second half of the past century NERNST could also name fundamental advances for the chemical science: the periodic chart of the elements, the stoichiometry, the structure of the organic compounds, the mass action law, and the phase rule. So he came to the conclusion, that the statement of HERMANN VON HELMHOLTZ expressed repeatedly, *"that physics represents the theoretical foundation of all branches of science, certainly in the spirit of the great deceased"*, now must be supplemented, *"that this foundation is formed by the concept of nature created by the common action of physics and chemistry".* [Nernst (1896b): 12]. As in the letter of the Faculty in Göttingen quoted above, in which the establishment of the new Institute has been proposed, NERNST emphasized again the outstanding importance of electrochemistry: *"It appears that among the different branches of physics just the theory of electricity has taken the role of the connecting bond between physics and chemistry, and the field of electricity seems to experience a tremendous growth. ... The special science of physical chemistry, which occupies itself with these phenomena, is called electrochemistry; it is one of our most important tasks to look into its laws."* [Nernst (1896b): 12–13].

In the *Festschrift* NERNST had pointed out: *"Therefore, it would be a misconception about the development of physical chemistry, if one would specialize oneself completely and would dedicate a separate Institute to a particular branch of the boundary field between physics and chemistry, for example, to electrochemistry. However, without doubt just now it is exactly the latter field, which in many ways promises the largest returns, and where it represents a great national and also scientific contribution of our chemical industry, to have strongly recommended special attention to it many times. With these remarks at the same time I want to justify the name, which the new Institute has obtained."* [Nernst (1896a)].

NERNST stressed *"The Institute essentially is meant for advanced people."* [Nernst (1896b): 13]. And *"Scientific papers, ... printed matter ..., these are the visible fruits, which our Institute hopes to harvest; in this way at the same time we hope to be useful to the students of the pure as well as the applied science, technology, and industry."* [Nernst (1896b): 14]. The already existing and still to be expected *"large number of PhD-theses in the field of physical chemistry"* mean *"a strong and clear support of science and of the German national prosperity, which in this case means the same. The latter must be emphasized. It is a patriotic duty to maintain the relations between our Institute and the German industry; however, also a duty of gratitude to our industry."* [Nernst (1896b): 15].

However, by no means for NERNST these duties meant to ignore the primacy of basic research. Basic knowledge of the laws of nature can serve technology and industry. *"However, it would be wrong again, if in our research we would restrict ourselves to very special targets of the technology."* In this context he quoted a sentence of HELMHOLTZ: *"Who in the pursuit of science is hunting for immediate practical benefits can be quite sure that he is hunting in vain."* [Nernst (1896b): 15–16]. Being himself a clever business man, as it turned out soon, he added by commenting on the need of accepting a certain risk in basic research: *"A large Institute such as ours may not adopt the attitude of the little merchant, who must ask for the profit in interest at each and still very small capital investment, but instead that of the major capital investor, who does not hesitate to risk large sums à fonds perdu, if required by the large tasks of his company."* [Nernst (1896b): 16].

It was the intention of this somewhat more detailed presentation of the contents of the dedication speech, which also can be found partially in the *Festschrift* [Nernst (1896a)], to illustrate interesting aspects regarding the position of physical chemistry at the end of the 19th century and to provide some further insight into the nature and the personality of

NERNST and into his mental attitude in many questions important for science. Some of these ideas may be up-to-date even today.

At the end of this section we list the Directors of the Institute of Physical Chemistry founded officially in 1895, who were working in the building complex in *Bürgerstraße 50*:

Name	Term of Office	Remarks
WALTHER NERNST	1895 – 1905	
FRIEDRICH DOLEZALEK	1905 – 1907	Student of NERNST in Göttingen
GUSTAV TAMMANN	1908 – 1929	Since 1903 Director of the newly founded Institute of Organic Chemistry in Göttingen
ARNOLD EUCKEN	1929 – 1950	Student of NERNST in Berlin
EWALD WICKE	1950 – 1953	Student of EUCKEN in Göttingen
WILHELM JOST	1953 – 1971	Previously coworker of MAX BODENSTEIN in Berlin and Professor in Leipzig, Marburg, and Darmstadt

In 1971 the Institute moved into a new building in *Tammannstraße 6* in Göttingen-Weende. Then the old building served as Institute of Anthropology.

4.8 Studies and Members in the New Institute

In his report about the new Institute NERNST remarked: *"During this summer term [1895] in addition to the Director and his two assistants, Dr. Lorenz and Dr. Roloff, there are working in the Institute two lecturers, Dr. Abegg and Dr. Des Coudres, and further about 30 probationers, the larger part of whom is occupied with their own scientific investigations. – The maintenance of the Institute is handled by the mechanic of the Institute."* [Nernst (1894a)]. RICHARD LORENZ was the assistant for chemistry. When he accepted an offer from Zurich in 1896, FRIEDRICH WILHELM KÜSTER took his position. In 1894 MAX ROLOFF obtained his PhD with the thesis *"Contributions to Our Knowledge of the Photo-Chemical Effects in Solutions"* carried out with NERNST.

On February 5, 1894 in conjunction with an *"excursion to Berlin"* NERNST had informed WILHELM OSTWALD in a letter: *"... from there I brought along Dr. Abegg, who faithfully helps me."* [Zott (1996): 68]. In 1891 RICHARD ABEGG got his PhD with AUGUST WILHELM VON HOFMANN in Berlin, and subsequently working with WILHELM OSTWALD and SVANTE ARRHENIUS became a specialist in the field of physical chemistry. In 1897 he took the position of *Extraordinarius* of Physical Chemistry in Göttingen. However, in 1899 he left the *Georgia Augusta* and went to the University of Breslau, where he became Section Head and in 1909 *Ordinarius* at the Institute of Physical Chemistry. In 1910 he accepted the position of *Ordinarius* at the newly established *Technische Hochschule* in Breslau, but he suffered a fatal accident during the crash of a free balloon, i.e., while performing a kind of sport, to the development of which he had contributed a lot. The family of NERNST and that of ABEGG enjoyed friendly relations, which, for example, are testified by a common trip during spring of 1896.

Within the new facility in Göttingen, ABEGG played an important role in precision measurements of the freezing point of solutions, which had started already in 1894. On this subject there appeared several papers published together with NERNST [Nernst and Abegg (1894)]. There developed a brief controversy with HARRY C. JONES, who had also performed such measurements in the laboratory of OSTWALD. In a letter to OSTWALD from November 18, 1895 NERNST referred to the papers by JONES only as *"twaddle (it is really nothing better!)"* [Zott (1996): 98].

At the turn of the year 1896/97 NERNST wrote to ARRHENIUS about THEODOR DES COUDRES: *"Des Coudres does nice things with cathode rays, but he cannot decide to publish. However, I am very happy that he works here, one cannot imagine a more pleasant colleague."* [Zott (1996): 115]. At this time DES COUDRES had completed for publication an article about the construction principles and the capability of mirror galvanometers. In 1897 he became *Extraordinarius* in Göttingen, and in 1901 he obtained the position of *Extraordinarius* of Theoretical Physics in Wurzburg.

According to the specialty of the Institute many investigations were carried out on electrochemistry. These were also dealing with electrochemical measuring techniques, including the development of a method to determine the inner resistance of galvanic cells, in which ERNST HAAGN participated, or the construction of a new kind of electrometer. The latter had been built in 1896 as a highly sensitive instrument (5 μV) by FRIEDRICH DOLEZALEK, a student of NERNST at the time. NERNST's publications concerned methods for determining the dielectric constants.

On June 27, 1896 at the 3rd General Meeting of the German Electrochemical Society in Stuttgart NERNST wanted to *"allow himself to say something about the quantitative electrochemistry, especially from the experience I have gained during the installation of the Institute in Göttingen"* [Nernst (1896c): 52]. He pointed out that the electrochemistry cannot rely only on the measuring instruments, which have been developed originally for purely physical applications. *"Hence, we have tried in Göttingen to serve the special targets of electrochemistry by the construction and development of new measuring instruments."* [Nernst (1896c): 52]. He could name and demonstrate four such instruments: the apparatus for determining dielectric constants described already in 1894, an instrument for measuring large electrolytic resistances developed by his former PhD student MARGARET ELIZA MALTBY, a thermal column equipped with a sensitive regulating device for the gas supply developed by his student HEINRICH DANNEEL, which allowed *"to keep an electromotoric force, supplying current, constant for a long time"* [Nernst (1896c): 53], as well as the electrometer described by him and DOLEZALEK. If these measuring instruments referred to studies using direct current, NERNST pointed out that in the future *"in an unexpected way also alternating current could be utilized by the electrochemistry. I am convinced that the electrochemical measurement technique would be well advised to prepare for this in time."* [Nernst (1896c): 53]. Within this context the polarization capacity represents an important quantity. Already in 1894 NERNST had proposed a principle for its measurement, which is based on the comparison of the polarization capacity with a capacitor in the standard bridge circuit, and which his student CLARENCE GORDON now had transformed into a simple and exact method.

Regarding the electrochemical measurement technique, it should be mentioned, that it was NERNST who proposed in 1900 in a paper on electrode potentials, based on such measurements and calculations by his colleague NORMAN T.M. WILSMORE in London, for the determination of the electrochemical potential to use as zero potential that of the normal-hydrogen electrode [Nernst (1900/01)]. Subsequently, this has been generally accepted. The problem was that electrochemical potentials can be measured only as differences, as it is also the case with the electric potentials or with the geographic levels in height. Therefore, for practical reasons one must fix a zero-point for a comparison of such quantities. In the case of the electric potentials, the potential of the earth (ground potential) represents the zero-point. In the case of the geographic height levels, the average sea level is fixed as the zero-point. The normal-hydrogen electrode, which according to NERNST fixes the zero-point of the electrochemical potentials, in principle consists of a platinized platinum sheet within a solution of hydrogen-chloride (HCl) of the concentration 1 mole/l in water, which is hosed by hydrogen gas with a pressure of 101325 Pa (earlier: 1 atm). The temperature is 25°C. The equilibrium generating the potential of a hydrogen electrode can be written as $½H_2 + H_2O \rightleftharpoons H_3O^+ + e^-$.

Of course, for the studies in the field of electrochemistry, there existed already various electro-preparative and electro-analytic publications. Examples are the PhD thesis of 1897 by HEINRICH DANNEEL *"Studies of the Electrochemical Precipitation of Metals According to Faraday's Law"* and NERNST's publication on the chemical equilibrium, the electromotoric effectiveness, and the electrolytic precipitation [Nernst (1897a)].

Last but not least, at the time the importance of the electrochemical research can be seen from the disintegration voltage and its determination. Just in 1891 this subject and the related concept had been introduced into science as a material-specific quantity by MAX LE BLANC when he applied the ideas of NERNST's habilitation thesis to electrolysis. In this context around 1898 in NERNST's laboratory WILLIAM CASPARI created the concept of the *Überspannung* (overvoltage). From this and from other studies suggested by NERNST we see, that he recognized the value of the kinetic methods for the electrochemical analysis, however, that he did not work himself on such a subject. In the end it was JULIUS TAFEL, a student of EMIL FISCHER, who in 1905 could announce the relation between the overvoltage and the logarithm of the current density, named after him.

Further, investigations of the conductance were carried out. Those dealing with the solid-state conductance of oxides and mixtures of oxides must be looked at in the context of the research on the lamp of NERNST and will be discussed below in Section 4.9. However, at this point we can mention also ABEGG's measurements of the conductance of pure materials under standard conditions, corresponding studies within liquid ammonia, and extended series of measurements on compressed powders of oxides and sulfides.

Also the investigations on the lead cell must be mentioned, which NERNST had suggested to his student DOLEZALEK. It is interesting, that primarily these were not focused on the technological, but instead on the fundamental aspects, such as the application of the new theory of the potential generation to this system invented already in 1859 by GASTON PLANTÉ and to the gas polarization in it. By the application of the concept of the *Überspannung* introduced by CASPARI, in his PhD thesis of 1898 DOLEZALEK was able to explain the processes within the lead cell and to describe in detail its reaction equations.

After 1900 together with ERNST HERMANN RIESENFELD, NERNST studied in particular phenomena associated with boundaries. In this field in 1901 the former completed his PhD thesis *"On Electrolytic Phenomena and Electromotoric Forces at the Boundary between Two Solvents"*. During this time also another student, FRIEDRICH KRÜGER, worked on the problem of the double-layer, the electro-capillarity, and the polarization capacity. In this case one could start from the fundamental results, obtained already by HERMANN VON HELMHOLTZ and GABRIEL LIPPMANN, where, however, the experimental data could not be described in terms of the simple capacitor model of the former. In 1898 MAX WIEN had proposed a bridge circuit, which allowed the direct measurement of the double-layer capacity, and which had been applied to the measurements at electrodes by EMIL WARBURG already in 1899. NERNST very much appreciated the results of his student presented in 1903 [Krüger (1903)]. The corresponding remarks in a letter to WILHELM OSTWALD from February 19, 1903 also indicate the objectivity of his judgment demonstrated by NERNST, who was well convinced of his own power and who paid close attention to its acknowledgment with not a small degree of vanity: *"... although his paper turned out a bit long, Krüger in a relatively brief form in particular has yet developed and experimentally justified a new theory of the polarization capacity, in which the old idea of Helmholtz as well as that of Warburg represent very special cases to be defined exactly theoretically. For years I did not enjoy a paper as much as that of Krüger."* [Zott (1996): 155].

By no means was the work in the Institute of NERNST restricted to electrochemical problems. We have mentioned the cathode-ray experiments by DES COUDRES. The versatile ABEGG also occupied himself with scientific photography. Furthermore, the research covered the theory of solutions, vapor pressure, chemical equilibrium, and reaction kinetics in heterogeneous systems.

In 1904 NERNST had developed a theory, according to which the rate of heterogeneous chemical reactions is determined only by the diffusion of the reaction partners across an intermediate layer, which is generated in the region between the solution of constant concentration, established by the rapid transport by means of convection, and the surface of the solid [Nernst (1904)]. Today this layer is referred to as NERNST's diffusion layer. If c_i and c_0 denote the constant concentration of the reaction partner in the solution and at the surface of the solid, respectively, and δ the thickness of the diffusion layer, then according to NERNST one finds, that in this layer the concentration increases with the slope

$\frac{\partial c}{\partial x} = \frac{c_i - c_o}{\delta} = $ const. If we connect this result of the concentration gradient with the first diffusion law by ADOLF FICK $\frac{dn}{dt} = -Dq\frac{\partial c}{\partial x}$ for the rate of the change of the amount of material n with the time t, where D is the diffusion coefficient and q denotes the surface area, then after division by the volume V one obtains the rate equation

$$\frac{dc}{dt} = -\frac{Dq}{\delta V}(c_i - c_o) = -k(c_i - c_o)$$

for the decrease of the molar concentration $c = n/V$. Hence, the rate constant k is directly proportional to the diffusion coefficient and the surface area, and inversely proportional to the thickness of the diffusion layer. This theory, which only has a limited validity, was worked out by NERNST together with HEINRICH DANNEEL and ERICH BRUNNER. In 1903 the latter obtained his PhD with a thesis based on this theory.

Within this period of NERNST's activities there were three further subjects of important research: the development of his lamp, the law of the electrical nerve stimulus threshold (*Reizschwellengesetz*) named after him, and the investigations of the chemical equilibria of the reactions of gases. These subjects will be treated separately in the following sections.

Before doing so, at this point we wish to turn to the first female student of NERNST, the American MARGARET ELIZA MALTBY mentioned already with the listing of the electrochemical measuring instruments.

Fig. 4.9 MARGARET ELIZA MALTBY.

On July 17, 1893 NERNST started a letter to WILHELM OSTWALD with the information: *"A Miss M.E. Maltby, A.M.S.B., Professor at a Women's-College in Wallasey has come to Europe, to study in particular physical chemistry."* [Zott (1996): 62]. She would have made a good impression on him, and would be recommended by ARTHUR NOYES among others. Since his laboratory would be overcrowded, he asked OSTWALD, if he could take her. However, the open-mindedness of the University of Göttingen in the case of the study of women and the personal nature of NERNST and also of his superior at the time, EDUARD RIECKE, to find the other sex quite attractive, two days later allowed to inform OSTWALD: *"I am quite willing to see that Miss Maltby becomes acquainted with the physical-chemical methods, and since Riecke as the Institute Director not only is not opposed, but even would welcome with joy the use of our instruments by female hands (recently the Philosophical Faculty here has very energetically declared itself in favor of the admittance of women to study), so there should not be any obstacle."* [Zott (1996): 63].

In fact, in 1893 MARGARET E. MALTBY was admitted to the lecture courses and to perform research in the Physical Institute. Nearly two years later, on July 11, 1895, NERNST reported to his Swedish friend ARRHENIUS: *"Yesterday Ms. Maltby got her PhD; fortunately, now her anxiety because of the exam is over a. in the most beautiful way."* [Zott (1996): 93]. For this success Miss MALTBY had to thank *"Professor Riecke for his many supports, and Mr. Professor Dr. Nernst for the interest with which he followed my work"*, as she wrote in her *vita* for the thesis. During the same year the English women GRACE CHISHOLM obtained her PhD with a mathematical thesis with FELIX KLEIN. The remark by NERNST from July 24, 1895, that the thesis of MARGARET E. MALTBY *"is the first experimental PhD thesis 'femininum generis' in the German language"* [Zott (1996): 96], is not quite correct, since before her JULIA LERMONTOFF (LERMONTOVA) had already obtained her PhD in Germany with an experimental thesis.

The thesis by Miss MALTBY is entitled *"Method for Determining Large Electrolytic Resistances"* and was published in the *Zeitschrift* [Maltby (1895)], *"although it was a bit long"* [Zott (1996): 95], as observed by NERNST. The latter emphasized the exact method as well as the very large electric resistances. Furthermore, Miss MALTBY had performed experiments at the critical temperature, and she had utilized the electromotor of HANKEL in a skilful way for measuring the resistance.

RIECKE was the main referee of the dissertation, while the documents of the University of Göttingen do not mention NERNST nor do they contain a report by him, although he was the actual advisor.

After the completion of her PhD, Miss MALTBY still worked in Göttingen for some time, where she studied high-frequency oscillations obtaining interesting results [Maltby (1897)], a subject, which *"also we cannot ignore in the long run"* [Zott (1996): 93], as NERNST had stated already in 1895. At the turn of the century she was occupied at the *Physikalisch-Technische Reichsanstalt* in Berlin. There under FRIEDRICH KOHLRAUSCH, its president at the time, she investigated the conductance of aqueous solutions of alkaline chlorides and nitrates [Kohlrausch and Maltby (1899)].

In the USA she became known because of her effective effort to obtain a scholarship for graduate and postdoctoral education for women. From 1912 until 1929 she was a member of the *Fellowship Committee of the American Association of University Women*, the Chair of which she occupied during the period 1913 – 1924, and she published a *"History of the Fellowship Awarded by the American Association of University Women, 1888 – 1929"*. In 1926 the *Association* established the *"Margaret E. Maltby Fellowship"* in her honor [Ogilvie (2000): 835].

4.9 The Nernst Lamp

Among the research performed by NERNST at the end of the 19th century, that part dealing with the construction and development of his lamp and the investigations of solid-state electrolytes associated with it has taken up an important fraction. It represents the first example of his interest in obtaining fundamental scientific information and at the same time in utilizing it for practical and profitable inventions, for which there exists a current requirement. It also shows his excellent capability to turn this intention into reality.

In order to correctly classify the NERNST lamp temporally and also with respect to its value, we present a brief overview of the history of the generation of light in the 19th century (see [Bartel *et al.* (1983)] and further references therein). In its second half new kinds of artificial light sources came into use. To these belong the petroleum lamp, the proper invention of which can be traced back to IGNACY ŁUKASIEWICZ in Galicia in 1852 and the first useful form of which had been developed by BENJAMIN SILLIMAN senior in the USA in 1855. To this must be added the gas and the electric illumination. Electric light sources existed in the form of arc lamps and filament bulbs. Among the three kinds of light generation just mentioned the gas illumination was the oldest. So WILHELM AUGUST LAMPADIUS had introduced the first German street illumination with gas lamps already in 1811 in Freiberg in Saxony.

By means of intensive research, with time it was possible to improve the application potential, the form of the burner, the gas mixture, etc. Within this context one must mention in particular also the development of the principle of the incandescent light, because this principle led to an important advancement and it is closely connected with the NERNST lamp. The history of the incandescent light extends from the lime light of DRUMMOND described in 1826 for the observation of distant stations during the survey of the earth, and in which calcium oxide is heated in a flame of detonating gas, up to the incandescent mantle covered by rare-earth oxides, which was invented by AUER VON WELSBACH in 1885. The latter led NERNST to the idea of his light bulb.

Starting with the 19th century, several researchers had turned to the problem of the generation of light by means of electricity. At this point, as representatives we can name the early pioneers of electrochemistry, namely JOHANN WILHELM RITTER and Sir HUMPHRY DAVY, as well as ROBERT WILHELM BUNSEN. In 1884 the development of the carbon-arc lamp started, the improvement of which occupied FRIEDRICH VON HEFNER-ALTENECK and PIERRE JABLOKOFF, among others. Both scientists played a role in conjunction with the NERNST lamp.

However, a meaningful development of electric lamps could start only after 1867, when in January of that year WERNER VON SIEMENS had announced the electrodynamic principle discovered by him, which subsequently became the foundation of the large scale generation of electric current. So in 1879 HEFNER-ALTENECK could develop the differential arc lamp, which could replace the JABLOKOFF candles being only three years older.

Within the first light bulbs a carbon filament acts as the glowing material. Based on this approach, in the USA in 1854 HEINRICH GOEBEL constructed a light source, which he used for the illuminetion of his watch-maker shop in New York. However, only the carbon-filament lamps constructed in 1879 by THOMAS ALVA EDISON gained any importance. From the many competitors being also active in this field we mention JOSEPH WILSON SWAN, who produced such a lamp in the same year in 1881, when EDISON opened the first filament bulb factory in Menlopark. One year later EMIL RATHENAU in Germany founded a Company with the goal of spreading the invention of EDISON, and out of which in 1887 the *Allgemeine Elektrizitätsgesellschaft* (AEG) developed. Since the Siemens Company was betting exclusively on the production of arc lamps, the first German factory for the production of glow lamps was opened by the AEG in 1884 in the *Schlegelstraße* in Berlin.

We must not overlook the fact that in the last two decades of the 19th century the production of electric light sources represented a major part of the growing electric industry, and that this activity resulted in becoming the origin of two very important developments for the industry in general: the vacuum technology and the automatization of a production process.

During this time, in the artificial generation of light there broke out a tough competitive fight, the complexity of which is indicated in Fig. 4.10.

In the past much had been invested already into research, development, and the installed facilities in particular for the gas illumination. In order to survive in the competition within the light-generation industry, all competitors had to work very hard to improve their lamps by means of new physical and technical solutions, and to eliminate existing deficiencies in the products already on the market. The advantages and profits expected from the mastering of this task initiated the corresponding investigation of the laws of nature fundamental to the process of light generation. The discovery of the radiation laws by WILHELM WIEN,

JAMES H. JEANS, and Lord RAYLEIGH, the generalization of which led MAX PLANCK to the formulation of his quantum hypothesis, are highly impressive examples for this.

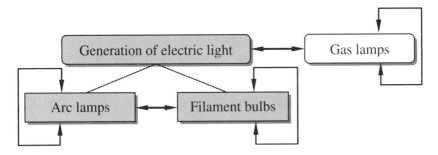

Fig. 4.10   Schematic presentation of the competitors (↔) in the fight for the artificial generation of light.

Corresponding to his personality, in a nearly natural way WALTHER NERNST must have been attracted into this exciting field of fundamental science and technical-industrial innovation. The theoretical origin of his ideas for the development of a novel light bulb was the radiation law found by JOSEPH STEFAN in 1879 and theoretically justified by LUDWIG BOLTZMANN in 1894, according to which the total radiation is proportional to the fourth power of the temperature. Hence, it must be advantageous to increase the amount of radiation or light emitted by a body by means of a strong increase in its temperature. In the case of an electric lamp the high temperature was obtained by means of JOULE heating, where the integral heat power $\dot{Q}$ for a finite volume is proportional to the resistance R and to the square of the current: $\dot{Q} = RI^2$. A pin made of an oxide mixture appeared suitable to NERNST as the glowing material. It was similar to the mantle used by AUER and consisted of about 85 % zirconia (ZrO_2) and 15 % yttria (Y_2O_3). In addition to the problem of the technical development, perhaps even primarily, NERNST was interested in the fundamental electrochemical problem of the conductance of solid-state electrolytes, closely connected with this light source.

On July 6, 1897 NERNST would present the first result of his work on the development of his lamp in a patent, which was issued on July 8, 1899 [Nernst (1897b)]. Here as glowing material he presented the use of a conductor of the second type, as it is given by the oxide mixture of his glowing pin. However, in order to start the passage of current, this had to be heated by a heating device separated from the electrodes. The use of an electric heater is the subject of a second patent of 1897 [Nernst (1897c)].

Fig. 4.11 A NERNST lamp (now owned by the Humboldt University in Berlin). On the right photograph we see the glowing NERNST pin and the opening in the frosted glass container (top), which allows the access of air to the glowing material.

Compared to the carbon-arc lamps, the NERNST lamp had several advantages. The complicated and expensive evacuation, required in the production process of the former in order to keep the filament from burning, was not necessary any more. The NERNST lamp could and even had to burn within air. Furthermore, it had a few better characteristics, to which belonged for example the light output and the light quality.

The important disadvantage of the invention of NERNST was connected with the use of the oxide mixture as the glowing material (Figs. 4.11 and 4.12). The need for heating this separately until the material became conducting led to problems, for the solution of which much research effort was spent. The shadow generated by the heating devices could be reduced, and the heating current could be switched off upon the start of the glowing process. However, the main deficiency caused by the relatively long time needed for heating from the switching on of the lamp until its light emission sets in, in spite of all attempts to find a suitable configuration of the glowing material and other possibilities for optimization, could not be eliminated satisfactorily. This fact appears to be the main reason, which led to the termination of the operation of this lamp within a relatively short time.

Fig. 4.12   Inner configuration of the NERNST lamp: (1) frosted glass container, (2) the glowing device with the NERNST pin (top), (3) the iron-hydrogen resistance.

The material of the NERNST bulb is a conductor of the second kind (ionic conductor), the electric resistance of which decreases with increasing temperature, in contrast to the carbon filaments and the metal filaments, which are conductors of the first kind (electronic conductors) and which show an increase of the resistance with increasing temperature. As a consequence, during permanent current flow the NERNST pin would be heated up until it melts (at about 2600 °C). For solving this problem NERNST found a solution by using the iron-hydrogen resistance (Fig. 4.12). As the name indicates, in this case one deals with an iron wire, located within a small tube filled with hydrogen and electrically connected to the outside. Here NERNST utilized the fact, that in this system at a certain value I_1 the current I remains practically constant within a larger interval ΔU_1 of the voltage U (Fig. 4.13).

Aside from their use within the NERNST lamps, subsequently under the notation '*variators*' the iron-hydrogen resistors were used as starting resistors for small motors. They survived the NERNST lamps for a longer period also in radio receivers. Only in 1920 an explanation of its voltage-current curve (Fig. 4.13) was given in the habilitation thesis of HANS BUSCH [Busch (1920)]. This thesis had been prepared in Göttingen at the *Radioelektrische Versuchsanstalt für Marine und Heer* (Radioelectric Laboratory for the Navy and the Army) and after its termination at the Institute of Applied Electricity of the *Georgia Augusta*.

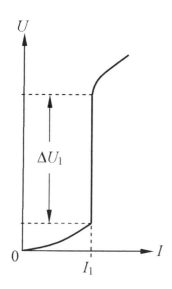

Fig. 4.13 Schematics of the voltage-current curve in the system iron (or nickel) / (dilute) hydrogen.

Since NERNST was well acquainted with the founder and chief officer of the AEG, EMIL RATHENAU, he was able to sell his patents to this Company. On May 9, 1899 in the meeting hall of the AEG in the *Luisenstraße* in Berlin the NERNST lamp was publicly demonstrated for the first time. At this time there existed already 14 German and 100 foreign patents, and within the next five years the first number increased up to about 80. In London the *"Nernst Electric Light Limited"* also contributed to the development of the lamp. In 1900 at the World Exhibition in Paris the pavilion of the AEG was illuminated using NERNST lamps.

For about half a decade the production rate of the lamp was rather high. It is estimated to have amounted to several thousand units per day. This means that about four million NERNST lamps or burners were produced.

Of course, as in the case of most highly promising inventions, the introduction of the NERNST lamp was preceded by not a small number of legal obstacles. Already the two patents of 1897 [Nernst (1897b), (1897c)] met 13 objections and 5 complaints. For example, it was argued against NERNST, that he had used an invention by PIERRE JABLOKOFF patented in 1877, in which kaolin platelets were made to glow by means of an electric current generated by an arc inductor. NERNST claimed that he did not know about this. FRIEDRICH VON HEFNER-ALTENECK supported NERNST by emphasizing that the latter has produced a useful invention out of his own creative activity, whereas the patent of JABLOKOFF remained without any application.

In other cases NERNST had been called to the Imperial Court of Justice in Leipzig as a consultant in patent lawsuits. Just in 1897 he appeared as a consultant in a case dealing with lead cells. Later on he was chief witness in the lawsuit dealing with the technical production of ammonia conducted against FRITZ HABER.

NERNST looked positively at the patent issue and its different aspects. In principle, he valued the mental effort connected with the formulation of a patent and its defense. According to him, in addition to the research results which can be applied technically, such results with a purely scientific character should also be protected by a patent, if they can be used by industry. Such developments would be interesting only for large companies and institutions, which have the necessary financial means and which can expect a future profit from the further development of the results. Then the high investments would cause a rapid scientific and technical progress.

Although NERNST was convinced of the high importance of his invention to such an extent, that he put it in line with the HELMHOLTZ eye mirror, the AUER glowing mantle, and even the rays discovered by RÖNTGEN, the period of its production was only relatively short. The osmium-filament bulb invented in 1898 by AUER VON WELSBACH did not represent a serious competition, since it was designed for 75 V, whereas the existing power network had a voltage of 110 V. Therefore, this lamp was not accepted. The situation was different in the case of the glow lamps with tantalum filaments marketed in 1904 by the Siemens & Halske Company. At about this time ALEKSANDR LODYGIN in Russia and several scientists in other countries worked on the use of tungsten as the glowing material, since this metal showed the highest melting point known at the time. In the end it was IRVING LANGMUIR, a student of NERNST in Göttingen (PhD in 1906), since 1909 working in the Research Laboratory of the General Electric Company in Schenactady (New York/USA), recipient of the Nobel Prize in Chemistry in 1932, who caused the final breakthrough of the tungsten lamps and practically completely pushed aside the invention of his German teacher. His ingenious idea was the use of a chemically inert protective gas instead of vacuum and a tungsten filament having many windings.

If the yield of the carbon-filament bulb was 3 lm/W, which was improved by the NERNST lamp up to 6 lm/W, then aside from all other advantages the metal-filament bulbs even reached about 12 lm/W. As quoted by JOHN EGGERT, the apparent defeat has been commented by NERNST with the words: *"There are inventions, which are too beautiful for this world. ... At any rate, I have shown the way for the light technology, how it can achieve progress, namely by the increase of the temperature of the glowing material."* [Eggert (1964): 447]. Some of the patents prepared in the 20th century, indeed, testify that the advantages of the NERNST lamp had not been forgotten [Birghall et al. (1973)].

The NERNST pin still remained the light source in infrared spectrometers until the second half of the 20th century.

Also NERNST himself continued to feel being under an obligation to the light technology. The mercury-discharge lamp owes its development also to suggestions by him. Its principle had been proposed shortly before the NERNST lamp, and it was realized for the first time by PETER COOPER-HEWITT. In 1913 NERNST could work on patents, which improved the spectral properties of the mercury-vapor lamp by means of the addition of metal halogenides (see, for example, [Nernst (1913a)]). Here we must mention also the suggestion of NERNST from 1906, in which he proposed to use for the primary standard of the light intensity, instead of the HEFNER candle, a black body with a distinct temperature [Nernst (1906a)]. In some sense, in 1948 it was realized in terms of the introduction of the international unit *'Candela'* (cd) (from Latin: *candela* = wax candle), which is connected with the solidification temperature of platinum (2042.5 °C).

At the beginning of this Chapter we had mentioned, that NERNST had closely connected his research on the development of his lamp with the problem of the solid-state electrolytic conductance. Here he had started from the assumption that the light output of the EDISON lamp can be improved in two ways: by increasing the radiation temperature and by utilizing glow elements, the coefficient for light emission of which being large in the visible spectral range and small in the infrared. Already in his first patent document of 1897 NERNST pointed out, that all conductors of the first kind (electron conductors) such as graphite, metals, and several oxides of the heavy metals, are unsuitable as a material for a glow lamp, since the large number of the conduction electrons causes a large absorption coefficient and, hence, at the same time a large emission coefficient in the infrared range. Therefore, as an alternative he proposed the use of solid-state conductors of the second kind (ionic conductors).

At the General Convention of the German Electrochemical Society in 1899 NERNST could point out already regarding this matter: *"Based on experiments aimed at the use of solid-state electrolytes as glowing elements in electric lamps, I was able to state that at high temperatures the electrolytic conduction of solids can reach surprisingly high values. ... Already soon the general result became evident, that the conductance of pure oxides, which are air-resistant only in their glowing state, and which, hence, primarily recommend themselves for investigation, increases very slowly with the temperature and remains relatively low, whereas mixtures show an immensely much higher conductance, a result in perfect agreement with the well-known behavior of the liquid electrolytes."* [Nernst (1899/1900): 41].

Initially NERNST had used alternating current in his experiments. During the subsequent experiments using direct current it became apparent that the employed oxide mixtures could be kept glowing for several hundred hours without any detectable decomposition. From this NERNST concluded, that in the stationary state oxygen enters the glowing element at the cathode and exits again at the anode. Based on the simple experimental arrangement, he had recognized the essence of the underlying conduction mechanism.

On the subject of the *"electrolytic glowing elements"* NERNST's paper from 1899 [Nernst (1899/1900)] was followed only by a second one published soon afterwards [Nernst and Wild (1900/01)]. Scientifically, they left a few questions open. These concerned the unequivocal demonstration of the electrolytic conductance of current, the nature of the ions responsible for the charge transport, the origin of the conductance difference between the pure and the mixed oxides, and the details of the mechanism of the ion transport. The advances achieved in solid-state physics in the 20th century made it possible, that in its first half detailed ideas about the conductance of solid-state electrolytes could be developed, which answered these questions. In the 1920s and the 1930s important contributions to this subject have been provided among others by JAKOV I. FRENKEL, WILHELM JOST, WALTER SCHOTTKY, and CARL WAGNER, whereas NERNST himself did not participate any more directly in this research.

However, the Soviet physicist ABRAM F. JOFFE reports, that NERNST, whom he had met personally only late and according to his description whom he had apparently almost completely misunderstood, still in Berlin had not lost sight of the problem of the solid-state conductance: *"Nernst's ideas about the electric breakdown of dielectric materials by means of an electron avalanche were connected with the nature of the electric currents in metal oxides, out of which the pins of his lamp were made."* [Joffe (1967): 83].

4.10 Nernst Law of Electrical Nerve Stimulus Threshold (*Reizschwellengesetz*)

In 1899 NERNST published his first paper about the theory of the electric stimulation [Nernst (1899)], which deserves a special mentioning since it represents an interesting contribution to the electricity of animals. About 100 or 50 years earlier two highly important works had appeared in this field: by JOHANN WILHELM RITTER *"Proof that a Steady Galvanism Accompanies the Life Process within the animal kingdom"* in 1798 in Weimar [Ritter (1798)], which became the foundation of the electrochemistry and for RITTER supplied the base for further research and philosophical considerations, extending also to animal and plant electricity, and by EMIL DU BOIS-REYMOND *"Studies on Animal Electricity"* in 1848/1849 in Berlin. In the 447th thesis of his *"Fragments of the Assets of a Young Physicist"* [Ritter (1810)] RITTER had formulated the remarkable statement *"The galvanic chain will be the image of life."*

KARL FRIEDRICH BONHOEFFER has expressed the merits of his teacher in Berlin with the words: *"The first quantitative attempt to understand the laws of the threshold of the electric stimulation in terms of physical chemistry was due to Walther Nernst. The suggestive power of his proposed interpretation originated from the simplicity of the underlying ideas, as is typical of many theories of Nernst."* [Bonhoeffer (1943): 270]. The starting point of these ideas was the occupation of NERNST with high-frequency alternating currents. In this context he noticed, that the explanation of the human insensitivity against these currents given at the time could not be correct. Because of their high frequency such alternating currents should not reach the interior of the conductor and, hence, should be unable to stimulate the nerves. NERNST recognized that this effect only occurs in the case of good conductors such as the metals, however, not in the case of organic tissue because of its very low conductance. Furthermore, in 1897 he had been able to prove experimentally, that such currents fill out the total cross section of salt solutions. Therefore, the stimulation threshold had to be explained in a different way.

For this purpose NERNST started from the following considerations: Within a living tissue an electric current causes specific changes, which can be traced back to those of concentrations. These will appear at the membranes existing within the tissue liquid. This process due to the electric current is counteracted by diffusion, which favors the balancing of the concentrations. Under certain simplifying assumptions the temporal development of the concentration changes at the membranes can be determined quantitatively. With the supposition, that the stimulation threshold is characterized by attaining a distinct value of the concentration change, the laws regulating this behavior can be indicated.

Just in 1899, EMIL WARBURG had published a paper on the concentration polarization in the presence of alternating currents, in which NERNST found the complete theory needed by him. Under the assumption mentioned before, that at the threshold a critical ion concentration must be reached, he could extract from it his law of the constancy of $i/\sqrt{f}$ for the stimulation threshold, where i and f denote the current and the frequency, respectively. Checking this result using experimental data obtained by JOHANNES VON KRIES in the range $100 - 1000$ Hz, NERNST found that his relation $i/\sqrt{f} = $ const. is well satisfied. Additional measurements suggested by NERNST or carried out by himself [Nernst (1904)] at least in the studied frequency intervals also confirmed the validity of NERNST's theory.

Up until his Berlin period NERNST worked on the problem of the electric stimulation. In 1908 in the *"Archives of Physiology"* of EDUARD PFLÜGER he published a longer paper, in which *"(essentially verbally)"* he repeated his theory of 1899, *"in order to attach to it a few further remarks."* [Nernst (1908): 276]. In addition to other points, to the latter belonged the discussion of the measurements using alternating currents, of the range of validity, and of the effect of pulses of direct current. In the case of constant stimulation by direct current, from NERNST's theory it follows immediately, that for reaching the threshold in the case of smaller currents a larger amount of electric charge is needed. Under the assumption of a sufficiently large distance between the membranes, the product of the current i and of the square root of the duration t of the current must be constant: $i\sqrt{t} = $ const. However, NERNST had to emphasize, that this law is valid only in the case of short stimulations: *"In addition, it was found, that the theory must be restricted to momentary stimulations, i.e., to currents alternating sufficiently rapidly or to current pulses of sufficiently short duration."* [Nernst (1908): 313]. Apparently, the stimulation effect decreases in the case of longer duration of the current flow. NERNST described this phenomenon using the concept of the *"accommodation"* (Latin *accomodatio*: adjustment). In this case he assumed that the current effects a change of the object, and that its stimulation threshold increases. Regarding the accommodation, NERNST applied simple physical-chemical arguments. However, he avoided a quantitative treatment. The latter had been carried out successfully only in the 1930s, for example, within the formal theory created in 1936 by ARCHIBALD V. HILL.

At the end of his paper of 1908 NERNST could summarize: *"If one keeps the assumption introduced by myself in 1899, that a stimulation by means of an electric current is caused by concentration changes, which are generated by the corresponding current at the boundary between protoplasm and cell liquid, then an exact theory of the stimulation phenomena can be developed, such that the stimulation threshold with its dependence upon the nature of the current can be calculated. ... As could be demonstrated by a large number of experimental observations, outside of this accommodation regime the theory is exactly valid."* [Nernst (1908): 312 – 313].

NERNST's $i\sqrt{t}$ -law in the case of pulses of direct current turned out to be valid only within an intermediate current interval. In the case of small currents there appears a fundamental phenomenon of the stimulation threshold, which could not be explained by NERNST's theory: Below a certain magnitude of the direct current (rheobase), independent of the duration of the stimulation, no excitation appears anymore. On the other hand, in the case of sufficiently short pulse times t or large currents i, the product $i \cdot t$, i.e., the amount of charge, becomes important.

In spite of this and also likely in view of the theories created and of the knowledge gained during the lifetime of NERNST, regarding the electric stimulation KARL FRIEDRICH BONHOEFFER could state: *"The ideas generated by Nernst have influenced immensely fruitfully experiments and theory."* [Bonhoeffer (1943): 271].

4.11 The Construction of Instruments

At the end of his creative period in Göttingen NERNST had turned to the problems of the chemical equilibria of gases at high temperatures, the study of which looked highly rewarding both from a theoretical and a practical point of view. In his lecture at the X. General Assembly of the German Bunsen Society in Berlin he indicated on July 3, 1903: *"For investigating these equilibria it appears beneficial, since one will deal primarily with reactions within gaseous systems, to concentrate upon the volume measurement or, what means the same, upon the study of the vapor densities of chemical systems at very high temperatures."* [Nernst (1903): 622].

In this case NERNST referred to the well-known and effective air-displacement method by VICTOR MEYER of 1878. This method for determining the molar mass M is characterized by the fact, that a known mass m of the investigated substance is gasified within a suitable apparatus under a given pressure p and temperature T, and that the air volume V displaced by the generated vapor is measured. The principle of the method can be seen from the ideal-gas equation, from which one obtains

$$M = \frac{m}{V}\frac{RT}{p} = \rho\frac{RT}{p},$$ where R is the universal gas constant and $\rho = m/V$ the vapor density.

During the application of this method, for NERNST an instrumental difficulty could have arisen from the fact that one had to operate at very high temperatures, which would have been impossible above 1800°C because of the melting points of the available container materials. However, the Company of WILHELM CARL HERAEUS in Hanau, *"which always had the warmest and most effective interest in all these studies"*, had developed already iridium products, and *"surprised us with a real 'iridium bulb', which represents a first-rate work of art and was supplied to us most graciously by the Heraeus Company."* [Nernst (1903): 625]. Since the melting point of iridium amounts to 2410°C, now temperatures above 2000°C could be applied.

The upper part of the bulb employed by NERNST was shaped as a capillary and water-cooled using a copper spiral. It carried a fall arrangement and the glass capillary filled with mercury, at which the displaced air volume could be measured. An iridium beaker served as the falling body. The scheme of the apparatus for determining the vapor density is shown in Fig. 4.14. The heating device also consisted of an iridium tube, which was heated at low voltage with alternating current passing through platinum contacts and copper ribbons serving as conductors. For heat protection the iridium tube was surrounded by a cover of magnesia and one of asbestos (Fig. 4.14). The heating power of 2000 – 3000 W was taken by NERNST from a self-made oil transformer. For measuring the temperature he used the mixed-oxide glowing pin from his lamp by raising its temperature and thereby its brightness by means of the applied current until it agreed with that of the iridium beaker. Based on the current value measured in this way, the emitted light intensity per unit area of the bulb could be determined. From the latter, however, under the bold assumption that the iridium tube would behave like a black body, its temperature could be estimated from the radiation laws which had just been formulated.

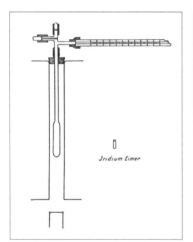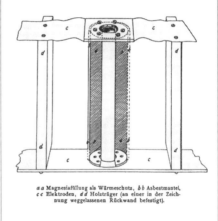

Fig. 4.14 Apparatus for measuring the vapor density according to the "VICTOR MEYER-NERNST method" (left, from [Nernst (1903): 624]) and its heating parts (right, from [Nernst (1903): 623]).

In order to guarantee a constant temperature, and because of the high price of iridium, the apparatus had to have small dimensions. However, this made it necessary to determine with sufficient accuracy the mass of only a small amount of substance. Therefore, together with his coworker

ERNST HERMANN RIESENFELD, NERNST constructed a microbalance, the so-called NERNST-balance [Nernst and Riesenfeld (1903)] (Fig. 4.15], and he asked OTTO BRILL, also working with him at the time, to test it carefully regarding its application potential [Brill (1905)]. Its description and operation principle can be taken from the following quotation: *"[In the case of the balance] the quartz beam of about 5 cm length, in the middle of which a balance beam made from a thick quartz thread with a sharp bend is perpendicularly attached, serves as the middle axis. The horizontal end of the beam carries the light scale of only 20 mg weight, fabricated from a thin platinum foil. The other end of the beam is drawn into a fine pointer moving across a sensitive mirror scale. The sensitivity amounts to about $^1/_{100}$ mg at a maximum load of a few milligrams. The balance represents a combination of a torsion and an inclination balance, since the torsion of the thin quartz thread in a small part leads to a restoring force."* [Eder (1952): 68].

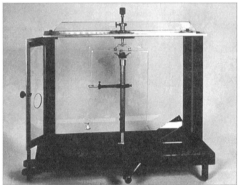

Fig. 4.15 The NERNST-microbalance and its principle.

MAX BODENSTEIN has properly summarized these achievements of his great colleague characterized by his genius: *"The bulb fabricated from iridium, just brought into an applicable form by Heraeus, Hanau, for determining the vapor density according to Viktor Meyer, heated within a short-circuit furnace made from iridium, together with the temperature measurement by means of a simple pyrometer ... and together with the ingeniously simple microbalance and with the measurement of the displaced gas volume by means of the shift of a mercury drop within a capillary -, this combination is a typical example of the elegance, with which – basically using very simple means – experimental difficulties appearing unsolvable were overcome in tiny instruments."* [Bodenstein (1942a): 85].

The NERNST-balance was produced and offered for sale by the Spindler & Hoyer Company in Göttingen, such that NERNST could announce: *"Said Company delivers the balance for about 70 Mk."* [Nernst and Riesenfeld (1903): 2093]. Actually, in 1906 the *"microbalance according to Nernst"* was offered for *"Mk. 80,00"* with the remark *"It allows to determine weights down to about 2 mg with an accuracy of 1 – 2 thousandth mg."* [Blücher (1906): 1176]. Due to the inflation, the price increased to 145 *Mark* and then to 212 *Mark* (1928).

The importance of NERNST's microbalance can be seen from the fact, that within two decades there were several improvements while the basic principle remained the same. A disadvantage of the basic principle was the small weighing range. In 1906 WILHELM KUHLMANN could increase the range up to 20 g at a sensitivity of 0.01 mg. In Graz, JULIUS DONAU and FRIEDRICH EMICH were working on the application and improvement of the construction by NERNST. So, for example, it was possible, to obtain from EMICH the NERNST-balance equipped with a mounting arrangement and with a read-off microscope. In 1915 RIESENFELD announced a *"new microbalance"*, which could measure masses down to $3.3 \cdot 10^{-8}$ g at a maximum load up to $5 \cdot 10^{-3}$ g, and where the improvement mainly consisted of the kind of mounting. In 1923 FRITZ PREGL increased the sensitivity of the balance of KUHLBAUM by a factor of ten. At any rate, even in 1928 the basic type of NERNST was offered together with the version developed by PREGL.

The described apparatus for determining the vapor density was used later in the laboratory of NERNST in Berlin, in addition to that in Göttingen. So in 1905 during his studies in Göttingen on the decomposition of CO_2 into CO and O_2, LEO LÖWENSTEIN replaced the iridium parts of the measuring device by those made of platinum. In Berlin, for measuring the temperature the WANNER-pyrometer introduced in 1900 was employed, in which NERNST had a general interest [Nernst and Wartenberg (1906)]. During his experiments on the dissociation of water vapor and for the determination of the mole mass of silver, HANS VON WARTENBERG used the apparatus in Berlin.

As will be discussed later, the research of NERNST in 1905 in Berlin led to the discovery of his thermal law, the Third Law of Thermodynamics. For the experimental demonstration of this law of nature, measuring techniques were needed, which NERNST improved again ingeniously or even developed. At this point we wish to describe briefly some examples of this.

During the research connected with the thermal theorem, the measurement of the specific heat at low temperatures played a central role. In this connection NERNST remarked: *"Together with my coworkers I constructed two kinds of calorimeters, the copper calorimeter and the vacuum calorimeter"*, and he emphasized, that *"for our purpose in particular the latter apparatus became important."* [Nernst (1918a): 21].

Starting from previous instruments, such as the calorimeter built by his PhD student HERMANN SCHOTTKY, in 1910 NERNST, FRITZ KOREF, and FREDERICK A. LINDEMANN described the copper calorimeter [Nernst et al. (1910)] (Fig. 4.16). In this case, instead of a calorimeter liquid, a thermally insulated block of copper, having an excellent thermal conductivity and, hence, providing a steady equilibration of temperature, was used carrying a hole, into which the heated or cooled substance to be studied was inserted. The temperature change of the block was measured using thermocouples, where the solder joints at the bottom were located within the calorimeter block, and the others within a ring-shaped copper block at constant temperature. In order to provide thermal insulation, the copper block was placed within a double-wall vessel, which was evacuated and silvered. The apparatus including the copper ring was inserted into a bath kept at constant temperature by means of ice or solid carbondioxide. The substances to be studied were taken up by thin-wall silver vessels and were inserted into the exactly fitting hole of the calorimeter block after the temperature had been measured exactly by means of a thermocouple attached to the center of this vessel.

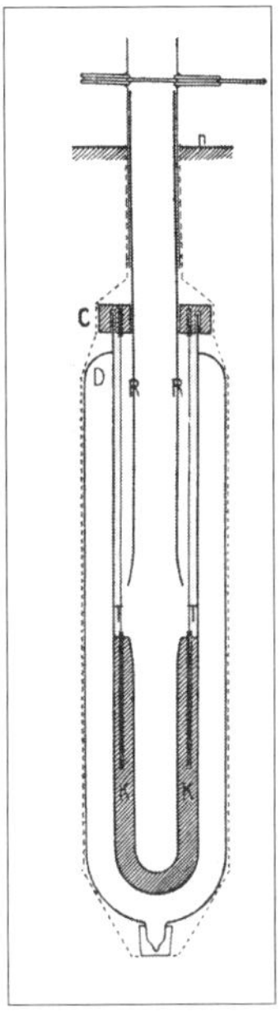

Fig. 4.16 The copper calorimeter of NERNST (from [Nernst (1918a): 22]) (K calorimeter copper block, T thermocouple, D vacuum container, C ring-shaped copper block, R tube for inserting the substance to be studied).

The copper calorimeter had the strong disadvantage that it allowed to measure the specific heat only as a value averaged over a large temperature interval, and that it would be unsuitable for its operation at lower temperatures such as that of liquid hydrogen, as presumed by NERNST without an experimental test.

In order to overcome these deficiencies, NERNST developed the vacuum calorimeter in two versions. In this case the basic principle consisted of the idea, that the substance itself to be studied was utilized as the calorimeter. In the first version, the substance was heated by adding to it a known amount of energy by means of a platinum wire. At the same time, this wire served as a resistance thermometer for measuring the temperature change. It is likely, that NERNST got this idea from the studies of the temperature dependence of the specific heat of metals, with which WOLFGANG GAEDE obtained his PhD in 1902.

Since GAEDE had restricted himself to the metals with their excellent thermal conductivity, an influence of the environment, leading to a temperature gradient within the calorimeter and, hence, to an inaccurate measurement, did not play any role. On the other hand, NERNST, who had to study a large variety of substances, placed the calorimeter within a good vacuum in order to obtain a practically perfect thermal insulation (Fig. 4.17a). So he could remark: *"By the way, Gaede already had worked with a similar method, however, without using a vacuum, without which the experiments described here could not have been carried out."* [Nernst (1918a): 24].

In 1909 NERNST suggested to his private assistant ARNOLD EUCKEN, to realize the technique invented by him, which the latter achieved perfectly. As a result, NERNST and FRANZ POLLITZER then developed the method further especially for the range of very low temperatures.

The vacuum was generated by means of the mercury rotation pump constructed by GAEDE only in 1905 and was improved in addition using charcoal from coconut strongly degassed by heating it in vacuum. For the actual calorimeter three types were developed, in order to obtain optimum experimental conditions corresponding to the particular kind of substance and temperature range. For the experiments with metals the calorimeter was fabricated from them as a cylinder block with a hole, into which a small rod of the same material covered by windings of a platinum wire was inserted (Fig. 4.17b). Nonmetallic substances were filled into a silver vessel, where the windings of the platinum wire were located either in its interior (Fig. 4.17c) or in the case of experiments with liquid hydrogen on the outside (Fig. 4.17d).

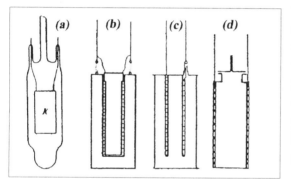

Fig. 4.17 First version of the vacuum calorimeter: The actual calorimeter K in the vessel to be evacuated (a), for the study of metallic (b) and of nonmetallic substances using liquid air (c) or liquid hydrogen (d) (from [Nernst (1918a): 25–26]).

The glass vessel with the calorimeter was surrounded by a cooling bath of liquid air or liquid hydrogen. Its filling with the corresponding cooling liquid provided the bath temperature of the calorimeter, which was kept also, if the pressure above the liquid air or hydrogen was reduced by pumping. In this way, NERNST was able to reach temperatures of 60 K in the case of liquid air and of 22 K in the case of liquid hydrogen.

The use of platinum as resistance thermometer had the disadvantage, that in the case of this metal at low temperatures the variation of the resistance with the temperature shows an unfavorable behavior, leading to uncertainties of the measurements in spite of a careful calibration. Furthermore, during the absorption of energy the platinum wire changes its resistance, such that the exact determination of the absorbed electric energy becomes very complicated.

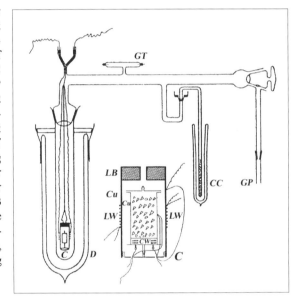

Fig. 4.18 Apparatus for measuring the specific heat at low temperatures showing the second version of the vacuum calorimeter C (D DEWAR vessel, GP connection to the GAEDE pump, CC thermally degassed charcoal of coconut within the cooling bath, GT GEISSLER tube for checking the vacuum, Cu copper jacket or copper vessel for taking up the substance, LB lead block, LW lead wire for measuring the temperature of the copper jacket, CW constantan heating wire).

In order to avoid these disadvantages, instead of the platinum wire, NERNST used wires of lead and constantan. In the case of the second version of the vacuum calorimeter, which NERNST developed together with FRÉDÉRIC SCHWERS, in order to simplify the procedure the temperature was measured by means of a thermocouple with one solder joint at the calorimeter and the other at a lead block, the temperature of which could be considered as being constant. For generating the vacuum the molecular pump or the mercury rotation pump of GAEDE was used, as well as thermally degassed charcoal of coconut cooled with liquid hydrogen. In Fig. 4.18 we show the scheme of the utilized measuring apparatus and the second version of the calorimeter.

With the help of the second version of the vacuum calorimeter, after the clarification of many details not discussed here, the temperature dependence of the specific heat at low temperatures of a large number of substances could be studied conveniently in NERNST's laboratory. The overall goal of this effort was explained by NERNST as follows: *"The task to know exactly the behavior of the specific heat of a large number of substances as completely as possible, is clearly important, both regarding every future theory of the solid state, which apparently will become further developed based on the behavior at low temperatures, and also regarding the applications of our Thermal Law and the study of the chemical affinity."* [Nernst (1918a): 32].

4.12 Mathematics and Chemistry

We have pointed out already several times, that as an experimental physicist and physical chemist WALTHER NERNST had good relations with mathematics. Its foundation had been laid by the excellent education along this direction at the *Gymnasium* (high school) in Graudenz and by his studies with important mathematicians. His outstanding capabilities in this field, beside those in certain parts of his papers and from his textbook on *"Theoretical Chemistry"*, can be seen from the fact, that in 1894 LUDWIG BOLTZMANN proposed his former student to become his successor as the Chair of Mathematical Physics in Munich, and that there was a desire to follow this recommendation.

Even from the standpoint as a physicist, during his occupation with physical chemistry NERNST had also turned to chemistry. Its relation to mathematics suffered from two aspects. On the one hand, IMMANUEL KANT had coined the word, which turned out to be invalid already during his time: *"... so 'Chymie' can never develop into more than a systematic art, or experimental rule, but never into a separate science, because its principles are only empirical and do not allow a representation a priori in the perception, hence, it can never make comprehensible even in a minimum way the basics of chemical phenomena, because it is unable to apply mathematics."* [Kant (1786): X]. On the other hand, the majority of the chemists, in particular those not representing physical chemistry, but also even a few physical chemists, were by no means convinced of the necessity to introduce mathematics into their discipline.

However, already during the first half of the 18th century the Russian scientist MIKHAIL V. LOMONOSOV had demanded: *"Anybody who wants to enter more deeply into the facts of chemistry must study mechanics without any question. However, since the knowledge about mechanics requires familiarity with pure mathematics, any person interested in a detailed study of chemistry must also be well acquainted with mathematics."* [Lomonosov (1961): 72]. Around the middle of the same century the encyclopedia edited by DENIS DIDEROT and JEAN LE ROND D'ALEMBERT under the entry «*Chymie ou Chimie*» had indicated different from KANT: *"Since chemistry had taken up the special form of a science, i.e., since it had adopted the dominating physical systems and step by step had turned into the Cartesian, corpuscular, Newtonian, academic, and experimental chemistry, different chemists have told us ideas from it, which are more clear and comprehensible, because they were guided by the standard logic of science."* (quoted from [Naumann (1984): 177]).

Two works, indirectly or directly important for chemistry, were connected with the town of Konigsberg (now Kaliningrad, Russia) of the 18th century, where KANT had created his «*Metaphysische Anfangsgründe*» (*"Metaphysical Origins"*) appearing in 1786 [Kant (1786)].

The first work had been published by LEONHARD EULER under the title «*Solutio problematis ad geometriam situs pertinentis*» (*"Solution of a Problem Concerning the Geometry of the Location"*) exactly 50 years prior to KANT's publication. It contains the solution of the so-called Konigsberg bridge problem and must be looked at as the hour of birth of the graph theory. The concept of the *"geometry of the location"* originating from GOTTFRIED WILHELM LEIBNIZ implicates a kind of mathematics, which only deals with qualities by ignoring the treatment of quantities (*quanta*). According to KANT the qualities can only be represented by the empirical perception. Since chemistry is a qualitative science in the sense, that the changes essential in its case are of a qualitative nature, i.e., in the sense of HEGEL they are changes of the object itself, the graph theory is extremely important for the description of its foundations. At any rate, the notion of 'graph' is a short version of 'chemicograph' customary in the mathematical treatment of the chemical theory of structure of the 19th century, which was introduced into science by JAMES JOSEPH SYLVESTER. Since these developments do not play any role in the following, these remarks may be sufficient, and we may refer to the expositions in [Bartel (1996): 11–26].

The paper, which was immediately important for chemistry, is the PhD thesis «*De Usu Matheseos in Chymia*» (*"About the Use of Mathematics in Chemistry"*), presented by JEREMIAS BENJAMIN RICHTER in Konigsberg in 1789, i. e., three years after the appearance of the «*Metaphysische Anfangsgründe*» by KANT. Prior to this, RICHTER had studied in Konigsberg mathematics and with KANT philosophy. This thesis contains the theory of the mass ratios of chemical reactions fundamental to chemistry, which RICHTER called '*Stöchiometrie*' (stoichiometry) derived from τὸ στοιχεῖον, the Latin *elementum*.

In particular because of the atomic theory developed by JOHN DALTON at the beginning of the 19th century, chemistry could be quantified and, hence, became accessible to a mathematical treatment. This did not only concern the now available theoretical foundation of stoichiometry, but also a possible newly provided connection with mechanics and, hence, with mathematics in the sense of the mentioned demand by LOMONOSOV and of the quoted statement in the encyclopedia of DIDEROT and D'ALEMBERT.

These developments contributed to the fact that with increasing degree the chemistry of the 19th century teamed up with the physics in generating Physical Chemistry in the form of electrochemistry, chemical thermodynamics, and chemical kinetics, understanding itself as 'theoretical chemistry'. The mathematical discipline primarily dominating these fields was the analysis, the foundations of which were created by NEWTON and LEIBNIZ, and which since experienced a continuing further development. Therefore, with respect to the entropy principle of thermodynamics HANS JAHN has observed: *"With this theory for the first time the cutting weapon of the mathematical analysis was made available also to the theoretical chemistry."* [Jahn (1892): 141].

In spite of the apparent necessity to penetrate into chemistry in the form of analysis and other mathematical calculating schemes, around 1900 many chemists were still opposed to mathematics. As a typical example we mention EMIL FISCHER, about whom WALTER ADOLF ROTH reported in connection with his habilitation in Berlin in 1903: *"In order to demonstrate to him as a co-referee the advantage of physical chemistry, I had prepared a lecture about physical-chemical methods of analysis. Fischer declined at the last moment: 'In the thesis there are even differentials and Intejrale[sic]! Planck must do this!'"* [Roth (1949): 226 – 227].

Of course, researchers, teachers, and students were open-minded in the case of becoming exposed to physical chemistry, if they looked at the latter from the standpoint of a physicist. This was clearly the case with WALTHER NERNST, who had to represent physical chemistry in Göttingen anyhow as a special discipline of physics. So it is not surprising, that around him there developed the desire for a textbook of mathematics written for scientists. For the task of writing such a book NERNST could win over his colleague ARTHUR SCHÖNFLIES, who had been offered the position as Professor of Applied Mathematics, which had been established in Göttingen in 1892. So on June 26, 1895 he reported to WILHELM OSTWALD: *"Within a few weeks there will appear a mathematics for scientists (ca. 20 sheets) by Schönflies and myself; the book will turn out quite well, I can say this myself, since I have written only little (a number of examples) and Schönflies is an excellent pedagogue. Since years I had been approached about such a book, in the end I decided to do it, after Schönflies had been won over for this. ... Now we tell each other, S. and I, that it is mainly due to your activity, if among chemists there is a requirement to differentiate and to integrate, and accordingly we want to document this also to the outside, most preferably by dedicating the book to you."* [Zott (1996): 94].

Indeed, the textbook with the complete title *"Introduction to the Mathematical Treatment of the Natural Sciences – Compact Textbook of the Differential and Integral Calculus with Particular Emphasis on Chemistry"* [Nernst and Schoenflies (1895)] was *"most friendly dedicated by the authors to Herr Professor Dr. Wilhelm Ostwald in Leipzig"*. The dedication and already its announcement in the letter just quoted represent a hidden critique, since on the one hand OSTWALD deserved extraordinary credit for the establishment of Physical Chemistry, but on the other hand he cared only very little for the implicit need to strengthen this new discipline with a mathematical backing.

In contrast to this, HANS JAHN, who had started his scientific career as a pure chemist like WILHELM OSTWALD and being of the same age, was a glowing fighter for mathematics in chemistry. Therefore, in the preface of their textbook NERNST and SCHÖNFLIES quote a passage [Nernst and Schoenflies (1895): V], which JAHN had placed at the beginning of his *"Compendium of Electrochemistry"*, which had also appeared in 1895. In spite of its length we quote it here completely, since it characterizes very well the situation at the end of the 19th century: *"Not infrequently one can hear the reproach against Physical Chemistry, that it had become too mathematical. In the interest of Theoretical Chemistry, to use the weapon of the mathematical analysis also for the solution of its problems as much as possible, I can only see a progress, and measured according to the successes even an eminently healthy progress. Similarly as there does not exist a physicist any more today, who does not try to obtain an understanding of the theoretical part of his science by means of detailed mathematical studies, also the chemists gradually must get used to the idea, that theoretical chemistry will remain for them a 'book with seven seals' without the familiarity with the elements of higher analysis. A differential or an integral symbol must not remain an unintelligible hieroglyphic for the chemist, the notation of these symbols must become to him as familiar as that of his stoichiometric formulas, if he does not want to expose himself to the risk of losing any understanding of the development of theoretical chemistry. Therefore, it is a hopeless task to try to explain halfway in a discussion of many pages what is indicated to the expert by a single line of an equation, and to lead to the final answer along bumpy and impassable roads, to which the mathematical analysis has paved already a royal path."* [Jahn (1895): IV].

NERNST and SCHÖNFLIES expressed the aim of their book with the words: *"It is the aim of the present book to make the study of the higher mathematics easier to the students of the natural sciences. ... The selection of the material followed mainly from the principle to provide an access to the study of physical chemistry as well as of the elements of theoretical physics. Also the physiologist, the botanist, the mineralogist, etc. will be able to orient himself sufficiently from it regarding the mathematical needs of his field."* [Nernst and Schoenflies (1895): V–VI]. The quality and popularity of this textbook is indicated by the fact, that after the first edition nine further editions followed until 1923, while during the same time the total number of pages increased from slightly more than 300 up to 502. Finally, in 1931 NERNST and WILHELM ORTHMANN prepared a revised edition with 478 pages. Already in 1900 there appeared an English translation [Young and Linebarger (1900)].

Chapter 5

Professor of Physical Chemistry in Berlin (1905 – 1922)

5.1 The Friedrich-Wilhelm University and Other Academic Institutions in Berlin and Charlottenburg

WALTHER NERNST was active in Göttingen as an academic teacher and scientist up to 15 years. In Berlin it was almost twice as long. Before we look more closely at his scientific activity and his actions dealing with the organization of science in the German capital, we discuss briefly the most important academic institutions, which existed in Berlin and in Charlottenburg located nearby, where NERNST continued his scientific career.

The first educational institution in Berlin was the *Collegium medico-chirurgicum* originating in 1713 from the *Theatrum anatomicum*. Here the chemist, medical doctor, and main founder of the phlogiston theory, GEORG ERNST STAHL introduced lectures on chemistry for physicians. Since 1716 he was physician in ordinary of the Prussian King FRIEDRICH WILHELM I. The practical instructions were held in the laboratory of the court pharmacy in Berlin.

Shortly before the end of the 18th century GOTTFRIED WILHELM LEIBNIZ presented to FRIEDRICH III, Elector of Brandenburg at the time, his ideas on the foundation of an Academy of Science in Berlin. Only a little later FRIEDRICH III became King of Prussia as FRIEDRICH I. In order to achieve a practical profit (*«realen Nutz»*), this association of scholars needed to *theoria cum praxi coniungere*. By then the institution founded in 1700 as a partnership of the sciences with LEIBNIZ being its first president had retained this motto during three centuries, although its name had been changed several times: "Royal Prussian Academy of the Sciences", ... , "Academy of the Sciences of the GDR", "Berlin-Brandenburg Academy of the Sciences". All important representatives of the humanities and of the natural sciences were elected to Ordinary Members of the Academy. This immediate circle was supplemented by outstanding scholars as Corresponding Members.

In addition to the actual partnership of scholars, within the Academy selected research facilities and project-oriented research groups were installed. For example, one of these was the Chemical Laboratory. It was established after the French mathematician, astronomer, and biologist PIERRE LOUIS MOREAU DE MAUPERTUIS, who was President of the Academy at the time, indicated to his King FRIEDRICH II in 1748: *"Our chemists outrank all chemists in Europe."* Its first director was ANDREAS SIGISMUND MARGGRAF, who in 1747 had discovered the raw sugar in the beet (*Beta vulgaris var. crassa*). His successor was FRANZ CARL ACHARD, who had worked in the fields of optics, acoustics, and electricity, and who had studied gases under chemical, physical, and physiological aspects, but who is also known as the founder of the beet sugar-fabrication. Since 1800 HEINRICH MARTIN KLAPROTH directed the Laboratory. He belonged to the group of founders of analytical chemistry as well as of archaeometry, and had discovered several chemical elements such as, for example, uranium (in 1789 in the form of UO_2).

Among the research projects of the Academy we mention one in particular, which was directed by HENRICUS JACOBUS VAN'T HOFF within the decade 1897 – 1908, and which was entitled *"Studies of the Generation Aspects of the Oceanic Salt Deposits and in Particular of the Stassfurt Salt Deposit"*. In 1895 VAN'T HOFF had accepted the offer of the position as a Research Professor at the Academy of the Sciences in Berlin, which was realized in 1896. Simultaneously, the position as an Honorary Professor at the University of Berlin was established for him with a budget of an *Extraordinariat*. The results of the studies were presented in 52 reports, which were prepared by VAN'T HOFF together with a total of 29 coworkers, to which belonged mainly WILHELM MEYERHOFFER, but also JEAN D'ANS, FREDERICK G. COTTRELL, and HANS VON EULER-CHELPIN.

In his inaugural speech VAN'T HOFF explained the motivation for turning to this research subject with the following remarks, the quotation of which is also interesting in view of our discussion in Section 4.12: *"It was quite correctly just recognized, especially by OSTWALD, that for the complete connection between chemistry and mathematics just the second joining element, physical chemistry, becomes necessary, since the combination of chemistry and mathematics bears fruit mainly due to physics, and since this physics is connected already with mathematics. ... It is clear in which direction I will work: the connection between chemistry and mathematics remains my main task, and each aspect in a new environment will be welcome. So at first I want to concentrate on that part of physical chemistry, which deals with the so-called conversion phenomena, the formation of the double salts, and the double exchange; the application of mathematics is possible also in that case, and the prospect of the additional connection with the Stassfurt industry and geology is particularly attractive."* [van't Hoff (1896): 746–747].

When in 1807 Prussia lost its famous University in Halle at the Saale river, since NAPOLEON I had attached this town to the new Kingdom of Westphalia, in 1809/10 the Prussian King FRIEDRICH WILHELM III had the Royal Friedrich-Wilhelm University being established in Berlin according to the plans of JOHANN GOTTLIEB FICHTE, FRIEDRICH DANIEL SCHLEIERMACHER, and WILHELM VON HUMBOLDT. In this case WILHELM VON HUMBOLDT can be looked at as the actual founder of this *Alma Mater*: *"Humboldt, Wilhelm Frhr. v. ... since 1809 in charge of the Prussian Ministry of Culture and founder of the University of Berlin."* [Asen (1955): 87]. To his brother ALEXANDER we owe, that soon this new educational and research establishment gained world renown in the humanities as well as in the natural sciences, because of his famous "Cosmos Lectures", his appointment strategy, the support of talented young scientists, and other accomplishments in the organization of science. Therefore, it is well justified, that 140 years after its opening the oldest University of Berlin in honor of the brothers HUMBOLDT is named "Humboldt University of Berlin".

In order to illustrate the importance of the Friedrich-Wilhelm University, we list a small selection of scientists, who were appointed within the first 50 years after its foundation. Here we must note also, that many of the people listed, prior to their appointment as Full Professor, had been active already as *Privatdozent* or as *ausserordentlicher* Professor. In addition to SCHLEIERMACHER (appointed in 1809 for Theology and Philosophy, Rector during 1815/16) and FICHTE (1810 for Philosophy), the first *Wahlrektor* (elected rector) during 1811/12, we mention the following: for the humanities FRIEDRICH AUGUST WOLF (1809 for Classical Philology), the main representative of the historical law school FRIEDRICH CARL VON SAVIGNY (1810 for Roman Law, Rector during 1812/13), AUGUST BOECKH (1810 for Classical Philology, Rector during 1825/26, 1837/38, 1846/47, and 1859/60), GEORG WILHELM FRIEDRICH HEGEL (1818 for Philosophy, Rector during 1829/30), FRANZ BOPP (1825 for Comparative Philology and Sanskrit), CARL RICHARD LEPSIUS (1846 appointed to the first German Chair of Egyptology), LEOPOLD VON RANKE (1833 for History), GUSTAV PETER LEJEUNE-DIRICHLET (1839 for Mathematics), ERNST EDUARD KUMMER (1855 for Mathematics), as well as THEODOR MOMMSEN (1861 for ancient History, Rector during 1874/75), who in 1902 as the first German received the second Nobel Prize of Literature, and for the natural sciences PAUL ERMAN (1810 for Physics and Meteorology), JOHANN GEORG TRALLES (1810 for Physics and Higher Mathematics), MARTIN HEINRICH KLAPROTH (1810 for Pharmaceutical Chemistry), SIGISMUND FRIEDRICH HERMBSTAEDT (1810 for Chemistry and Technology), CHRISTIAN SAMUEL WEISS (1810 for Mineralogy, Rector during 1818/19, and 1832/33), EILHARD MITSCHERLICH (1825 for Chemistry, Rector during 1854/55), HEINRICH ROSE (1835 for Analytical Chemistry), GUSTAV ROSE (1839 for Mineralogy), HEINRICH WILHELM DOVE (1844 for Physics, Rector during 1871/72), GUSTAV MAGNUS (1844 for Physics, Rector during 1861/62), JOHANN FRANZ ENCKE (1844 for Astronomy, Rector during 1853/54), RUDOLF VIRCHOW (1856 for Pathology and Anatomy, Rector during

1892/93, and EMIL DU BOIS-REYMOND (1858 for Physiology, Rector during 1869/70 and 1882/83).

We must mention that also distinguished scientists belonged to the teaching members of the University, which exercised this function as lecturing members of the Academy. Among these people we find, for example, ALEXANDER VON HUMBOLDT and the brothers JACOB and WILHELM GRIMM, who took up this task since 1827 for geography, since 1841 for German literature and mythology, and since 1842 for *Germanistik* (German language and literature studies).

In 1806 ALBRECHT THAER, the founder of the agricultural sciences, in the village of Möglin located about 50 km north-east of Berlin had created the first European educational Institution of Agriculture, the "Royal Prussian Academic Educational College of Agriculture". In February of 1881 it changed into the "Royal Agricultural *Hochschule*" located in the *Invalidenstraße* in Berlin. The establishment of this Agricultural College with the right of awarding the PhD was due in large part to the sugar industry, which also was heavily involved in the appointment of HANS LANDOLT in 1880 as Full Professor of Chemistry, who then became its first Rector. Previously, this well-known physical chemist had studied successfully at the Polytechnic Institute in Aix-la-Chapelle the behavior of solutions of optically active substances in polarized light, and in 1879 had published an extended monograph on this subject. Since 1881 at the Agricultural College, there were working as teachers and researchers in addition to many others: RICHARD BÖRNSTEIN as Professor of Physics (Rector 1908 – 1910), EDUARD BUCHNER, recipient of the Nobel Prize in Chemistry of 1907, during 1898 – 1909 as Professor of Chemistry, and since 1899 MAX EMIL JULIUS DELBRÜCK as Professor of Technical Chemistry (Rector 1898 – 1900).

Similar as the Institution established in 1798 and raised to the College of Veterinary Medicine in 1887, during October of 1934 also the Agricultural College was incorporated into the University of Berlin each as a special Faculty, respectively.

We must mention also the Trade School of Berlin, which was established in 1821 as "Technical School" at the initiative of the high-level

Prussian state official CHRISTIAN PETER WILHELM BEUTH. Since 1827 it carried the name "Trade Institute", and in 1971 it became the Technical College (*Technische Fachhochschule*) of Berlin. The subjects, which were taught at the Technical College to the future technicians, were physics, chemistry, technical drawing, and mathematics. One of the most famous teachers was FRIEDRICH WÖHLER, who taught chemistry here during 1825 – 1831, and who was promoted to Professor in 1828. During this time there happened, among other things, WÖHLER's synthesis of oxalic acid from cyanogen (NC–CN) (1824) and of urea by means of the transformation of ammonium cyanide (1828), on the basis of which the doctrine of the *vis vitalis* could be disproved. This *vis vitalis* had been considered to represent the necessary driving force (*Lebenskraft*, vital force) for the generation of organic substances, existing only within a living organism. The organic compounds oxalic acid and urea had been prepared from the inorganic substances cyanogen and ammonium cyanide, respectively, outside of an organism.

In Charlottenburg, which had received its city rights in 1705 and which belonged to "Greater Berlin" since 1920, in 1879 the Building Academy established in 1799 and the Trade Academy founded in 1821 were united to form the Royal Technical College (*Königlich Technische Hochschule*) of Berlin, to which in 1916 was added also the Mining College (*Bergakademie*) existing since 1770. After the First World War this educational institution was renamed to become the Technical College of Berlin (*Berliner Technische Hochschule*). As one of the first such institutions it was ranked equal to the classical Universities. When in 1946 it was opened again, as the first Technical College in Germany it was allowed to carry the name "Technical University". The importance of this teaching and research institution in Charlottenburg for the natural sciences can be seen, for example, from the fact, that here the physicists CARL ADOLF PAALZOW (1885–1904), HEINRICH RUBENS (1904–1906), FERDINAND KURLBAUM (1906–1927), and GUSTAV HERTZ (1928 to 1936), recipient of the Nobel Prize in Physics of 1925, were working.

In a subsequent Chapter we will cover the *Physikalisch-Technische Reichsanstalt* established also in Charlottenburg in 1887 at the initiative of the engineer and industrialist WERNER VON SIEMENS, the physicist HERMANN VON HELMHOLTZ, and the astronomer WILHELM FOERSTER.

5.2 The Famous Year 1905

For ALBERT EINSTEIN the year 1905 was an *annus mirabilis*, since in this year among other things there appeared his famous papers revolutionizing physics because of their unconventional concepts, dealing with Special Relativity, Quantum Theory (hypothesis of the light quanta), and the BROWN molecular motion. In some sense it was such an exceptional year also for WALTHER NERNST.

In this context we must mention his transfer from Göttingen to the German Capital of Berlin that perfectly met the intentions and wishes of the professor who was just 40 years old. NERNST felt strongly, that his scientific power was approaching its summit. To reach this top level, the scientific environment, based on an excellent tradition in science and technology, promised the most effective possibilities. For illustration we mention a selection of scholars in the fields of physics and chemistry, who were active as researchers and teachers in scientific institutions in Berlin and Charlottenburg since 1871 or still were in 1905, when NERNST started his position as the Chair of Physical Chemistry. The year 1871 was chosen, since in this year HERMANN VON HELMHOLTZ received an offer from Berlin, and since this event represented the beginning of the epoch of the Berlin Physics, in which NERNST then played an important role. At the Friedrich-Wilhelm University in Berlin in the field of experimental physics in addition to HELMHOLTZ we can mention AUGUST KUNDT, EMIL WARBURG, and PAUL DRUDE, in the field of theoretical physics GUSTAV ROBERT KIRCHHOFF and MAX PLANCK, in the field of chemistry AUGUST WILHELM VON HOFMANN, CARL FRIEDRICH RAMMELSBERG, and EMIL FISCHER, and in the field of physical chemistry HANS LANDOLT. During this period at the *Technische Hochschule* in Charlottenburg physics was taught by CARL PAALZOW and HEINRICH RUBENS. At the *Physikalisch-Technische Reichsanstalt* in 1894 FRIEDRICH KOHLRAUSCH had followed HELMHOLTZ as its president, and just in 1905 he was superseded in this office by EMIL WARBURG. In addition to LANDOLT, for the Agricultural Academy in Berlin we mention RICHARD BÖRNSTEIN and EDUARD BUCHNER. Last but not least there was also VAN'T HOFF, who occupied the position of a professor at the Academy and was teaching at the University.

In addition to the sciences, the medical field and the humanities in Berlin also enjoyed international prestige. Regarding the latter, we think in particular of the philosophical, the philological, the historical, and the archaeological disciplines, as well as in particular of the mathematics, which similar to physics had established a typical Berlin School. This school had been influenced for example by PETER GUSTAV LEJEUNE-DIRICHLET, ERNST EDUARD KUMMER, KARL WEIERSTRASS, LEOPOLD KRONECKER, and HERMANN AMANDUS SCHWARZ.

The offer to NERNST from Berlin was triggered by the intention of LANDOLT to retire from his chair at the University of Berlin on March 31, 1905 because of age. The selection committee formed by the order of FRIEDRICH ALTHOFF consisted of its chairman LANDOLT and among others of EMIL FISCHER, MAX PLANCK, and EMIL WARBURG. Of course, the professor they were looking for was to represent the field of Physical Chemistry in an excellent way. This field had been represented at the University of Berlin for the first time by LANDOLT since 1891, and had well established itself such that it definitely had to be continued because of its importance. Last but not least, HANS JAHN heading a section in LANDOLT's institute would have also been selected as a candidate due to his merits in the field of electro-chemistry, if he was not severely hindered by his almost complete deafness. However, primarily one had to consider WILHELM OSTWALD, JACOBUS HENRICUS VAN'T HOFF, SVANTE ARRHENIUS, and, of course, NERNST. The first was not suitable, since at this time he devoted himself almost totally to philosophy. Furthermore, in Berlin it was a necessary custom, that the professor of physical chemistry had to also present the lectures in inorganic chemistry. One did not want to burden OSTWALD and VAN'T HOFF with this load. For 1905 ARRHENIUS had been promised the position of the director of the section on physical chemistry at the Nobel Institute in Stockholm, since he did not want to leave Sweden. Therefore, EMIL FISCHER proposed to put those three physical chemists on the selection list only honorarily and to give NERNST the first place. This plan was supported by MAX PLANCK, and all other committee members then accepted it. The list further included the names of PAUL WALDEN, GUSTAV TAMMANN, and MAX LE BLANC. Based on the corresponding note of LANDOLT to the Secretary, on December 2, 1904 the University of Göttingen was told: *"Effective from April 1, 1905 I have transferred Professor Dr. Walther Nernst to the Philosophical Faculty of the University of Berlin with identical duties as before."* [UAHUB: I, 92]. In fact LANDOLT retired from his office on the day of Easter, i.e., on PLANCK's birthday on April 23, as NERNST always emphasized.

During the year 1905 some events took place in the organization of science, which became highly important to the future. Following a suggestion by WILHELM OSTWALD, NERNST and EMIL FISCHER tried to establish a *Chemisch-Technische Reichsanstalt*, which whould allow intensive and expensive experiments in the field of chemistry without perturbation by teaching duties, following the example of the *Physikalisch-Technische Reichsanstalt* established in 1887. Apparently, NERNST was highly interested in this subject, since his letters to OSTWALD from the year 1905 were concerned mainly with this projected research facility and its organization. A corresponding outline by OSTWALD was complemented and returned in this completed form on August 7. On September 8, NERNST presented to his two colleagues his proposal for the organization of the governing board. The subsequent events will be discussed later in conjunction with NERNST's activities in the organization of science.

Prior to his departure for Berlin, NERNST received a special recognition of his scientific achievements. Already in 1896 WILHELM II had presented to him the Order of the Red Eagle, class IV. However, now he was told, *"His Majesty the Emperor and King has given to Professor Dr. Nernst the title of 'Geheimer Regierungsrat'. The patent is dated from March 20, 1905."* [UAHUB: I, 98]. Still in this year VICKY ZAESLEIN-BENDA created the oil painting of *"Geh. Rat Prof. Dr. Nernst"* (Fig. 5.1), which existed for some time within the trade in works of art, but seems to be lost presently. NERNST was very proud of this title and was quite displeased, when somebody without a proper reason addressed him only with *"Herr Professor"*.

NERNST received another honor after he had started his position as Professor in Berlin: On November 24, 1905 he was elected as a full member of the Royal Prussian Academy of Science of Berlin. The corresponding nomination had been presented to the Physical-Mathematical Class on June 22. It was prepared by VAN'T HOFF and was signed together with HANS LANDOLT, EMIL FISCHER, MAX PLANCK, and EMIL WARBURG.

Fig. 5.1 Painting by VICKY ZAESLEIN-BENDA of «*Geh. Rat Prof. Dr. Nernst*» (1905).

All these honors in the year 1905, University Chair in the Capital, Title of *Geheimrat*, and membership of the Academy, were based on the past achievements in science. However, they also expressed expectations for the future. Already during the summer of this eventful year NERNST produced the foundation for these expectations: the discovery of NERNST's thermal theorem, which already soon became known as the Third Law of Thermodynamics.

5.3 The Institute of Physical Chemistry at the University of Berlin

Before we discuss in detail the discovery and the subsequent examination of this extremely important law of nature, we will describe briefly the Institute entered by NERNST during April of 1905, and where during August of the same year he first publicly announced his thermal law. It was in this Institute, where within a little more than a decade he and his co-workers investigated this theorem experimentally and theoretically.

During the last decades of the 19th century the required new institute buildings for the research in the field of science and technology were built in a highly elaborate way corresponding to the high esteem these fields enjoyed in Germany. In a particular way this concerned also the complex of the Institute of Physics at the University of Berlin and the attached residence of the director. During the time 1873 – 1878 along the river Spree near the *Reichstag* building, this building had been erected for HERMANN VON HELMHOLTZ under the direction of the architects FRITZ ZASTRAU and MORITZ HELLWIG. HELMHOLTZ still desired an extension of the building, which together with the residence of the director was to symmetrically complete the institute building. Again, this task was given to ZASTRAU. The additional construction had started already in 1879.

During this time the chemical industry experienced a tremendous growth, the number of students of chemistry increased strongly. Therefore, the Institute of Chemistry in the *Georgenstraße*, built by AUGUST WILHELM VON HOFMANN and ZASTRAU between 1865 and 1867, became too small, and even more so, since in addition to that of HOFMANN it also accommodated the second Chair of Chemistry occupied by CARL FRIEDRICH RAMMELSBERG. When in 1881 HOFMANN became Rector of the University, he was able to effect that the extension of the building planned for Physics could be occupied by Chemistry. That is the reason, why the year 1881 and not 1879 is mentioned by NERNST as the start of the construction of the Second Chemical Institute [Nernst and Sand (1910): 306].

The new institute building at the corner of *Schlachtgasse* and *Reichstagufer* was opened in 1883. The ensemble of the buildings at the *Reichstagufer*, which all became the working locations or the residence of NERNST, then looked as can be seen in Fig. 5.2. The ground floor of the new building became occupied by the Technological Institute, directed by CARL HERMANN WICHELHAUS during the time when NERNST came to

Berlin. The new Chemical Institute, referred to as "2nd Chemical Laboratory", was located in the two upper floors. Here under the direction of RAMMELSBERG people worked in the field of inorganic chemistry and in particular of the chemistry of minerals.

Fig. 5.2 The Technological Institute and the 2nd Chemical Laboratory (the later Institute of Physical Chemistry) (left), the Institute of Physics (middle), and the residence of its director (right) at the river Spree in Berlin.

When in 1891 HANS LANDOLT left the Agricultural Academy in Berlin, where he had worked as a scientist and teacher since 1881 and where he had become its first Rector, in order to occupy the first Chair of Physical Chemistry at the University of Berlin as the successor of RAMMELSBERG, the name of the Institute remained unchanged. However, the *Schlachtgasse* has been renamed after ROBERT WILHELM BUNSEN, LANDOLT's teacher from Heidelberg, who also had contributed significantly to the development of the field of physical chemistry. Only on April 13, 1905, following an order of the Secretary, the 2nd Chemical Laboratory was changed to "Physical-Chemical Institute", because *"under the new director the main research goals of the Institute should also become visible to the outside."* [Nernst and Sand (1910): 308].

If one compares the building costs of 377 000 Marks for the Technological Institute/2nd Chemical Laboratory with the costs for the Institute of Physics and for the residence of the director, which amounted to 1 264 000 and 315 000 Marks, respectively, [Diestel (1896): 265, 268] one can understand, that in his *"Memoirs of an Old Thermal Chemist"* WALTHER ADOLF ROTH, a student of LANDOLT and HANS JAHN, referred to the 2nd Chemical Laboratory as a *"little Institute"*, however, with a *"cozy"* atmosphere [Roth (1949): 226]. This was the reason, why already during the winter 1891/92 LANDOLT had arranged for modifications in the building. Then from March 1907 until August 1908 NERNST even had the building extended. He had been able to convince the Ministry of Education and that of Finances, as well as that of Public Works, that *"the available space was quite insufficient for the specific requirements of physicochemical research"*, clearly leading to *"a state of emergency, making the need for an expansion most urgent"* [Nernst and Sand (1909): 229]. The building extension, which also concerned the Technological Institute, consists of two parts, which along the *Bunsenstraße* closed the gap to the Pharmacological Institute. GEORG THÜR participated in the preparation of the plans for this new construction. MAX BODENSTEIN, at the time head of a Section, was the representative of NERNST's Institute for the construction project. On January 9, 1909 NERNST and WICHELHAUS presented the new rooms to the public. NERNST felt that a description of his enlarged Institute should be published. Therefore, together with his new Section Head JULIUS SAND, who had taken up this position from BODENSTEIN in 1908, he published such a report in the *«Zeitschrift für Elektrochemie»* [Nernst and Sand (1909)]. The ground plan with the different rooms of the Physical-Chemical Institute of 1909 is shown in Fig. 5.3. Here the laboratory space of NERNST and the lecture hall, where NERNST had announced for the first time the Third Law of Thermodynamics, are marked by D and LR, respectively. The new rooms (shown in bright color in Fig. 5.3) are joining the lecture hall and the laboratory hall P underneath on the left hand.

An impression of the Institute building with its entrance in the *Bunsenstraße* and the extension on the left hand is given in Fig. 5.4.

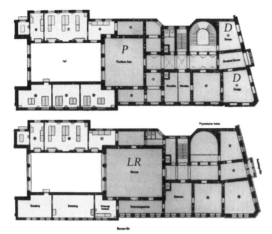

Fig. 5.3 Ground plan of the Physical-Chemical Institute in the *Bunsenstraße* in Berlin in 1909 [Nernst and Sand (1909): 231]. Upper part: First floor. Lower part: Second floor (*D* laboratory space of the director, *LR* lecture hall, *P* laboratory hall). Bright color on the left hand: extension part of 1907/08. Gray color on the right hand: original building.

Fig. 5.4 The Institutes of Physical Chemistry and of Technology in the *Bunsenstraße* in Berlin in 1909.

5.4 The First Lecture in Berlin – Announcement of a Fundamental Law of Nature

Prior to his transfer to Berlin at the end of April 1905, at the 46th Main Assembly of the «*Verein deutscher Ingenieure*» NERNST gave a lecture on the combustion processes in gas engines [Nernst (1905)]. In order to obtain physicochemical data on the equilibrium of these specific reactions between gases, one needs the corresponding equilibrium constant K as a function of the temperature T. Then from the equation $A = RT \cdot \ln K$ the maximum work to be gained from this process, or the affinity A identical with it, can be calculated and vice versa (R universal gas constant). However, the affinity cannot be found easily, if K is unknown. For some important fuels such as methane, alcohol, and benzene neither K nor A were known. Therefore, in his lecture NERNST had to point out: *"However, in these and in similar cases we will not produce large errors, if we simply set A equal to the heat of combustion. ... However, one does not always keep in mind, that sometimes this can only be a rough approximation, since A can be either smaller or larger than the heat of combustion. However, in the absence of more exact data, one cannot argue much against the generally accepted procedure to relate the efficiency to the heat of combustion."* [Nernst (1905): 9]. In principle, these statements, which analogously apply to all chemical equilibria, characterize the difficulty very soon to be solved by NERNST in terms of the Third Law of Thermodynamics, as will be discussed in detail below.

Again the study of the processes in gas engines indicates a peculiarity of NERNST, to combine his personal interests with his scientific research. Also it corresponds to his character, that already at his arrival in the Capital he wanted to demonstrate himself as a modern scientist being at the top in his field. Therefore, he undertook his transfer from Göttingen to Berlin in his own automobile. This event being unusual at the time was well documented in a photo taken from the start in Göttingen (Fig. 5.5). However, due to a breakdown of his vehicle on the road, an unexpected delay affected the trip.

Fig. 5.5 At the departure from Göttingen to Berlin.

A few days after the arrival of NERNST in Berlin, on April 26 the summer term started, ending only on August 15. NERNST was supposed to lecture on inorganic chemistry. However, because of his insufficient time for preparation, he was excused from giving this course. Hence, his course *"Introduction to the Thermodynamic Treatment of Chemical Processes"* allowed him to present himself and his previous research results from this topic to the academic community of Berlin.

In order to demonstrate again the situation at the University of Berlin regarding the lectures in the field of physics and chemistry, we mention a few courses given in 1905 in addition to those of NERNST, FISCHER, and WICHELHAUS: JACOBUS HENRICUS VAN'T HOFF: *"Selected Chapters of Physical Chemistry"*, HANS LANDOLT: *"Inorganic Experimental Chemistry"* (supposed to be covered by NERNST), LUDWIG CLAISEN: *"Organic Experimental Chemistry"*, OTTO DIELS: *"Introduction to the Methods of Organic Chemistry"*, PAUL DRUDE: *"Experimental Physics"*, EMIL WARBURG: *"Electron Theory"*, EDUARD GRÜNEISEN: *"Acoustics"*, ERNST PRINGSHEIM: *"Physics of the Sun"*, MAX PLANCK: *"Theory of Heat, especially Heat Radiation"*, and EMIL ABDERHALDEN: *"Protein Chemistry"*. HANS JAHN, whom NERNST appointed as Section Head, had not announced any course for the summer term in 1905. In the previous winter term he had given a course on thermochemistry and mathematics. It is very likely, that during the winter term 1885/86 in Graz NERNST had attended the course of JAHN *"On the Fundamentals of Theoretical Chemistry"*. In this course NERNST became aware of the problem, the solution of which is the Third Law of Thermodynamics.

We must mention also WILLY MARCKWALD, who was working at the University as a lecturer since 1889 similar to JAHN. In addition to his contributions to the experimental and theoretical organic chemistry, the stereochemistry and physical chemistry, in particular we must mention his contributions since 1902 to radiochemistry, thus ensuring a place next to MARIE and PIERRE CURIE as a pioneer in this field.

Within this environment, with his teaching and research activities NERNST continued in an excellent way a tradition of Berlin: the *"Relations between Chemistry and Physics"*, as a lecture course given by the lecturer FRIEDRICH NEESEN during the winter term 1880/81 had been called.

Fig. 5.6 NERNST (on the right), JAHN (next), and further coworkers in the lecture hall of the Institute in the *Bunsenstraße* during the summer term 1906.

His first lecture in Berlin was given *privatim* on Tuesdays from 11 a.m. until 1 p.m. in the lecture hall of the Institute in the *Bunsenstraße* (Fig. 5.6). Later ERNST HERMANN RIESENFELD recalled: *"During the last lectures it became clear to him, that in order to calculate chemical equilibria from thermal data, one must extend the so-called classical thermodynamics by adding the hypothesis, that near the zero value of the absolute temperature the change of the free energy with the temperature vanishes."* [Riesenfeld (1924): 438]. NERNST had found the solution of a problem in chemistry in the form of a theorem, which became known as "NERNST's Thermal Law" and subsequently as the "Third Law of Thermodynamics". In the year 1905, August 1, 8, and 15 fell on a Tuesday. Hence, the first public announcement of this fundamental law of nature can be placed within the noon hours of these days, and perhaps even of the last day, if we take RIESENFELD's remark quite literally. So in addition to the location, also at least the approximate time of this important event in the history of science can be specified. NERNST himself is said to have claimed, that he discovered the theorem actually during his lecture.

This fact is also indicated on a bronze plate mounted in 1964 in the historic lecture hall: *"In the year 1905 during his lecture given in this lecture hall Walther Nernst discovered the 3rd Law of Thermodynamics"* (Fig. 5.7).

Fig. 5.7 Bronze plate mounted in the lecture hall of the Institute of Physical Chemistry in Berlin.

However, it is more likely, that NERNST kept his announcement until the end of his first presentation in Berlin. Because of his talent as an actor, which he demonstrated quite often since his youth, it is possible that he only enacted his "discovery" during the lecture. After all, it was a great, perhaps the greatest event for him. It is possible, that even at the meeting of the German engineers he did not disclose the solution of the problem of the exact caloric determination of the affinity, already known to him, in order to announce it in Berlin with great fanfare.

At any rate, RIESENFELD's opinion, which he expressed along with his report quoted above, can be accepted: *"It required the nearly unique intuition of Nernst, to find such a hypothesis, and the conviction of being on the right track, as it is given only to great men, for daring such a bold hypothesis."* [Riesenfeld (1924):438].

5.5 The Nernst Law of Heat or the Third Law of Thermodynamics

5.5.1 Remarks on the First and Second Law of Thermodynamics

It turned out, that this *"hypothesis"* is even of such a fundamental nature, that it can be placed next to the two already well known laws of the classical or phenomenological thermodynamics, i.e., being included in the axioms of this field of physics. Here, under an axiom we understand a fundamental statement, the truth of which is only based on experience, but which cannot be demonstrated by deduction due to its fundamental character. Furthermore, it can be evident so clearly, that a proof is not necessary. NERNST recognized the high importance of his thermal theorem from the very beginning. Of course, the theorem was not so obvious, that it did not require an experimental test. Instead, the opposite was the case, since consequences could be derived conflicting with views well supported experimentally and even by the classic theory.

For a better understanding of the following and in order to emphasize the importance of the NERNST thermal theorem as the Third Law of Thermodynamics, at this point we include short remarks on the thermal laws already known before 1905.

The empirical foundation of the First Law of Thermodynamics originates in particular from the experiments performed by JAMES PRESCOTT JOULE during the time 1840 – 1845. These experiments demonstrated the so called equivalence principle, according to which the heating of a specific amount of material by the same temperature increment each time requires exactly the same amount of mechanical work. WILLIAM THOMSON (LORD KELVIN) has expressed this principle in the following way: *"If the same amount of mechanical work W is generated from thermal sources or is dissipated by means of thermal effects, then the same amount of heat Q disappears or is generated"*. According to the subsequent understanding, W can stand for any kind of work.

In addition to JOULE, other scientists of the 19th century independent of each other have looked into the connections between energy, work, and heat, and in this way have contributed significantly to the recognition of the First Law of Thermodynamics. Here we must mention in particular HERMANN VON HELMHOLTZ, JULIUS ROBERT MAYER, and LUDWIG AUGUST COLDING. Furthermore, it is worth mentioning, that in the notes left by SADI CARNOT after his death in 1832 the equivalence principle already had been clearly formulated.

If one defines a quantity U, for which RUDOLF CLAUSIUS has proposed the denotation inner energy of the system, satisfying the equation $\Delta U = \Delta W + \Delta Q$, then from the consideration of a cycle process it follows, that the value of the inner energy depends only on the corresponding state of the system and not on the path leading to this state. Therefore, U is referred to as a state variable, which is generally not true for W and Q.

Furthermore, keeping in mind that in principle extensive quantities, depending on the amount or the total mass of material, such as, for example, U and the entropy S, for a specific system can change internally and externally ($dU = d_{int}U + d_{ext}U$), the First Law of Thermodynamics can be formulated as follows: "• The inner energy U is a state variable; • There exist no internal changes of U (law of energy conservation): $d_{int}U = 0$; • U can change externally by generating or by absorbing work in the system, or by delivering heat from the system to its environment or by absorbing heat: $d_{ext}U = dU = đW + đQ$."

Based on the equivalence principle, MAX PLANCK expressed the First Law of Thermodynamics using a more practical but negative statement: "It is impossible, to perform or to consume work continuously using a periodically operating machine (impossibility of a *perpetuum mobile* of 2nd kind), as it is also impossible, to absorb or to generate heat continuously with such a machine." Here a periodically operating machine is a system, which again and again returns to a specific state.

Highly important for the history of the Second Law of Thermodynamics are CARNOT's treatise «*Réflexions sur la puissance motrice du feu*» from 1824 and the paper by CLAUSIUS «*Über die bewegende Kraft der Wärme und die Gesetze, welche sich daraus für die Wärmelehre selbst ableiten lassen*» (*About the Driving Force of the Heat and the Laws, which can be derived from it for the Theory of Heat itself*) from 1850. In the former CARNOT formulated the theorem, that the work produced by a heat engine only depends on the temperature difference and not on the working medium. In the latter, to which JOSIAH WILLARD GIBBS referred to as the beginning of thermodynamics as a science, CLAUSIUS combined CARNOT's theorem with the equivalence principle and in this way generated the fundamental idea of the Second Law. It was formulated as follows: "It is impossible, that in a system having no exchange with its environment, heat is transported from an object at lower temperature to an object at higher temperature." This formulation is awkward, since it is only by means of the Second Law itself that it becomes clear, in which case a temperature is larger or smaller than another.

Based on these papers and those of JOULE, in 1851 WILLIAM THOMSON presented a formulation for both Thermal Laws. For the Second Law it said: "It is impossible to generate mechanical work using a periodically operating machine only by cooling a heat reservoir." (impossibility of a *perpetuum mobile* of 2nd kind). Subsequently, this formulation was used also by PLANCK. As can be shown easily, it is equivalent to that of CLAUSIUS. We note that the terms *perpetuum mobile* of 1st and of 2nd kind have been introduced into science by WILHELM OSTWALD.

Using the term 'entropy' (from Greek ἐντρέπω: 'turning to'), created by him in 1854 and supplied with a caloric character, CLAUSIUS could formulate the Second Law in a new way. Analogous to the formulation of the First Law given above, today we can say: "• The entropy S is a state variable; • Its internal change, the entropy production, cannot be negative: $d_{int}S \geq 0$; • The external entropy change, the entropy current, is given by the heat current multiplied with an integrating factor a^{-1}: $d_{ext}S = a^{-1}dQ$." One can show, that a can be set equal to the temperature T: $a = T$.

Already in 1848, THOMSON could define this absolute temperature being independent of the thermometer substance, based on the «*Réflexions*» by CARNOT. Even up to the time after the discovery of the Third Law one had overlooked the fact that an explicit axiomatic definition of the concept of temperature did not exist. This was accomplished only by RALPH HOWARD FOWLER in form of the so-called Zeroth Law, which must be placed in front of the axiomatics of thermodynamics. Therefore, and since the numbering of the other Laws of Thermodynamics had already been established, one had chosen this curious denotation. The Zeroth Law indicates that the temperature ϑ is a state variable, and that for two systems A and B the statements "The temperatures of both systems are the same (relation =).", and "Both systems are in thermal equilibrium (relation Γ_{th})." are equivalent ($\vartheta_A = \vartheta_B \Leftrightarrow A\,\Gamma_{th}\,B$), and that the relation Γ_{th} is transitive ($A\,\Gamma_{th}\,B$ & $B\,\Gamma_{th}\,C \Rightarrow A\,\Gamma_{th}\,C$).

Of course, the content of these relations was known already in the 19th century. However, when the theory of thermodynamics was developed with the intention to study the efficiency of heat engines, the content of the Zeroth Law regarding the temperature looked so trivial, that its summary in terms of a law with axiomatic character seemed unnecessary. Assuming this implicit knowledge of the Zeroth Law, we can say, that since about the middle of the 19th century the axiomatics of the phenomenological thermodynamics had reached its preliminary completion.

This preliminary character can be seen, for example, by the fact, that according to the First and Second Laws the inner energy as well as the entropy is given only in the form of differences and not absolutely. However, this presents a problem only regarding the entropy. Although starting from a different original question, it was WALTHER NERNST who achieved the axiomatic completion of the phenomenological thermodynamics.

5.5.2 The problem and its solution given by Nernst

The problem, the solution of which NERNST could present for the first time in his lecture hall in Berlin, had been posed originally by chemistry. In order to describe thermodynamically exactly a chemical reaction in terms of its direction and the location of the equilibrium it reaches finally under the given conditions, we must know the "affinity" A as a function of the (absolute) temperature and perhaps of other variables. Under isochore conditions (no generated power and constant volume, $dV = 0$, if only the work by volume change would be possible) the affinity is equal to the change ΔF of the free energy caused by the reaction, and under isobaric conditions (constant pressure, $dp = 0$) it is equal to the change ΔG of the free enthalpy due to the reaction. In the following we only deal with the first case which can easily be generalized, in this way taking up the arguments by NERNST.

From the First and Second Laws of thermodynamics one can derive the equation

$$A = U + T\frac{\partial A}{\partial T} \quad (1)$$

usually connected today with the names JOSIAH WILLARD GIBBS and HERMANN VON HELMHOLTZ. Here constant temperature ($dT = 0$) is assumed. The quantity U denotes the (inner) energy, where U is taken for the change ΔU, as it can be found in the papers by the NERNST school. The partial derivative $\partial A/\partial T$ indicates the constancy of all extensive variables, in particular of the volume. Under these conditions the change in energy corresponds to the heat of formation Q.

Equation (1) indicates, that the knowledge of the function $A(T)$ allows the calculation of $U(T)$. However, the opposite is not true. In fact, the general solution of the differential equation (1)

$$A = -T\int \frac{U}{T^2} dT + JT \quad (2)$$

contains the integration constant J, for which a specific value could not be given prior to the discovery of the third law. Hence, in principle, for each $U(T)$-curve there exist infinitely many $A(T)$-curves, as illustrated in Fig. 5.8. However, it was clearly impossible to find the true solid $A(T)$-curve of Fig. 5.8, based on a well known rule of nature.

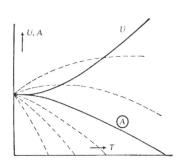

Fig. 5.8 A single given $U(T)$-curve can correspond to infinitely many $A(T)$-curves, only one of which (the solid line) is correct. (from [Eggert (1941): 422]).

Obviously, this circumstance is quite unsatisfactory in particular for chemistry. In 1888 HENRY LE CHATELIER has emphasized this fact in his book «*Les équilibres chimique*» by stating: "*It is very likely that the integration constant [J] ... is a specific function of certain physical properties of the reacting substances. The determination of the nature of the* function [$A(T)$] *would lead to a complete knowledge of the laws for the equilibrium. It would determine a priori the complete conditions for the equilibrium; up to now the exact nature of this constant could not yet be determined."*

The problem *"to base chemistry ... on the same mechanical principles, which dominate already the different branches of physics"*, had been addressed since 1864 also by MARCELIN BERTHELOT in his thermochemical studies. An important result of these studies was his *«principe de travail maximum»*, which in his *«Essai de mécanique chimique fondée sur la thermochimie»* from 1879 is formulated as follows: *"Each chemical change occurring without the input of an external energy tends toward the production of the material or of systems of materials, which sets free the maximum amount of heat."* [Berthelot (1879): 421]. Here he had started from the knowledge gained in mechanics, that a system resides in a stable equilibrium, if it contains a minimum amount of energy which can be transformed into work, i.e., of free energy F. Under the (valid) assumption, that the equilibrium condition $F \to minimum$ also applies to chemistry, and assuming that the heat production of a chemical reaction is equal to the total loss of energy ($Q = U$) and the latter being equal to the loss of free energy, one obtains BERTHELOT's law of maximum work. In a similar way JULIUS THOMSON had already postulated before: *"The heat production provides us with a measure of the chemical force developed during the action."*

Following these remarks on the affinity, the THOMSON-BERTHELOT principle can be written as

$$A = U \qquad (3)$$

Assuming its validity, we can calculate the affinity and, hence, the chemical equilibrium only from thermal measurements. However, from the comparison with the GIBBS-HELMHOLTZ equation (1) we see that it only applies in a limiting case.

HANS JAHN characterized the situation as follows: *"As a first approximation for the orientation in the immense number of the relations* (between materials), *Berthelot's principle of the maximum work can be highly useful, since it provides welcome information on the possibility and perhaps also on the probability of certain reactions, as long as one deals with the balancing of strong affinities. However, this principle does not concern the necessity of a reaction, since its derivation is based on assumptions, the validity of which remains quite uncertain."* [Jahn (1892): 122]. He had considered a CARNOT cycle, in which a chemical compound is expanded isothermally ($dT = 0$) close to its temperature of disintegration Θ up to the separation into its components. The amount of heat Q, which must be added in this case, corresponds to the heat developed during the formation of the compound. Hence, for the gained work we have

$$W = Q \cdot \Theta^{-1}(\Theta - T).$$

We see that BERTHELOT's assumption $W = Q \to$ *maximum* would be satisfied only in the limiting cases, in which $T = 0$, or Θ is very large compared to the temperature T.

During the time, in which NERNST studied in Graz, HANS JAHN worked at this university as an instructor, it is very likely, that here at least NERNST had heard about the problem of calculating the affinity. The problem confronted him again, when in Göttingen he dealt with the equilibria of the reactions of gases. He had turned to this subject, since to him it appeared *"of high value experimentally, to apply the basic formula to a temperature interval as large as possible."* [Nernst (1918a): 9]. Here with the *"formula"* the GIBBS-HELMHOLTZ equation (1) was meant. However, an extended temperature range could only be investigated using systems in the gaseous state. On the other hand, NERNST later emphasized, that the results obtained in this work only contributed indirectly to the solution of the problem.

Already in 1898, regarding the relation (1), NERNST had distinguished between two cases:

1. $U = 0$ and $A = T \dfrac{\partial A}{\partial T}$, realized, for example, during the expansion of ideal gases, and

2. $A = U$ or $\dfrac{\partial A}{\partial T} = 0$.

The temperature independence of A and, hence, also of U given in the last case is observed exactly, if only gravitation and electromagnetic forces are acting on the system. However, one could also mention examples of chemical and electro-chemical processes, for which the second case is only approximately valid. Already in 1894 NERNST could show, that during the mixing of concentrated solutions sometimes this limiting case can be observed.

In some way *"guided by a lucky chance"*, NERNST arrived at *"the following opinion"* [Nernst (1918a): 9]: The first case occurs at very high temperatures, since there all systems approach the state of an ideal gas, whereas the second case should be realized close to the zero value of the absolute temperature.

For the discovery of the new thermal law, also the critique of NERNST of an ansatz of VAN'T HOFF from the year 1904 certainly played some role. Starting from the experience, that during chemical processes the produced heat and, hence, U changes only little with the temperature, he had proposed the most simple ansatz $U = U_0 + aT$, which together with (2) leads to $A = U_0 - aT\ln T + JT$. However, from this one finds $\frac{\partial A}{\partial T} = J - a(\ln T + 1)$, such that at $T = 0$ the tangent of the $A(T)$ curve would coincide with the ordinate, i.e., we have $\lim_{T \to 0} \frac{\partial A}{\partial T} \to \infty$ (Fig. 5.10).

NERNST had observed *"Although ... in general the quantities A and U are not equal to each other, it is highly striking, that at least at not too high temperatures, as a rule the difference between both quantities remains within modest limits."* [Nernst (1906b):5]. As a result, when approaching the zero value of the absolute temperature A and U should become identical:

$$\lim_{T \to 0} A = \lim_{T \to 0} U.$$

However, in this case one could claim that for $T \to 0$ both curves should join together with the same slope. Therefore, in the ansatz of VAN'T

HOFF one must have $a = 0$, from which follows $\lim_{T \to 0} \dfrac{\partial U}{\partial T} = 0$ even in the case, if it would be expanded by terms with higher powers of T such as $U = U_0 + aT + \sum_{\alpha \geq 2} b_\alpha T^\alpha$. The postulated coincidence of the tangents for $T \to 0$ then leads to

$$\lim_{T \to 0} \frac{\partial A}{\partial T} = 0. \tag{4}$$

This statement about the temperature independence of the affinity upon approaching the zero value of the absolute temperature represents the actual content of the thermal law discovered by NERNST, i.e., of the Third Law. Originally he wrote it in the form

$$\lim_{T \to 0} \frac{\partial A}{\partial T} = \lim_{T \to 0} \frac{\partial U}{\partial T} = 0. \tag{5}$$

ALBERT EINSTEIN characterized this discovery with the words: *"This assumption is simply that A becomes temperature-independent under low temperatures. The introduction of this assumption as a hypothesis (third main principle of the theory of heat) is Nernst's greatest contribution to theoretical science."* [Einstein (1942): 196].

Of course, at first the new thermal law was only valid for condensed, i.e., for solid or (supercooled) liquid systems.

If we apply the assumption (4) to the solution (2) of the GIBBS-HELMHOLTZ equation (1), we obtain the important result, that the integration constant J becomes zero:

$$J = 0. \tag{6}$$

Now the function $A(T)$ can be found only by means of thermal measurements. Hence, in the sense of the above quotation of LE CHATELIER, *"the complete equilibrium conditions can be determined a priori."*

The still remaining determination of the function $U(T)$ can be reduced to the measurement of (the differences of) specific heats C_α by means of KIRCHHOFF's law

$$\frac{\partial U}{\partial T} = \Delta C = \sum_\alpha \nu_\alpha C_\alpha \tag{7}$$

leading to $U = U_0 + \sum_\alpha v_\alpha \int_0^T C_\alpha \, dT$. Here v_α are the stoichiometric coefficients. By differentiating the GIBBS-HELMHOLTZ equation (1) one obtains

$$\frac{\partial^2 A}{\partial^2 T} = -T^{-1}\frac{\partial U}{\partial T} = -T^{-1}\sum_\alpha v_\alpha C_\alpha$$

and by integrating this relation and by application of the third thermal law (4)

$$\frac{\partial A}{\partial T} - \lim_{T \to 0}\frac{\partial A}{\partial T} = -\sum_\alpha v_\alpha \int_0^T \frac{C_\alpha}{T} dT = \frac{\partial A}{\partial T}.$$

We see that according to

$$A = U_0 + \sum_\alpha v_\alpha (\int_0^T C_\alpha \, dT - T \int_0^T \frac{C_\alpha}{T} dT) = U_0 + \sum_\alpha v_\alpha \int_0^T \frac{dT}{T^2} \int_0^T C_\alpha \, dT \quad (8)$$

the affinity can be found from the specific heats. Therefore, NERNST was confronted with the task to measure these quantities down to very low temperatures.

From KIRCHHOFF's law (7) together with the thermal theorem (5) it follows, that for $T \to 0$ the differences of the specific heats also must approach zero:

$$\lim_{T \to 0} \Delta C = 0.$$

However, it was known that this behavior is not observed in ideal gases. Hence, the new thermal law could not be applied directly to pure gases. Being convinced of the general validity of his theorem, NERNST postulated that at very low temperatures the gases assume a particular state, which he called *"degeneration of gases"*. Here among other things the energy should take up a fixed value, the so-called *"zero point energy"*. Then also the specific heats would show the required behavior.

For the case of equilibria, in which gases participate and which had interested NERNST primarily, he derived the following practical procedure: by integration of the VAN'T HOFF isochore reaction curve $\frac{\partial \ln K}{\partial T} = \frac{U}{RT^2}$ (R is the universal gas constant) one obtains for the equilibrium constant K

$$\ln K = \int \frac{U}{RT^2} dT + I = E(T) + I \tag{9}$$

with the integration constant I remaining completely undetermined at first. The affinity is connected with the equilibrium constant K according to

$$A = -RT(\ln K - \sum_\alpha v_\alpha \ln p_\alpha) = -RT(E(T) + I - \sum_\alpha v_\alpha \ln p_\alpha), \tag{10}$$

where p_α denotes the equilibrium partial pressures of the gaseous partners of the reaction. The CLAUSIUS-CLAPEYRON equation $\dfrac{\partial \ln p_\alpha}{\partial T} = \dfrac{L_\alpha}{RT^2}$ yields the relation

$$\ln p_\alpha = \int \frac{L_\alpha}{RT^2} dT + i_\alpha = \Lambda_\alpha(T) + i_\alpha, \tag{11}$$

where L_α indicates the heat of condensation or sublimation. The integration constant i_α represents a value characteristic for the particular material, which is independent of the physical state and of the modification of the material. Hence, NERNST referred to the quantities i_α as *"chemical constants"*. By means of the relation (11) they can be found from measurements.

From (10) and (11) one obtains

$$\frac{\partial A}{\partial T} = -R(E(T) - \sum_\alpha v_\alpha \Lambda_\alpha(T) - T(\frac{\partial E(T)}{\partial T} - \frac{\partial \Lambda_\alpha(T)}{\partial T}) + I - \sum_\alpha v_\alpha i_\alpha)$$

yielding

$$I = \sum_\alpha v_\alpha i_\alpha$$

together with the thermal law (4). We see that the integration constant I can be found from the experimentally determined chemical constants of the gaseous components participating in the reaction. If the temperature dependence of the reaction heat is known, we see that from equation (9) one can also calculate the dependence of the equilibrium constant.

On December 23, 1905 NERNST reported his discovery to the Royal Society of Science in Göttingen. His chosen title *"On the Calculation of Chemical Equilibria from Thermal Measurements"* emphasizes both the problem resulting in the discovery of a new fundamental law of nature and also the first applications of his thermal law. In addition to the determination of the integration constant we have just described, among other things NERNST treated chemical equilibria in homogeneous gaseous and heterogeneous systems, the issue of the stability of chemical compounds, and the sublimation equilibria.

Fig. 5.9 First publications of the Third Law of Thermodynamics.

Summarizing, he could announce, *"that the final goal of thermochemistry, namely the exact calculation of chemical equilibria from the heats of reaction, appears within reach"*, if for this purpose *"one uses his new hypothesis."* And further *"The relations between heat energy and chemical affinity essentially appear to be clarified."* [Nernst (1906b): 39]. However, an exact test of the formulae derived by him will only be possible, if better data on the specific heats at low temperatures become available.

The lecture by NERNST appeared in the Bulletin of the Mathematical-Physical Class of the Society of Science in Göttingen in the first issue of the year 1906 [Nernst (1906b)] (Fig. 5.9). Since this represents the first publication of the Third Law of Thermodynamics, often this year is listed as the year of its discovery.

Also in 1906 NERNST presented his new thermal theorem and its application to a considerably increased number of examples on the occasion of the "Mrs. Hepsa Ely Silliman Memorial Lectures" at the Yale University in New Haven, Connecticut. The corresponding publication of 123 pages appeared in 1907 [Nernst (1907a)]. Only on December 20, 1906 the Royal Prussian Academy of Science in Berlin in the Session of its Physical-Mathematical Class was informed *"On the Relations between the Heat Generation and the Maximum Work in Condensed Systems"* [Nernst (1906c)] (Fig. 5.9). In this report NERNST compared the incorrect $A(T)$ curve of his certainly present colleague VAN'T HOFF (which had been discussed in a similar way in 1906 also by JOHANNES BRØNSTED) with his own theoretical curve obtained for the case of the transition of prismatic to octahedral sulfur (Fig. 5.10). As additional examples NERNST treated the transition of optical isomers, the formation of salts containing crystals (salt mH$_2$O + nH$_2$O (ice) $\rightleftharpoons$ salt·$(m+n)$H$_2$O), and the CLARK element with its potential-generating process

$$Zn + Hg_2SO_4 + 7H_2O \rightleftharpoons ZnSO_4 \cdot 7H_2O + 2Hg$$

playing a role in telegraphy.

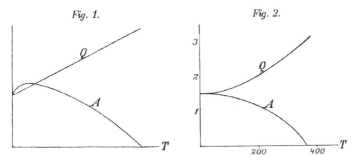

Fig. 5.10 The incorrect $A(T)$ curve (*"Fig. 1"*) according to VAN'T HOFF and BRØNSTED and the correct curve according to the thermal law of NERNST for the transition of the modifications of sulfur (*"Fig. 2"*) [Nernst (1906c): 935]. We must note $Q = U$.

Responding to an inquiry of WALTER OSTWALD, the son of his teacher in Leipzig, about one year before his death NERNST, being already very ill, summarized the essential part of his discovery of 1905, i.e., the content of the Third Law of Thermodynamics in a letter (Fig. 5.11).

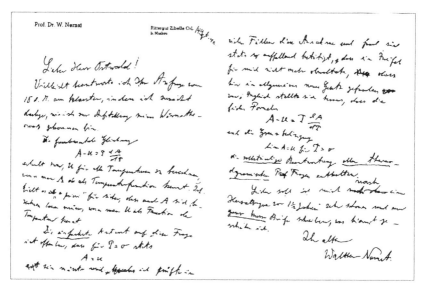

Fig. 5.11 Letter of September 14, 1940 from NERNST to WALTER OSTWALD:

Manor Zibelle O/L Sept. 14, 1940
near Muskau

Dear Herr Ostwald!

Perhaps I answer your inquiry from 10th of the month most clearly by explaining first how I arrived at the formulation of my thermal theorem.

The fundamental equation $A - U = T \dfrac{dA}{dT}$ *allows to calculate U for all temperatures, if one knows A as a function of temperature. 'A priori' I felt certain, that also A can be calculated, if U is known as a function of the temperature.*

Apparently, the <u>most simple</u> answer to this question is, that for $T = 0$ we must have always $A = U$ and I checked this assumption in many cases and found it always so well confirmed, that without any doubt for me a new general law had been found. At the same time it became clear that the two formulas $A - U = T \dfrac{dA}{dT}$ *and the limiting condition $\lim A = U$ for $T = 0$ contains the <u>complete</u> answer to <u>all thermodynamical</u> questions.*

Unfortunately, following a heart attack 1½ years ago I must be careful about myself and must write only short letters, which has been done.
Your old
Walther Nernst."

In his memorial speech about WALTHER NERNST, MAX BODENSTEIN expressed the importance of the Thermal Law with the words: *"With the Thermal Theorem of Nernst a Third Law of Thermodynamics had been found, which, in its universal validity extending far beyond the field of chemistry, is placed at an equal level next to the other two Laws. This implies its theoretical importance as well as its practical usefulness, where the calculation of the location of unknown chemical equilibria is of the highest value also for technology and the inventive work."* [Bodenstein (1942b): 141].

5.5.3 The calculation of chemical equilibria

During a careful check of his thermal theorem in conjunction with chemical equilibria, NERNST could note that *"the theorem is numerically well confirmed in a very large number of cases"* [Nernst (1907b): 521]. However, the equilibrium of ammonia $3H_2 + N_2 \rightleftharpoons 2NH_3$ provided difficulties, since strong deviations appeared between the calculated result and the experimental data obtained by FRITZ HABER and GABRIEL VAN OORDT. During the summer of 1906 NERNST and KARL JELLINEK could prove experimentally, that the deviations were caused by the very small yield of ammonia, which could only be obtained by HABER and his co-workers by operating at high temperatures. In this way errors in the analyses could not be avoided resulting in an incorrect determination of the equilibrium.

According to the relation $\left(\dfrac{\partial \ln K}{\partial p}\right)_T = -\dfrac{\Delta v}{RT}$ given by MAX PLANCK and JOHANNES JACOBUS VAN LAAR for the dependence of the equilibrium constant K upon pressure p at constant temperature T, indicating the proportionality to the change of the partial molar volumes Δv, during an increase of the pressure the equilibrium of the formation of ammonia must be shifted toward the product, since here we have $\Delta v < 0$ and $K = \dfrac{p_{NH_3}^2}{p_{H_2}^3 p_{N_2}}$. NERNST utilized this fact by constructing an *"electrical high-pressure furnace"*, which allowed the formation of NH_3 under *"suitable pressure"*.

In Fig. 5.12 we show the operation of this apparatus. The copper capillary *a* is connected to a high-pressure vessel, containing the mixture of the starting cases under the pressure at which the synthesis within the porcelain tube *p* should be performed. The platinum wire *h* serves for heating, and the temperature is measured using the thermocouple *T*. By means of the adjusting screw *s* at the valve in the cap k_2 the exit velocity of the gases could be well regulated.

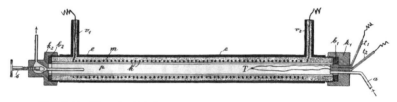

Fig. 5.12 Electrical high-pressure furnace, from [Jost, F (1908): 6].

During the end of September of 1906 the results achieved with this high-pressure furnace caused NERNST to set up a consulting contract with the nitric acid factory Griesheim-Elektron. In this contract he agreed to provide his expertise at the return charge of his participation with 20% of the net profit for 15 years in the case of success [Hoechst (1966)]. However, the experiments of Griesheim never have lead to results which were industrially competitive. On the other hand, this industrial relevance became apparent more strongly experimentally and theoretically in the experiments performed by FRITZ JOST, a student of NERNST, during the winter of 1906 and reported in his dissertation of 1908 [Jost, F (1908)]. In his dissertation, in addition to the equilibrium of ammonia, he also treated the reaction between hydrogen and coal. In the former case the experiments performed at temperatures between 958 and 1313 K and pressures from 50 to 70 atm yielded results in good agreement with the theory based on the new thermal theorem.

Therefore, at the 14th Main Assembly of the *Bunsengesellschaft* on May 12, 1907 in a ruthless way NERNST could criticize HABER: *"In my opinion, if one operates with a yield of only fractions of a milligram over long periods, indeed, one cannot gain very much. I would recommend increasing the amount of ammonia by one hundred."* He proposed that *"instead of his method applied earlier and yielding only highly questionable results, Prof. Haber now should apply also a method, which must result in highly reliable values due to its large yield."* [Nernst (1907b): 524]. On the other hand, the new results have also demonstrated, *"that the equilibrium is shifted more strongly toward a much reduced production rate than expected from the highly incorrect data of Haber."* In 1916 NERNST stated regretfully *"unfortunately, at the time I felt that technically an improvement would be quite difficult."* [Mittasch (1951): 84]. However, with his results he has accomplished *"a milestone in the history of the catalytic ammonia production"*, as it was referred to by AL-WIN MITTASCH [Mittasch (1951): 69]. Actually, in 1908 HABER and ROBERT LE ROSSIGNOL took up the suggestion by NERNST to operate at an elevated pressure. This is demonstrated in the comparison of Fig. 5.13. Following this principle, in 1913 the process for the large scale production of ammonia from nitrogen and hydrogen taken from the air and from water, respectively, could be realized at the *Badische Anilin- und Sodafabrik* (BASF) in Oppau near Mannheim under the direction of CARL BOSCH and MITTASCH.

On the occasion of the 100th anniversary of the foundation of the Friedrich-Wilhelm University in 1910, in his report covering the Physical-Chemical Institute NERNST could summarize: *"In one of its main groups the scientific projects of the Institute dealt with the measurement of chemical equilibria at high temperatures. Very often these experimental projects were closely related to the theoretical work of the director in the field of chemical thermodynamics. They included the dissociation of water vapor and carbon dioxide (Nernst and v. Wartenberg), the formation and decomposition of nitrogen oxide (Jellinek), ammonia (Jost), phosgene gas (Bodenstein* [and DUNANT]*); the measurement of vapor densities at the highest available temperatures (Nernst and v. Wartenberg); the exact determination of the maximum pressures during the explosion of gases (Pier); the measurement of vapor pressure curves within large temperature regimes.*" [Nernst and Sand (1910): 309].

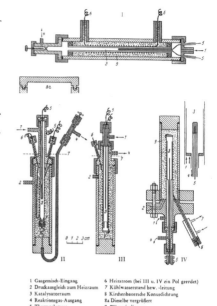

Fig. 5.13 Furnace constructions for the synthesis of ammonia following NERNST, HABER, and the BASF (from [Mittasch (1951): 95]).

1 Gasgemisch-Eingang
2 Druckausgleich zum Heizraum
3 Katalysatorraum
4 Reaktionsgas-Ausgang
5 Thermoelement
6 Heizstrom (bei III u. IV ein Pol geerdet)
7 Kühlwasserstand bzw. -leitung
8 Kirchenbauersche Konusdichtung
8a Dieselbe vergrößert
9 Wärmeisolierung

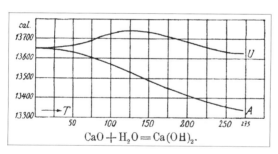

Fig. 5.14 Confirmation of the thermal law for the equilibrium indicated (from [Nernst (1918a): 92]).

However, this report did not list all projects dealing with chemical reactions up to this time. For example, we can mention also the calculations of the hydrocarbon equilibria by HANS VON WARTENBERG, and the dissertations of FRANZ HORAK on the dissociation of phosgene gas, of FRANZ POLLITZER on the equilibrium $H_2S + I_2 \rightleftharpoons 2HI + S$ and on the dissociation of hydrogen sulfide, and of MARIA WASJUCHNONOWA on the equilibrium between copper(II)- and copper(I)-sulfide. Also after 1910 NERNST sometimes asked his PhD students to investigate chemical equilibria, which were expected to confirm his thermal theorem for condensed matter systems. In this context among others we must mention the studies by ALFRED SIGGEL on the formation of $CuSO_4 \cdot H_2O$ (1913) and by WILLY DRÄGERT on the equilibrium $CaO + H_2O \rightleftharpoons Ca(OH)_2$ (1914). The data from the latter are shown in Fig. 5.14. Summarizing these studies, NERNST could state: *"In all cases studied so far there never appeared the smallest occasion to doubt the validity of the thermal law."* [Nernst (1918a): 98].

5.5.4 Specific heats and low-temperature physics

Above we have already discussed that the specific or the atomic and molar heats C_α in some sense were important from the beginning regarding the new thermal theorem. On the one hand, this concerned the need for its measurement at low temperatures, in order to determine the affinities. On the other hand, this resulted from the conclusion, that during the approach of the zero value of the absolute temperature its differences must vanish: $\lim_{T \to 0} \Delta C = 0$, where as above C and C_α stand for the corresponding quantities at constant volume.

For solid-state chemical reactions this means that at low temperatures the difference of the molar heats of the reaction partners must vanish. For example, this could be satisfied, if the rule established by FRANZ ERNST NEUMANN (1831) and HERMANN KOPP (1864), according to which the molar heats can be approximately given by the sum of the atomic heats, becomes exactly valid at low temperatures. In 1906 NERNST had suggested in the SILLIMAN lecture, that the atomic heat of each element approaches a limiting value between 0 and 2 cal/(g-atom·K), which he approximated for his calculations by 1.5 cal/(g-atom·K) [Nernst (1907a): 63].

This assumption of NERNST demonstrates a certain lack of knowledge existing at the beginning of the 20th century. However, this was soon overcome by NERNST himself, and by EINSTEIN and PLANCK. Only few and insufficient data were known regarding the temperature dependence of the specific heats, such as those of, say, ULRICH BEHN, WILLIAM AUGUSTUS TILDEN, and JAMES DEWAR. The latter pointed out, that at low temperatures the atomic heat of diamond decreases. However, based on his studies he was certain, that the atomic heat of lead remains constant down to the temperature of liquid hydrogen. This highly regarded scientist arrived at this opinion, since experimentally he still used the technique of the evaporation calorimeter yielding only very rough average values. In addition, theoretically he ignored the discovery of quantum theory in his work. The low-temperature laboratory of HEIKE KAMERLING ONNES in Leiden only published its important results on the low-temperature behavior of the energy content of the chemical elements since 1914.

The formulation of NERNST's Thermal Law by MAX PLANCK in 1910/11, which will be discussed below, holds the sharper statement, that not only the differences but also the specific heats themselves must vanish at low temperatures. Already the observations by HEINRICH WEBER, whom NERNST must have met during his studies in Zurich, could have pointed in this direction. In 1875 WEBER had found, that a few elements such as boron, calcium, and silicon show very low values of the heat capacity, which increase, however, with increasing temperature. It remained the objective of NERNST and his school to provide experimental and theoretical clarity within about a decade.

In this context the paper by EINSTEIN *"Planck's Theory of Radiation and the Theory of the Specific Heat"* from 1907 represents an important milestone [Einstein (1907)]. In this paper for the first time PLANCK's quantum hypothesis is applied to crystals and not just to radiation problems. It *"could be shown, that the theory of radiation ... leads to a modification of the molecular-kinetic theory of heat, by which some difficulties could be removed, which up to now obstructed the implementation of the theory."* [Einstein (1907): 180]. The essential idea of EINSTEIN concerned the fact, that not only the radiation energy but also the energy of the oscillators (constituting the crystals) must be quantized, if one wanted to derive PLANCK's radiation law from MAXWELL's theory. Hence, the energy of the oscillators can assume only integer values of $h\nu$ (h

PLANCK's constant, ν frequency). Now one could expect that this quantization would remove certain contradictions between experiment and theory.

This concerned in particular the rule of DULONG and PETIT, found empirically in 1819, which had been generalized by NEUMANN and KOPP, and according to which the atomic heat of all solid crystals of the elements amounts to about 6 cal/(g-atom·K). Based on the classical kinetic theory of heat, starting from the lattice vibrations of the atoms in the crystal, indeed, a value of $3R$ (R universal gas constant), i.e., about 5.96 cal/(g-atom·K) could be derived. However, especially at low temperatures, at which the best agreement should exist with the theory based on equal distribution of the energy, the experiments yielded strong deviations the observed values being much too small. By treating the atoms of the crystal as quantized oscillators elastically bound to each other in the crystal lattice, EINSTEIN could show that *"each vibrating elementary object"* must contribute *"per gram-equivalent"* the amount

$$C_V = 3R \cdot \frac{e^{\frac{\beta \nu}{T}} \cdot \left(\frac{\beta \nu}{T}\right)^2}{\left(e^{\frac{\beta \nu}{T}} - 1\right)^2} \tag{1}$$

to the specific heat (at constant volume V). Here $\beta = h \cdot N_L / R$, where N_L is the LOSCHMIDT number (N_L = number of atoms per g-atom). Figure 5.15 shows the temperature variation of equation (1). In the range $T/\beta \nu > 0.9$ the value of (1) differs only little from the DULONG-PETIT rule. However, in the range $T/\beta \nu < 0.1$ it approaches zero, such that in the limit $T \to 0$ the specific heat should vanish contrary to the classical theory. Qualitatively the temperature dependence of (1) corresponded to the experimental observations.

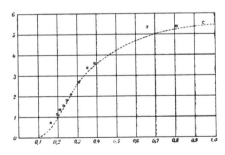

Fig. 5.15 Expression (1) plotted versus $T/\beta \nu$ (from [Einstein (1907): 186]).

Already in his publication of the Thermal Law NERNST had pointed out, that at very low temperatures the atomic heats of all non-gaseous substances should take up extremely small values which are independent of the nature of the material. Initially, NERNST proceeded with the limiting values from the SILLIMAN lecture in 1906 mentioned above. Of course, this had to be tested experimentally. On the one hand, because of the development of the copper block calorimeter, NERNST had been able to improve the accuracy of the measurements of the average specific heats. On the other hand, he had recognized that for reaching his goals the absolute specific heats must be measured with high precision, instead of the average values. Only in this way the complete curve of the temperature dependence of this quantity could be determined, the knowledge of which was needed anyway for calculating the equilibria of gases. For this purpose in an ingenious way NERNST developed the measuring technique based on the vacuum calorimeter, which allowed to determine easily the true specific heats at low temperatures, and which belongs to the classical methods of low-temperature research. The vacuum calorimeter will be described in the section dealing with the measuring devices developed by NERNST. At this point we only quote the following report by ARNOLD EUCKEN: *"Recently, the knowledge of the specific heats in particular at low temperatures seems important regarding the theoretical implications. Since the experimental data available up to now are still only few, Professor Nernst was kind enough to suggest some experiments for testing the following method ...: the substance itself represents the calorimeter."* [Eucken (1909): 586].

EINSTEIN's paper of 1907, in which based on the quantum theory of solids, showing that *"at the lowest temperatures the atomic heats always not only become zero but even zero of infinite order"*, must have been for NERNST extremely interesting in several respects. Hence, NERNST emphasized: *"At this point my measurements with the vacuum calorimeter started, which ... experimentally were shedding new light."* [Nernst (1918a): 46]. Subsequently, NERNST in fact could present curves such as shown in Fig. 5.16 as solid lines to be compared with the dotted lines calculated from EINSTEIN's theory. (Regarding the dashed lines from the theory of PETER DEBYE see below).

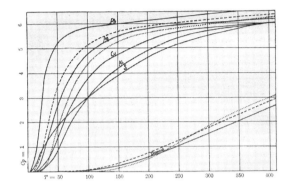

Fig. 5.16 Experimental results on the temperature dependence of the specific heat C_p (at constant pressure) from NERNST's laboratory (solid lines), theoretical curves according to EINSTEIN (dotted curves) and DEBYE (dashed curves), (from [Nernst (1918a): 46]).

In order to reach temperatures as low as possible, an apparatus had to be constructed, which could liquefy hydrogen. In Berlin, the always impatient NERNST did not want to invest long periods for development, as had been spent by the laboratory of KAMERLING ONNES in Leiden. Therefore, without extended calculations NERNST outlined the apparatus, which was soon afterwards built by his Institute mechanic HÖHNOW, *"an expert in his field ..., who was difficult to persuade for doing trivial tasks"* [Cremer (1987): 191]. In 1911, NERNST was able to report on this apparatus to the scientific community [Nernst (1911b)]. The previous history can be well extracted from letters, which at the end of the year 1910 NERNST had sent to his colleague OTTO WIENER in Leipzig, who was operating himself a hydrogen liquefier which NERNST wanted to look at. So in October 1910 he told WIENER: *"I have procured a small apparatus for the liquefaction of hydrogen, with which one can use commercial hydrogen bombs. However, I have no experience yet with it. I plan to measure the true specific heats at temperatures as low as possible for testing Einstein's theory and, hence, also indirectly Planck's radiation law."* [Nernst (1910a)] and one month later: *"By the way, in the meantime I have built for myself an apparatus for the liquefaction of hydrogen, which does not yield larger quantities, but which allows to cool small glass vessels down to $T = 21°K$ and to keep any temperature* *between liquid air and liquid hydrogen fairly constant."* [Nernst (1910b)]. Initially they operated at the pressure of the bomb, and later at 150 – 200 atm using a compressor. In this way the total content of the bomb could be liquefied instead of only 5%. Per hour about 300 – 400 cm^3 liquid hydrogen were obtained. However, here we must note, that the facility was only functioning, when it was operated by HÖHNOW. However, due to his highly developed diabetes, sometimes the latter was not available for health reasons. Nevertheless, the smart NERNST had several units of it built and offered them for sale. Therefore, for some time one could see them in different laboratories. In spite of its obvious deficiencies, it can be considered being the prototype of the small liquefier for hydrogen and helium.

Also the following episode is worth mentioning: In the beginning of December 1910 NERNST told WIENER in a letter, that EMIL FISCHER would give a lecture in the presence of WILHELM II on the occasion of the establishment of the *Kaiser-Wilhelm-Gesellschaft*. In this lecture, experiments with liquid hydrogen was also to be performed and WIENER's liquefier was to be presented, in order to demonstrate how expensive modern research could be. Since his own liquefier could produce only too little liquid hydrogen for this purpose, NERNST asked if WIENER could supply about four liter from his laboratory for the lecture, however also for measurements at the Institute in Berlin. A few days later WIENER gave the answer: *"It will be an honor for the hydrogen from Leipzig to evaporate for the German Emperor and to perform various magic tricks."* [Wiener (1910)]. The task of the carefully planned transfer of the liquid hydrogen was given by the Institute in Leipzig to WIENER's assistant EDGAR LILIENFELD. FISCHER's lecture was given on January 11, 1911. Only eleven days later NERNST thanked WIENER for his friendly support, apologizing for the delay. Using the hydrogen from Leipzig, in addition one had carried out intensive measurements (*"With it we have worked nearly 16 hours without interruption and nearly without something to eat and the poor Politzer*[sic] *could not sleep for 2 nights."* [Nernst (1911a): 1]), and the numerical analysis still had to be done. Furthermore, the measurements had to be completed *"using the small amounts of liquid hydrogen supplied by my apparatus, which really does very well."* [Nernst (1911a): 1].

In addition to the generation of low temperatures and of the vacuum, more problems had to be solved for obtaining reliable and precise data from the measurements with the vacuum calorimeter. This concerned the establishment of a temperature scale in the interval between boiling hydrogen and liquid air, in which there existed only very few fixed points. This task required tedious investigations. Since essentially nothing was known initially about the heat conduction at low temperatures, it was even possible, that perhaps the program could not be carried out. In particular this would have been the case, if the thermal conductivity had been strongly reduced compared to higher temperatures, such that in principle, exact measurements would be impossible. However, in 1911 EUCKEN showed, that such behavior of the thermal conductivity was not observed.

These efforts and the success soon achieved in the Institute in Berlin were summarized by KLAUS CLUSIUS: *"From 1910 until the first years of the First World War NERNST, at the summit of his power and supported by a group of highly motivated coworkers, in a restless exertion clarified the basic facts about the behavior of the heat content at low temperatures."* [Clusius (1943): 399]. In a very short time NERNST had accomplished the establishment of a world-renowned low-temperature laboratory. For illustrating this point we can quote some lines from a letter written by MAX BORN to ALBERT EINSTEIN on August 21, 1921: *"A large paper of mine on thermodynamics is in print, ... The result ... is curious: The Grüneisen Theorem of the proportionality between energy and thermal expansion is not valid at low temperatures; ... This should be tested experimentally (Nernst?)."* [Born (1972): 66].

NERNST's studies of the specific heats of solids at low temperatures, in particular to confirm EINSTEIN's relation (1), had indicated, that quantitatively the latter relation is incorrect, as also seen for example in Fig. 5.16. Together with his favorite student FREDERICK ALEXANDER LINDEMANN, in 1911 NERNST empirically found a formula, which agreed better with the observations [Nernst and Lindemann (1911)]:

$$C_V = \frac{3}{2}R \cdot \left[\frac{e^{\frac{\beta\nu}{T}} \cdot \left(\frac{\beta\nu}{T}\right)^2}{\left(e^{\frac{\beta\nu}{T}} - 1\right)^2} + \frac{e^{\frac{\beta\nu}{2T}} \cdot \left(\frac{\beta\nu}{2T}\right)^2}{\left(e^{\frac{\beta\nu}{2T}} - 1\right)^2} \right]. \quad (2)$$

Theoretically, deviations from EINSTEIN's relation were expected, since in its derivation the interaction between the vibrating atoms had been neglected.

In 1912 based on the quantum theory the ansatz of EINSTEIN had been extended by the papers of MAX BORN and THEODORE VON KÁRMÁN as well as of PETER DEBYE. For the first time, the latter had treated the atomic structure of the solid as a continuum, and for the energy E had obtained the equation

$$E = 9RT\left(\frac{T}{\beta\nu}\right)^3 \int_0^{\frac{\beta\nu}{T}} \frac{x^3 \, dx}{e^x - 1}. \quad (3)$$

For very low temperatures with $C_V = dE/dT$ this equation yields the so-called DEBYE T^3-law for the decrease of the specific heat at low temperatures

$$C_V = 4aT^3 \text{ with } a = \frac{3}{5}\pi^4 \frac{R}{(\beta\nu)^3}.$$

After initially NERNST had written an angry and rejecting letter, later he noted in a laudatory way: *"Indeed, Debye's formula extremely well reproduces the curves of simple solids experimentally observed by myself.* [see the dashed lines in Fig. 5.16] ... *Except for very low temperatures, the behavior of the formula of Debye [(3)] differs only insignificantly from the formula"* for the energy, from which the empirical equation (2) of NERNST and LINDEMANN is found by differentiation [Nernst (1918a): 51]. For the first time, in 1913 NERNST's coworkers EUCKEN and FRÉDÉRIC SCHWERS have confirmed the T^3-law in pyrites (FeS_2) and in fluorite (CaF_2).

After NERNST had clarified sufficiently the problems concerning the decrease of the specific heat of solids at low temperatures, which had been of prime interest for him, he could turn also to the questions dealing with the temperature dependence of the molar heat of gases. This will be discussed in the following section along with the contributions of NERNST to quantum theory.

"The insight gained from the low-temperature research of NERNST cannot be overestimated at all in its importance for the theory." [Clusius (1943): 399]. These were the words of KLAUS CLUSIUS praising the achievements of NERNST and his school in this field up to about 1916. After the First World War NERNST left the low-temperature research in the hands of his younger coworkers, of whom FRANZ SIMON must be mentioned in the first place. Later on, FREDERICK ALEXANDER LINDEMANN had established a low-temperature laboratory in Oxford following the operating style he had learned from his teacher in Berlin.

5.5.5 Quantum Theory

As we have seen, NERNST became closely involved in the quantum theory, in particular because of EINSTEIN's paper of 1907, which was extremely important for him in connection with the Third Thermal Law. Certainly, this still young theory had caught his interest. However, its relevance to solids and their specific heat at low temperatures demonstrated by EINSTEIN was so extremely important to NERNST that afterwards his research began to concentrate on quantum theory. In fact, the vanishing of the specific heats at low temperatures, following directly from his thermal law, could not be explained with the classical theory.

Only after NERNST had explored sufficiently the behavior of the specific heats of solids at low temperatures, as we have discussed in the previous section, he turned to the molar heat of gases. Also in this case there existed an analogous problem: Based on the classical theory one could show that for each degree of freedom the molar heat of a gas should be equal to half of the universal gas constant R: $C_f = R/2$. Assuming rigid molecules, then for the molar heats at constant volume, C_V, one obtains three fixed values independent of temperature: $C_V = 3R/2$ (one-atomic gas), $C_V = 5R/2$ (linear molecules), and $C_V = 6R/2$ (nonlinear, multi-atomic molecules). In particular in the latter case, frequently the experiments yielded strong deviations toward higher values, which still increased with the temperature. If one allows inner-molecular vibrations, one obtains a larger number of the degrees of freedom and, hence, higher values of C_V. However, their increase with temperature remained unexplained. Also the idea of a retardation of the thermal equilibrium of the vibrational degrees of freedom exceeding the duration of the experiment, did not look satisfactory.

Also in this case NERNST was one of the first scientists recognizing the fundamental character of the problem. The origin of the observed deviations from the theory was the theory itself. The classical theory had to be replaced by the quantum theory. Similarly as for the temperature dependence of the atomic heat of a solid, which could be explained by EINSTEIN's assumption of quantized atomic vibrations, NERNST related the corresponding problem of the molar heat of a gas to the contribution arising from the inner-molecular vibrations. Since for gases the interaction between the molecules is negligible, in 1911 he could use directly PLANCK's formula for the energy of a harmonic oscillator. Then by its differentiation he obtained the contribution of the vibrations to the molar heat [Nernst (1911c)]. NERNST felt that in principle the problem had been solved in this way. The further detailed questions, including an accurate test of the applicability of PLANCK's formula, were taken up and clarified subsequently by his coworkers and students. Initially, here we must mention NIELS BJERRUM and then in particular ARNOLD EUCKEN.

NERNST proposed for the treatment of the temperature dependence of the molar heat of molecules, which is more complicated than that of gases, to describe the motion of the center of gravity of the molecule and the vibrations using a DEBYE function and PLANCK functions, respectively, and to superimpose these contributions. This ansatz then turned out to be suitable.

Because of the excellent applicability of quantum theory for explaining the contribution of the vibrations to the specific heat, NERNST presumed that the rotational energy of the molecules should also be quantized. In this case, also at low temperatures the so-called rotational heat should drop down to values below R or $3R/2$ derived from the classical theory. This could be easily detected for the molar heat of hydrogen because of its small moment of inertia. Actually in 1912 EUCKEN confirmed this expectation of his teacher.

Although NERNST had been dealing only relatively less with the molar heat of gases, ARNOLD EUCKEN came to the following conclusion: *"Without any doubt, among the contemporary scientists it was he to whom this field owes by far the largest advance."* [Eucken (1943): 321].

The strong interest of NERNST in the development of quantum theory is also demonstrated by his initiative for organizing a meeting, at which nearly all important scientists in this field could discuss the development and the results of the *"Theory of the Quanta and Radiation"* within the decade after PLANCK's discovery at the end of the year 1900. He succeeded to persuade the Belgian scientist and business-man ERNEST SOLVAY to finance this plan. Hence, from October 30 until November 3, 1911 the First Solvay Congress was held in Brussels in an attractive atmosphere. In his opening and welcoming remarks on the eve of the meeting SOLVAY told the participants: *"I also thank you in the name of Mr. Nernst, since it was he who first had the idea to organize this meeting. He has encouraged and guided this action like a real leader, which he is, indeed."* [Solvay (1914): 1].

In order to illustrate the international character and the intellectual potential of the congress, we list the scientists who participated in the discussions in addition to NERNST* (Fig. 5.17): HENDRIK ANTOON LORENTZ* (Leiden, as chairman), MARCEL BRILLOUIN (Paris), MARIE CURIE (Paris), ALBERT EINSTEIN* (Prague), FRIEDRICH HASENÖHRL (Vienna), JAMES H. JEANS* (Cambridge), HEIKE KAMERLINGH ONNES* (Leiden), MARTIN KNUDSEN* (Copenhagen), PAUL LANGEVIN* (Paris), JEAN PERRIN* (Paris), MAX PLANCK* (Berlin), HENRI POINCARÉ (Paris), HEINRICH RUBENS* (Berlin), ERNEST RUTHERFORD (Manchester), ARNOLD SOMMERFELD* (Munich), EMIL WARBURG* (Charlottenburg near Berlin), and WILHELM WIEN (Wurzburg). The participants marked with * presented a *"report"* during the meeting. JOHANNES DIDERIK VAN DER WAALS (Amsterdam) and LORD RAYLEIGH (London) did not accept the invitation. However, the latter had submitted a letter for discussion. MAURICE DE BROGLIE (Paris), ROBERT GOLDSCHMIDT (Brussels), and FREDERICK A. LINDEMANN (Berlin) were appointed as Secretaries of the Congress. Further, SOLVAY's coworkers ÉDOUARD HERZEN and G. HOSTELET attended the sessions. As noted later on by FRIEDRICH HUND, PAUL EHRENFEST, PETER DEBYE, and NIELS BOHR were missing [Hund (1967): 30].

Fig. 5.17   The participants of the First Solvay Congress 1911 in Brussels: (1) NERNST, (2) BRILLOUIN, (3) SOLVAY, (4) LORENTZ, (5) WARBURG, (6) PERRIN, (7) WIEN, (8) CURIE, (9) POINCARÉ, (10) GOLDSCHMIDT, (11) PLANCK, (12) RUBENS, (13) SOMMERFELD, (14) LINDEMANN, (15) DE BROGLIE, (16) KNUDSEN, (17) HASENÖHRL, (18) HOSTELET, (19) HERZEN, (20) JEANS, (21) RUTHERFORD, (22) KAMERLINGH ONNES (23) EINSTEIN, (24) LANGEVIN. (During the time of the photograph SOLVAY was absent. His head was inserted later by means of photo-montage.).

NERNST had entitled his contribution *"Application of the Quantum Theory to Various Physical-Chemical Problems"* [Nernst (1914a)]. It was focused mainly on the quantum-theoretical treatment of the specific heat of solids, which we have discussed in the previous section. In the end NERNST talked *"about a general law concerning the behavior of solids at low temperatures"*. Starting from the fact that his measurements of the specific heat agreed with the quantum theory and demonstrated the existence of a temperature interval beginning at the absolute zero of temperature, in which *"for all solids ... the concept of temperature becomes practically meaningless"* [Nernst (1914a): 229], he proposed a model of the solid state which could explain the behavior of many physical properties at low temperatures. In this temperature range only very few atoms of the solid had taken up a single energy quantum and the number of atoms with more than one energy quantum can be neglected. Therefore, the number of atoms not being completely at rest is directly proportional to the energy content. Hence, the situation is similar to that in highly dilute solutions: The atoms with a single energy quantum correspond to the molecules of the dissolved matter, and the energy content of the solid to the concentration of the solution. Since in the case of dilute solutions we know that each property change is proportional to the concentration, analogously at low temperatures each property of a solid should change proportional to the energy E.

In this way for the volume he obtained the linear relation $V = V_0 + aE$, from which the proportionality between the cubic thermal expansion and the specific heat can be derived: $\alpha = \left(\dfrac{\partial V}{\partial T}\right)_p = a\dfrac{\partial E}{\partial T} = aC_V$ (p pressure).

EDUARD GRÜNEISEN had obtained a similar relation being valid at even higher temperatures: $\alpha = aC_p$. This corresponds to the former relation at low temperatures, since there we have $C_p \approx C_V$. Using his Thermal Theorem and assuming the functional relation $C_p = f(v/T)$ (v frequency) following from his results, as well as the temperature independence of v and $\partial v/\partial p$, NERNST could derive thermodynamically the GRÜNEISEN equation: $\alpha = \dfrac{1}{v}\dfrac{\partial v}{\partial p}C_p = aC_p$. Since NERNST's student CHARLES LIONEL LINDEMANN, the brother of FREDERICK A. LINDEMANN, by means of measurements down to the temperature of liquid hydrogen, could demonstrate not only the strong decrease of α with temperature but also the proportionality between the temperature change of α and that of the atomic heat; the Third Thermal Law had also been confirmed again.

According to *"the general law"* the inner energy U and the free energy or the affinity A should become temperature independent at low temperatures. However, this is exactly the original content of the Third Thermal Law, such that *"in this case the thermal theorem formulated already six years ago ... appears as a special case of a more general theorem derived from quantum theory."* [Nernst (1914a): 231].

It was also the task of NERNST together with EUCKEN to edit the proceedings of the meeting (Fig. 5.18).

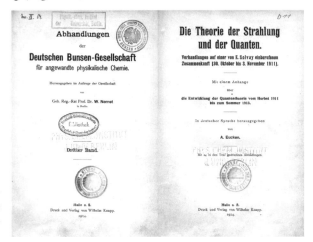

Fig. 5.18 The publication of the proceedings of the 1st Solvay Congress.

Of course, NERNST was interested in a paper by MICHAEL POLANYI from 1913, in which he applied quantum theory to the energy of a material under very high pressure and found that under these conditions also the specific heat must approach zero. NERNST found it interesting, that the limit of his law $\dfrac{\partial A}{\partial T} = \dfrac{\partial U}{\partial T} = 0$ under elevated pressure should still be valid at finite temperatures.

HENDRIK BRUGT GERHARD CASIMIR, a student of WOLFGANG PAULI, summarized the importance of the Thermal Theorem of NERNST in the following way: *"Initially Nernst's thermal law was a useful heuristic principle, then a firm guide which led from the classical to the quantum theoretical statistics; it remains a challenge for the mathematicians."* [Casimir (1964): 533]. On the other hand, as emphasized by MAX BODENSTEIN exactly thirty years prior to CASIMIR, just the *"development of quantum mechanics has made something in it more understandable, but it has left its core untouched in spite of certain restrictions."* [Bodenstein (1934): 438]. Although it has been discovered within and attached to the field of phenomenological thermodynamics, in fact the Third Thermal Law essentially is most strongly connected with quantum statistics.

In his *"History of Quantum Theory"* FRIEDRICH HUND demonstrated the importance of NERNST for the development of quantum theory during its early period by means of a diagram, of which in Fig. 5.19 we show a section in a slightly changed form.

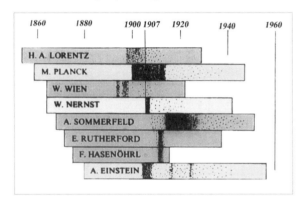

Fig. 5.19 Plot of the "density" of their research on quantum theory versus the time for some participants at the 1st Solvay Congress, from [Hund (1967): 233]. (The length and placement of the bars correspond to the lifetime of the different physicists).

The physical chemist WILHELM JOST, one of the successors on NERNST's chair in Göttingen, emphasized the close connection between the quantum theory and the thermal theorem in the following way: *"During the time of its discovery Nernst's thermal law could have been derived from quantum theory. Of course, on the other hand, this means that it could have led to the discovery of quantum theory."* [Jost, W (1964): 527].

5.5.6 The impossibility of reaching the absolute zero of temperature

On February 1, 1912 NERNST presented a report to the Physical-Mathematical Section of the Prussian Academy of Science in Berlin, which focused on the question of the impossibility of reaching the absolute zero of temperature [Nernst (1912a)] (Fig. 5.20). As NERNST himself emphasized later, the immediate occasion was a discussion at the Solvay Congress, in which EINSTEIN, being supported by LORENTZ, remained unconvinced that the thermal theorem of NERNST can be inferred from the vanishing of the specific heat near the absolute zero of temperature [Nernst (1914a): 243–244]. EINSTEIN based his opinion on the fact, that the reversible and isothermal transition of a system from one state A to another state B would be impossible sufficiently close to the absolute zero temperature. On the other hand, *"Nernst's theorem includes ... the assumption that the transition from A to B is always possible along a purely static path as seen from the point of view of molecular mechanics."* [Nernst (1914a): 243].

Fig. 5.20 The first publication of the principle of the impossibility of reaching the absolute zero temperature.

In his report NERNST started by noting that the question of reaching experimentally the absolute zero temperature has been discussed only little. In addition, the vanishing of the specific heat near the absolute zero temperature, confirmed by measurements and predicted by quantum theory, would represent a new situation for answering this question.

NERNST wanted to prove that the possibility of reaching the absolute zero of temperature by means of a finite process contradicts the Second

Law of Thermodynamics. For this purpose he considered the (reversible) CARNOT cycle process operating between the temperatures $T_1 > 0$ (isotherme $\overline{AB}$) and $T_2 = 0$ (isotherme $\overline{CD}$) (Fig. 5.21). Apparently, external work would be generated, if the process is possible, as seen from the area ABCD being different from zero. For this the system must have taken up thermal energy. Along the adiabatic curves $\overline{BC}$ and $\overline{DA}$, heat input is not possible. The experimental fact, that for $T \to 0$ the specific heat C of solids must approach zero, because of $A - U = -T\int \frac{C}{T} dT$ (see above) means the vanishing of the latent heat, i.e., of the difference between the maximum work A and the heat energy $U = Q$, $A - U = T\frac{\partial A}{\partial T}$.

We see that the statement $\lim\limits_{T \to 0} \frac{\partial A}{\partial T} = 0$ of the Third Law of Thermodynamics remains valid, and we conclude that also during the isothermal compression CD at $T_2 = 0$ heat exchange is not possible. The delivered work must be generated only by heat input during the isothermal expansion AB. Therefore, the system could generate work continuously by taking up heat energy from a reservoir without effecting anything else. However, this contradicts the Second Law of Thermodynamics, as we can see from the formulation by WILLIAM THOMSON mentioned above. Hence, the isotherme $\overline{CD}$ must be excluded, i.e., the absolute zero of temperature cannot be reached by a finite process such as $\overline{BC}$. Therefore, in his lecture for the first time NERNST formulates: *"It is impossible to have a process operating in finite dimensions, by means of which an object can be cooled down to the absolute zero of temperature."* [Nernst (1912a): 136].

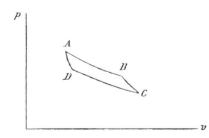

Fig. 5.21 Diagram to prove the impossibility of reaching the absolute zero temperature in [Nernst (1912a): 135] (p pressure, v volume).

This result can be obtained in a simpler way (however, without the relation with the specific heat important for NERNST) by noting

$$-\frac{\partial A}{\partial T} = \Delta S = \Delta S_{AB} + \Delta S_{BC} + \Delta S_{CD} + \Delta S_{DA} = 0$$

for the CARNOT cycle: For the two reversible adiabatic changes of the state we have $\Delta S_{BC} = \Delta S_{DA} = 0$. Because of the Third Law and since $\frac{\partial A}{\partial T} = -S$, we have also $\lim_{T \to 0} \Delta S_{CD} = 0$ and further $\Delta S_{AB} = \frac{Q}{T_1} = 0$, although an amount of heat $Q \neq 0$ is extracted from the reservoir mentioned above. Again, the principle of the impossibility of reaching the absolute zero of temperature follows from this contradiction.

At the end of his report in the session of the Academy of Science of Berlin NERNST said: *"Based on the experimental fact that at low temperatures the specific heat of solids approaches zero, it was shown that the thermal theorem claimed by the author can be formulated also in the following way:*

It is impossible to cool an object down to the absolute zero of temperature by means of processes operating within finite dimensions." [Nernst (1912a): 140].

Fig. 5.22 Portrait of WALTHER NERNST by MAX LIEBERMANN (1912).
(«*M. Liebermann pinxit. Nernst dedicavit. Frisch reproducit.* »)

This formulation of the thermal theorem, being equivalent to the original version, as NERNST has shown, was called by him the *"principle of the impossibility of reaching the absolute zero value"* of the temperature. In its negative statement it connects to the First and Second

Thermal Laws as they appear in the formulations by PLANCK or by CLAUSIUS and THOMSON mentioned above. In this way it further emphasizes its position within thermodynamics as its Third Thermal Law. This reiterated perfectly the overall importance which NERNST attributed to his theorem.

The year 1912, in which NERNST presented the most popular formulation of the Third Thermal Law in form of the impossibility principle, represents the summit of his creativity and fame. This situation is well expressed artistically by the portrait of NERNST created by MAX LIEBERMANN during this period (Fig. 5.22).

5.5.7 Formulation of the Third Thermal Law by Max Planck

In December of 1917 NERNST wrote in the preface of his monograph on the new thermal law: *"It is known that Planck, the first theoretician who has studied my thermal law in detail, has given an excellent presentation of it in the latest editions of his textbook on thermodynamics; the treatment presented here extends a bit further, since the theorem is assumed to be valid also for gases of finite density and, hence, also for solutions."* [Nernst (1918a): IV].

In fact, in November of 1910 in the preface of the 3rd edition of his *"Lectures on Thermodynamics"* MAX PLANCK had written: *"The introduction of the thermal theorem by W. NERNST in the year 1906 represents a technical extension of principal importance. If this theorem should turn out to be correct, as it appears presently, then thermodynamics would be enriched by a principle, the implications of which cannot be foreseen at all both from the practical and the molecular-theoretical point of view."* [Planck (1927): VII]. Since in his textbook PLANCK emphasized in particular the macroscopic view-point, he also treated NERNST's thermal law in this way, and the more so since at the time its importance for the atomistic theory did not appear to be established. Furthermore, he felt that he should formulate the theorem in the farthest possible way, and *"in this case he went a bit beyond Nernst himself not only in the form but also in the content."* [Planck (1927): VIII].

Of course, being an expert in thermodynamics, MAX PLANCK must have been strongly interested in the fundamental extension of the theory of heat discovered by NERNST. Furthermore, it was quite natural for PLANCK that he looked at this extension on the basis of the concept of entropy, created by RUDOLF CLAUSIUS, but disliked by the majority of the scientists. The formulation $\lim_{T \to 0} \frac{\partial A}{\partial T} = 0$ of the theorem by NERNST, because of $-\frac{\partial A}{\partial T} = \Delta S$, could be written perfectly equivalently using the entropy S in the form $\lim_{T \to 0} \Delta S = 0$. In this case we must note that this formulation of the theorem requires the vanishing of the entropy change ΔS in the limit $T \to 0$.

PLANCK discussed his proposed extension in the following way: As we have discussed above, the entropy difference can be connected with the change of the specific heat $\Delta C = C(T) - C_0$ according to $\Delta S = S(T) - S_0 = \int_0^T \Delta C \, d\ln T$. In the limit $T \to 0$, NERNST's formulation only requires that $C(T) = C_0$, but not the vanishing of the specific heat. However, if C remains finite in this limit, the entropy becomes negatively infinite. In the case that the vanishing of C during the approach of the absolute zero value of temperature is confirmed, the theorem could be extended beyond its version given by NERNST as follows: *"... that for an unlimited decrease in temperature the entropy of each chemically homogeneous object with finite density approaches a certain value, independent of pressure, state of aggregation, and of the special chemical modification."* [Planck (1927): 267]. Now not only for the difference ΔS, but also for the entropy S itself we have: $\lim_{T \to 0} S = S_0 = \text{const}$. Without any loss of general validity, this constant can be set equal to zero:

$$\lim_{T \to 0} S = S_0 = 0.$$

Therefore, PLANCK formulates: *"For an unlimited decrease in temperature the entropy of each chemically homogeneous object with finite density approaches the value zero. ... and from now on in this sense one can speak of an absolute value of the entropy."* [Planck (1927): 268].

If X stands for an arbitrary property of a thermodynamic system, PLANCK's formulation of the Third Thermal Law can also be written as $\lim_{T \to 0} \frac{\partial S}{\partial X} = 0$. Because of $\left(\frac{\partial S}{\partial V}\right)_T = \left(\frac{\partial p}{\partial T}\right)_V$ and $\left(\frac{\partial S}{\partial p}\right)_T = -\left(\frac{\partial V}{\partial T}\right)_T$ it follows that the pressure coefficient and the cubic expansion must vanish during the approach of the absolute zero value of temperature.

As mentioned above, PLANCK had conceded, that his formulation of the thermal theorem is only valid if the specific heat becomes vanishingly small for $T \to 0$. In later editions of his textbook he did not change this, although NERNST emphasized already in 1918, that this property of the specific heat is confirmed both theoretically and to a large extent also experimentally (even for gases). Therefore, *"of course, each difference between the interpretations* [of him and of PLANCK] *has disappeared; for the practical application of the thermal law, dealing always only with the changes in entropy, it has never existed."* [Nernst (1918a): 72]. However, PLANCK's formulation made the molecular-theoretical access easier.

5.5.8 Research between 1906 and 1916, the monograph, and the Nobel Prize in Chemistry

In the previous sections we have discussed the essential points of the Third Thermal Law and the related research within the relevant decade from 1906 to 1916. Next we want to present at first a general overview about the achievements of the NERNST school within this decade.

During this time, in conjunction with the Third Thermal Law, 117 papers were published in which in addition to NERNST 51 people (students and coworkers) participated. In the previous sections we have mentioned already a few people, of whom we recall in particular ARNOLD EUCKEN, the brothers LINDEMANN, HANS VON WARTENBERG, NIELS BJERRUM, MATHIAS PIER, FRANZ POLLITZER, and MAX BODENSTEIN. In addition to these and to the other scientists mentioned before, a few further names must be quoted.

JAMES RIDDICK PARTINGTON had worked with NERNST during 1911 – 1913 on the specific heat of gases. Later at the University of London he became known because of his papers on problems of physical chemistry as well as his voluminous book *"A History of Chemistry"*. To him we owe *"The Nernst Memorial Lecture"* [Partington (1953)].

Also NERNST's student HANS SCHIMANK later became known as a famous historian of science. He obtained his PhD in 1914 with a thesis on the behavior of the electrical resistance at low temperatures, and until the first days of the First World War he worked in the *Bunsenstraße*. His book *"Epochs of Science"* is dedicated to *"Dr. Walther Nernst ... my admired teacher."* [Schimank (1930)].

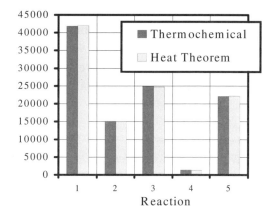

Fig. 5.23 Comparison between the thermo-chemically obtained data of F. KOREF and H. BRAUNE and the theoretical values of the reaction heat from [Nernst (1918a): 98] for the reactions
(1) $Pb + I_2 \rightleftharpoons PbI_2$
(2) $Ag + I \rightleftharpoons AgI$
(3) $Pb + 2AgCl \rightleftharpoons PbCl_2 + 2Ag$
(4) $Hg + AgCl \rightleftharpoons HgCl + Ag$
(5) $Pb + 2HgCl \rightleftharpoons PbCl_2 + 2Hg$

We should mention also FRITZ KOREF, who obtained his PhD in 1910 with a thesis on the equilibrium of the formation of carbon disulfide. Later he held leading positions in different companies of the light-bulb industry. Until about 1914 in NERNST's institute he participated in studies of the specific heat at low temperatures and in measurements of reaction heats. Together with HERMANN BRAUNE the latter were carried out on condensed systems (lead- and silver-halides) and served for testing the thermal theorem. KOREF's papers were valued highly by NERNST. Based on a summarizing table in [Nernst (1918a): 98], Fig. 5.23 shows the good agreement between the thermo-chemically obtained data of

KOREF and BRAUNE and the theoretical values of the reaction heat calculated from the thermal theorem.

Only at the end of the decade 1906 – 1916 PAUL GÜNTHER participated in the studies dealing with the Third Thermal Law. He also investigated the specific heat at low temperatures, and in 1917 he obtained his PhD. Between 1936 and 1945 he occupied the Chair of Physical Chemistry at the University of Berlin. PAUL GÜNTHER is one of the pioneers of radiation chemistry.

The following table presents a statistical overview of the publications which have appeared from NERNST's institute within the first decade after the discovery of the Third Thermal Law. The table only includes the scientists we have mentioned before. The contribution of a scientist to a paper with a total of n authors is given by $1/n$. We note that within this period on average we find 2.25 papers per scientist and 10.64 papers per year. The contribution of NERNST to the total volume of papers calculated from these numbers amounts to 26.4%. Of course, the impact of his ideas is much larger.

Name	Year											Σ
	1906	07	08	09	10	11	12	13	14	15	16	
BJERRUM, N.						1	2					3
BODENSTEIN, M.			$\frac{1}{2}$	$\frac{1}{2}$								1
BRAUNE, H.							$\frac{1}{2}$		$\frac{1}{2}$			1
DRÄGERT, W.								1				1
DUNANT, G.				$\frac{1}{2}$								$\frac{1}{2}$
EUCKEN, A.			1				1	$2\frac{1}{2}$	1		2	$7\frac{1}{2}$
GÜNTHER, P.										1		1
HORAK, F.				1								1
JOST, F.			2									2
KOREF, F.					$1\frac{1}{3}$	1	$1\frac{1}{2}$		$\frac{1}{2}$			$4\frac{1}{3}$
LINDEMANN, C.L.						1	1					2
LINDEMANN, F.A.					$1\frac{5}{6}$	3	$1\frac{1}{2}$					$6\frac{1}{3}$
NERNST, W.	$2\frac{1}{2}$	1		3	$6\frac{1}{3}$	7	$4\frac{1}{2}$	3	$1\frac{1}{2}$		2	$30\frac{5}{6}$
PARTINGTON, J.R.									1			1
PIER, M.					1	1						2

	Year											
Name	1906	07	08	09	10	11	12	13	14	15	16	Σ
POLLITZER, F.					1	2						3
SCHIMANK, H.								1		1		2
SCHWERS, F.							$\frac{1}{2}$	$\frac{1}{2}$				1
SIGGEL, A.						1						1
VASYUKHNOVA, M.				1								1
WARTENBERG, H. v.	$\frac{1}{2}$	2		1			1					$4\frac{1}{2}$
⋮					⋮							⋮
Σ (52 scientists)	6	6	7	13	15	19	18	14	12	1	6	117

Figure 5.24 shows the almost bell-shaped envelope of the number of papers plotted versus time with a maximum in the years 1911/1912. Furthermore, we see the important contributions of the coworkers of NERNST to the establishment of the Third Thermal Law.

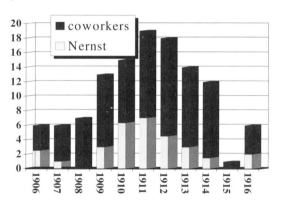

Fig. 5.24 Number of publications of NERNST's school dealing with the Third Thermal Law in the years 1906 – 1916.

At the beginning of the First World War NERNST became occupied with other problems from science and politics, which essentially he took up on his own. We will return to this subject below. Furthermore, NERNST felt that he had investigated the most important aspects of his thermal law, such that he could leave its further development and application to his younger colleagues. So the time had come, when he could take up the task to present in a *"little paper ... an attempt for the further development of one of the most important chapters of theoretical physics, of thermodynamics."* [Nernst (1918a): III]. Another sad cause for this decision is hinted at by NERNST in the words at the beginning of the preface written in December 1917: *"During times of sorrow and distress many among the ancient Greeks and Romans looked for and found comfort in philosophy; today we can say that no other science is as well suitable as theoretical physics to take the mind from the ... miserable present times and to move it into other spheres."* [Nernst (1918a): III]. Both his sons had been killed by the war, the older, RUDOLF, already in 1914, and the younger, GUSTAV, just then in 1917.

This *"little paper"* is NERNST's monograph *"The Theoretical and Experimental Foundation of the New Thermal Theorem"* [Nernst (1918a)], the first edition of which appeared in 1918, and which presents the research results of his school dealing with the Third Thermal Law during the period 1906 – 1916. It is interesting that the second edition appeared in 1924 and was produced by means of a printing technique invented by MAX ULLMANN. This reproduction technique for the fabrication of original printing plates for the reprinting of books and other printing matters operated on a photo-chemical principle. Therefore, all printing errors remained unchanged. However, the second edition contains some supplements covering the results on the Third Thermal Law obtained in the years after the First World War. As an example we mention an improvement of the vacuum calorimeter by FRANZ SIMON. The second edition served for the preparation of the English editions of 1926 [Nernst (1924a)].

On November 10, 1921 the Nobel Committee announced in Stockholm, that WALTHER NERNST would receive the Nobel Price in Chemistry of the year 1920. In this way NERNST was honored mainly because of his discovery of the Thermal Theorem of thermo-chemistry. The Swedish text justifying the award read «*Såsom ett erkännande för hans termokemiska arbeten*», and in his speech during the presentation of the award on December 10, 1921 the president of the Royal Swedish Academy of Sciences, the geologist GERARD DE GEER, said: *"In view of the great significance which Nernst's thermo-chemical work has for chemistry, a significance which may become more and more apparent with the course of time, the Academy of Sciences has decided to bestow on Professor Nernst the Nobel Prize for Chemistry. .... The Academy of Sciences has decided to hand you the Nobel Prize for Chemistry as recognition of the exceptional merit of your work on Thermochemistry."* On the same day at the banquet, responding to NERNST's speech focusing on the international character of science and art, the physiologist and Nobel Laureate of 1911, ALLVAR GULLSTRAND, said that NERNST had accomplished *"to fill his outstanding investigations in the interdisciplinary field between physics and chemistry with shining life. It is too early to say if your brilliant thermo-chemical works will bear more beautiful fruits in one or the other field. However, by the award of the Nobel Prize in Chemistry the Academy of Sciences has testified that in this field you have generated great benefits for humanity."* In this way the emphasis of thermochemistry, which may look restrictive from our present point of view, can be understood. The impact on quantum theory, solid state, and low-temperature physics still had to become apparent. Furthermore, it was closely connected with the research in the field of thermochemistry honored with the Nobel Prize.

Likely because of these reasons NERNST chose for the title of his Nobel Lecture on December 12, 1921 *"Studies on Chemical Thermodynamics"*, and he began his lecture with the words: *"I would fulfill my duty of presenting a lecture focused on my prize-winning papers in the most direct way, if I would develop here my thermal theorem completing the older thermodynamics, and if I would take into account especially its chemical applications and the experimental tests it has undergone in many studies performed in my laboratory."* [Nernst (1923): 1]. However, since these aspects had been treated already in detail recently, NERNST decided *"to treat the subject from the view-point of electro-chemistry."* [Nernst (1923): 1]. Nevertheless, he accomplished to point out the connections of his studies with the low-temperature and solid state physics as well as with the quantum theory and to discuss the resulting experimental and theoretical perceptions in these fields. At the end he mentioned *"that recently my thermal law has been used successfully by Eggert, Saha, and others for solving astrophysical problems."* [Nernst

(1923): 11]. This remark indicates one of his future research subjects, in which he was interested already for some time.

We want to mention the other scientists, who received a Nobel Prize in 1920 together with NERNST (Fig. 5.25). The Nobel Prize in Physics was given to CHARLES ÉDUARD GUILLAUME, at the time director of the *Bureau International des Poids et Mesures* in Sèvres, for his discovery of the anomalies in nickel-steel alloys and for his precision measurements. Based on his measurements the fabrication of "National Prototypes" of the original meter (made of a platinum-iridium alloy) using relatively cheap nickel-steel alloys became possible. The Danish physiologist SCHACK AUGUST KROGH received the Nobel Price in Medicine for the discovery of the regulating mechanism of the blood capillaries and for his studies of the gas exchange during breathing.

Fig. 5.25 CHARLES ÉDUARD GUILLAUME (left) and SCHACK AUGUST KROGH (right).

5.5.9 Critique and priority conflict

It remained inevitable, that NERNST's thermal theorem attracted criticism regarding its content and also the priority of its discovery. In the case of FRITZ HABER still another aspect appeared which JAMES FRANCK pointed out in an interview in 1958 with the words *"Haber believed that he was very close to discover the Third Law of Thermodynamics himself, and he blamed Nernst that he picked it just from his nose."* (quoted from [Szöllösi-Janze (1998): 169]). This might have been particularly painful for HABER, since in general the relation between NERNST and him was highly strained. For example, in this case we must remember the humiliation which NERNST inflicted upon his colleague in 1907 with the question of the equilibrium of ammonia, which was well documented for the scientific community. However, regarding the Third Law of Thermodynamics NERNST took a fair position. So we can read in his first paper dealing with the thermal theorem: *"Furthermore, I want to mention, that in his recent highly remarkable book 'Thermodynamics of Reactions of Technical Gases' Haber has also clearly formulated this problem and has attempted a solution, at least for some cases."* [Nernst (1906c): 8]. Obviously the results presented in HABER's book [Haber (1905)] are different from those obtained by NERNST, and they are by no means comprehensive. For example, HABER believed that the integration constant of the GIBBS-HELMHOLTZ equation is equal to zero only in the case of reactions of gases, which occur without any change in the number of molecules. Also in his monograph NERNST emphasized: *"At this point I want to direct attention in particular to the well-known excellent book by Haber 'Thermodynamics of Reactions of Technical Gases' (1905) which appeared however only after my preparatory work was essentially finished. In this book the importance of the integration constant for the equilibrium between gases is strongly emphasized, the nature of which was then cleared up by our thermal theorem."* [Nernst (1918a): 185].

At this point we want to mention also the curious relation between NERNST and his former lecturer in Graz and first section head in Berlin, HANS JAHN, in connection with the Third Thermal Law, although a conflict or something similar is not documented. Already in his first textbook on thermo-chemistry in its last chapters JAHN had emphasized the importance of the affinity and the *"application of the thermo-chemical data for solving the problems of chemical statistics and dynamics"*, interestingly with the subtitle *"Third Law of Thermo-Chemistry"* [Jahn (1882): 154]. The meaning of these words chosen by JAHN can be seen from his following remarks: *"Since we know from experience that each system of masses approaches the state in which the particles are as close to the stable equilibrium as possible, a system of chemical atoms will always tend toward the compound which shows the largest loss of its motional energy during its formation, i.e., the largest production of heat energy. This is the third law of thermo-chemistry, the so-called 'principle of maximum work', which has been formulated first by Berthelot in full generality."* [Jahn (1882): 169]. Avoiding the term 'Third Law of Thermodynamics', subsequently JAHN attached the last chapter (of his book) to the Second Law of Thermodynamics in the sense of the entropy theorem. In the *"second improved edition"* of his textbook he had pointed out: *"The last chapter, which in the first edition was based too one-sidedly on the principle of maximum work, had to be completely rewritten."* [Jahn (1892): VIII]. It is hard to understand, that in connection with his Thermal Theorem NERNST never mentioned his teacher and coworker.

From the fact that the Thermal Theorem is not satisfied in the case of the VAN DER WAALS law of gases, in 1911 PHILIPP KOHNSTAMM and LEONARD ORNSTEIN derived a principal objection against NERNST's Thermal Theorem. NERNST responded to this attack briefly, but very strongly [Nernst (1911d)]. He pointed out that the invalidity of the VAN DER WAALS equation in the case of liquids at low temperatures can be easily seen from the molecular-kinetic theory. For example, studies by GUSTAV TAMMANN had shown that strongly supercooled liquids assume a rigid amorphous state, in which there exists no more any molecular mobility. However, this was exactly the assumption required for the validity of the VAN DER WAALS equation. Hence, the latter is not meaningful any more at low temperatures, and the thermal law cannot be connected with it. NERNST illustrated this situation in his typical way: *"As is well known, when Tait set out to fight the Second Law of Thermodynamics with the assumption of 'demons', Clausius could justly point out that his formulae would not tell how heat would act with the help of demons, but instead what it would do by itself. Like-wise, here one might point out to Mr. Kohnstamm and Mr. Ornstein, that the equation expressing the thermal theorem cannot be applied to substances which only exist in their imagination. Instead, the real behavior of matter at low temperatures must be taken into account."* [Nernst (1911d): 67].

A real priority conflict developed between NERNST and the American THEODORE WILLIAM RICHARDS. Contrary to the case of VAN'T HOFF, a paper by this scientist from the year 1903 had no influence upon NERNST, *"because its thermodynamic contents did not influence my ideas due to their obscurity and incorrectness."* [Nernst (1918a): 185]. In his SILLIMAN-lecture in 1906 he had mentioned RICHARDS' paper and pointed out the partly qualitative, but not at all quantitative agreement. Although, together with EMIL FISCHER and VAN'T HOFF, NERNST had signed in 1909 the proposal for nominating RICHARDS to become a corresponding member of the Prussian Academy, the American stated in 1914, that all his ideas had been taken up unchanged by NERNST in the development of his thermal theorem. Therefore, in his monograph in 1918 NERNST felt obliged to point out all errors of RICHARDS and to prove his priority in the discovery of the Third Law of Thermodynamics.

The unique merit of NERNST regarding the Third Law of Thermodynamics has been expressed by the international scientific community not only for the priority question. The importance of his Thermal Law for thermodynamics is pointed out, in addition to many other places, in its discussion in the *"Handbook of Physics. Vol. IX: Theory of Heat"* of 1926 [Handbuch Physik (1926)]. Here NERNST's student KURT BENNEWITZ published an article on *"Nernst's Thermal Theorem"* [Handbuch Physik (1926): 141–174], the size of which amounts to about one quarter of the article by KARL FERDINAND HERZFELD on *"Classical Thermodynamics"* [Handbuch Physik (1926): 1–140], which covers the First and Second Laws of Thermodynamics.

5.6 Other Scientific Studies during this Period

In addition to the studies of chemical thermodynamics, low-temperature solid-state physics, and quantum theory, at the Institute of WALTHER NERNST, other subject areas were also investigated. Up until about 1910 these concerned *"the theory of the galvanic polarization, leading to laws about the nature of the electric nerve stimulation, which were experimentally verified (Nernst, Eucken), and several studies of the shape of current-voltage characteristics (Eucken), moreover, investigations of the kinetics of the reactions of gases with suitable catalysts (Bodenstein), photochemical, electrochemical, and organic chemical studies were published."* as reported by NERNST [Nernst and Sand (1910): 309].

Furthermore, NERNST paid attention also to the radiochemical research of his coworker WILLY MARCKWALD. Already in 1902 MARCKWALD became known because of his work dealing with radioactive chemical elements, and subsequently he had successfully investigated radium and polonium, the latter with higher accuracy than MARIE and PIERRE CURIE. Whereas the French couple had considered already, that polonium would follow a specific decay law, MARCKWALD could verify the general validity of this law also in this case, since for the first time he was able to prepare a pure polonium sample. In addition to radiochemical problems, MARCKWALD treated in particular the optical activity of organic compounds.

In Section 4.10 we have already discussed the investigations of his Law of the Electrical Nerve Stimulus Threshold mentioned by Nernst, which he had begun in Göttingen and continued up to 1908.

Among the studies dealing with electrochemistry, in addition to those connected with the Third Law of Thermodynamics [Nernst (1909b)], we must mention the construction of an apparatus by NERNST for determining the electrolytic conductance, the feasibility test of which he left to ALFRED MAGNUS in 1905. This instrument is *"superior to those used up to now mainly because of the convenient variation of its resistance capacity*[sic]*"*, since this was *"variable to a high degree, such that one could measure with it large as well as small conductances."* [Magnus (1906): 1–2]. In the case of a sufficient amount of sample material the measuring instrument turned out good enough, such that it was manufactured by the Keiser & Schmidt Company in Berlin and was offered for sale at a low cost of 30 marks together with a thermometer calibrated in $\frac{1}{10}$ degrees.

Among other electrochemical studies, three PhD theses suggested by NERNST are interesting to mention. These could utilize the results and the positive experience, which were achieved by a method invented by NERNST and performed and tested by EDMUND SAWYER MERRIAM in his PhD thesis carried out in Göttingen. NERNST had suggested eliminating the large error sources appearing during measurements of the residual currents by rapidly rotating one of the two electrodes and in this way achieving a strong constant mixing. Because of the rotation, around the electrode a homogeneous diffusion layer is generated, the thickness of which remains unchanged during the constant rotation. With this simple technique diffusion coefficients and reaction rates could be measured with high accuracy, for example.

This method was taken up by ARNOLD EUCKEN during his experimental and theoretical investigations carried out in the Physical-Chemical Institute in Berlin, which he published in 1907 in his PhD thesis [Eucken (1907)]. Initially, he had measured the diffusion coefficient of some organic acids, *"however, here I was confronted with a few peculiarities, which are only partly associated with that method and were providing a reason for a more general investigation."* [Eucken (1907): 73]. The experimental arrangement consisted of a small rotating cathode and a large fixed platinized anode constantly rinsed with hydrogen. In this way it was achieved that concentration changes only occurred near the cathode. For the theoretical discussions EUCKEN assumed that of all processes only diffusion takes place at finite velocity. Of the accomplished results we mention the demonstration of the agreement between theory and experiment in the case of the measurement of the current-voltage curves, which were obtained during the electrolysis of acid solutions without and with an equal-ionic admixture. EUCKEN was able to explain a partial break and a steep rise of the curves in terms of causes other than decomposition. Furthermore, in the case of purely acid solutions he could qualitatively verify that with increasing voltage the polarization and the resistance approach a constant limiting value. In addition to NERNST, his *"highly admired teacher, for his kind and helpful suggestion and for his continuous benevolence to me"*, EUCKEN had to thank also *"Prof. H. Jahn, who unfortunately passed away too early, and who supported me frequently most kindly with his rich experience."* [Eucken (1907): 117].

In 1909 KURT BENNEWITZ submitted a PhD thesis, which treated problems in connection with the decomposition voltage [Bennewitz (1909)]. It was customary to determine these quantities by graphical extrapolation from the discontinuities (breaks) of a measured current-voltage curve. However, in general this method was not unambiguous. No problems appeared in cases, where solutions of metal halides became electrolyzed. Therefore, BENNEWITZ primarily studied solutions of sulfuric acid. It was the task of his thesis to work out a method, which avoids the uncertainty and arbitrariness of the graphical extrapolation, and which allows to handle reliably such cases as the electrolysis of sulfuric acid presenting problems up to then. A discontinuity occurring during the gas separation process, which had nothing to do with the break at the decomposition point, could be eliminated by means of rotating electrodes. Of the essential results which BENNEWITZ could achieve we mention, that he was able to formulate an equation for the current-voltage curve above the decomposition point, which well reproduced the measured curves. He could show that close to this point the voltage does not depend on the logarithm of the current as in the relation by TAFEL. BENNEWITZ determined unambiguously the decomposition voltage of dilute sulfuric acid in the range 1.50 – 1.63 V as a function of increasing surface roughness of the used platinum electrode.

In his PhD thesis presented also in 1909 GEORG PFLEIDERER treated the generation of oxygen during the electrolysis of hydrochloric acid by using a platinum anode [Pfleiderer (1909)]. At this electrode Cl^--ions can be oxidized to Cl_2 as well as OH^-- ions to O_2. In the equilibrium case NERNST was able to show in which ratio both substances are generated. However, in the case of the electrolysis the processes at the electrode are not reversible, such that a relation for the thermodynamic equilibrium cannot be formulated. For the example of the electrolysis of the hydrochloric acid PFLEIDERER investigated *"by which rule under such circumstances the fraction of the current, which corresponds to one component of the secreted mixture, is determined."* [Pfleiderer (1909): 7]. Already in 1898 in this connection FRITZ HABER had considered the depletion of ions to be crucial, however, without determining its magnitude. In order to be able to test this assumption, which was considered unlikely in Berlin, PFLEIDERER used a rotating cylindrical platinum anode, since NERNST, MERRIAM, and EUCKEN had demonstrated that in this case the depletion of the ions can be calculated exactly. As one of its results this study yielded the finding, that during the electrolysis of hydrochloric acid, independent of the magnitude of the current, the platinum anode experiences a permanent change, which can be seen from a voltage rise of 0.5 V. Oxygen is generated appreciably only if the anode is covered with an oxide layer. Based on the observed behavior, *"a theory of the oxygen generation was proposed, by which the processes during the discharge were connected with the phenomena of the overvoltage [Überspannung], without presenting special hypotheses about its nature."* [Pfleiderer (1909): 37–38].

These investigations suggested and supported by NERNST, also regarding the electrochemical research, indicate a typical feature of him, which we had mentioned already in connection with his Thermal Law: In the cases when he could consider some fundamental advances achieved by him to be completed in some sense, he left the further clarification of special problems, which were interesting to him, to his students and younger coworkers. So within his working group, based on his ideas and because of his supporting advice, important advances were achieved, where his name did not explicitly appear in the corresponding publication.

In some sense one must include within these projects also the study carried out by KURT MOERS at the Institute of Physics in Berlin and dealing with the chemical nature of lithium hydride, LiH, unknown up to then and announced to the scientific community by NERNST [Nernst (1920)]. The comparison of a few properties of lithium hydride with those of lithium halides had indicated that the hydrogen in LiH perhaps behaves analogously to the halide anions. NERNST praised the achievement of MOERS with the words: *"However, the more superficial similari-*

ties between the lithium hydride on the one hand and the lithium halides on the other hand would hardly be crucial, if Mr. Moers would not have had the idea of the electrolytic splitting of the pure molten salt, and if he would not have succeeded, apparently, to prove unambiguously, that during the electrolysis of the molten hydride lithium metal is deposited at the cathode, hydrogen, however, at the anode like a halogen." [Nernst (1920): 323]. Therefore, in LiH and related salt-like hydrides the hydrogen is present in the form of H^--ions. In aqueous solution the hydrogen molecule H_2 can be treated as an extremely weak acid, and similarly the hydride ion H^- as an extremely strong corresponding base, such that we have in modern terminology

$$H_2 + H_2O \rightleftharpoons H_3O^+ + H^- \text{ or } H^- + H_2O \rightleftharpoons H_2 + OH^-.$$

In order to explain the demonstrated different kind of H_2-molecule, NERNST postulated its formula as *"H_2E_2, if E denotes the negative electron, which represents a chemical element as well as all others."* [Nernst (1920): 324]. Therefore, in addition to the usual dissociation *"$H_2E_2 = HE + HE$"* ($H_2 \to H\cdot + H\cdot$) there would exist another one of the form *"$H_2E_2 = H^\bullet + HE_2 [= H']$"* ($H_2 \to H^+ + H^-$).

After the First World War NERNST also turned to modern photochemistry. In 1912 ALBERT EINSTEIN had founded this field by means of his law of quantum equivalence. This law, which is named sometimes after JOHANNES STARK and EINSTEIN, says, that an absorbed light quantum $h\nu$ with frequency ν delivering enough energy for causing the reaction, exactly induces an elementary reaction on an atom or molecule of the absorbing substance, and that, on the other hand, an atom or molecule experiencing a photochemical change takes up exactly an energy quantum $h\nu$ from the irradiating light during the primary photochemical process. The products of the primary reaction can cause secondary photochemical processes in many different ways.

The law of the quantum equivalence has been confirmed in the case of many reactions. In this context just the reaction of the formation of hydrogen chloride, HCl, following the reaction formula $H_2 + Cl_2 \rightleftharpoons 2HCl$, with which ROBERT WILHELM BUNSEN and Sir HENRY ROSCOE had started their photochemical studies in 1856, still remained not understood. Because of the initial reaction $Cl_2 + h\nu \rightarrow 2Cl$ one expects the formation of two HCl-molecules per absorbed light quantum. However, actually this number is at least one million, as has been shown by MAX BODENSTEIN in 1913. In 1918 NERNST solved this problem in a simple but ingenious way [Nernst (1918b)]. The starting reaction of the formation of the two chlorine atoms corresponding to the law of equivalence of EINSTEIN represents only the primary photochemical process. In two reaction steps the two chlorine atoms mediate a cycle, in which hydrogen chloride is generated *via* hydrogen atoms as an intermediate product in such a way, that in the end one obtains the stoichiometric formula we have mentioned:

$$\begin{aligned} Cl + H_2 &\rightarrow HCl + H \\ H + Cl_2 &\rightarrow HCl + Cl \\ \hline H_2 + Cl_2 &\rightleftharpoons 2HCl \end{aligned}$$

According to NERNST the number of cycles is limited due to reactions of the free atoms such as $2Cl \rightarrow Cl_2$, $2H \rightarrow H_2$ and $H + Cl \rightarrow HCl$. However, he stated: *"One can understand immediately, that the free chlorine and hydrogen atoms can collide with the hydrogen molecules or with the chlorine molecules many million times before they neutralize themselves in the way mentioned due to a mutual collision."* [Nernst (1918b): 325] and *"It appears that Einstein's law is meant to play a similar role in the photolysis as Faraday's law in the electrolysis."* [Nernst (1918b): 326].

NERNST left it to his coworker and previous PhD student LOTTE PUSCH to pursue further and test his ideas on the law of equivalence. Regarding the work of NERNST and of herself she summarized: *"Basically the effect of light upon chlorine-hydrogen gas can be reduced to the idea, that the chlorine atoms are split due to the light, which process follows Einstein's law, and that the photochemical yield is immensely enhanced because of side reactions. This concept could be experimentally verified in different directions by investigating the effect of light upon bromine and upon substances adding bromine."* [Pusch (1918): 329].

Shortly before NERNST discussed the development of the reaction or chain cycle in the case of the formation of HCl by means of free atoms, MAX BODENSTEIN instead had proposed energetically excited Cl_2- and HCl-molecules. Later BODENSTEIN pointed out regarding NERNST's assumptions: *"However, the fine scent and the deliberate or unintended neglect of irrelevant circumstances were needed, that the ingenious man has shown so often in the case of his great discoveries, in order to make this assumption at the time. ... The termination of the chain happens differently from that proposed by Nernst; however, he was right regarding the beginning and the members of the chain."* [Bodenstein (1942c): 127–128]. Although BODENSTEIN had to mention also a few incorrect points of the argumentation, it remains the great achievement of NERNST, based on his concept of the free atoms, to have introduced into science the idea that radicals can function as members of a chain.

Also NERNST's coworkers carried out research in the field of photochemistry in connection with the law of quantum equivalence. Here we must mention primarily JOHN EGGERT and WALTER NODDACK. Since 1921 they have published several papers dealing with this law and with the photographic process. In 1922 KARL FRIEDRICH BONHOEFFER had also presented a PhD thesis about this problem [Bonhoeffer (1922)].

NERNST and NODDACK had already published a paper on photochemical reactions as a *"Report of the Physikalisch-Technische Reichsanstalt"* [Nernst and Noddack (1923)]. In their summary the authors made the following remarks: *"Based on our present knowledge, for the interpretation of photochemical processes the quantum theory is not needed in any different way than that the light absorption happens in the form of quanta. About the question, which chemical processes follow from this, reliable predictions cannot be made even in a single, still very simple case. ... As a rule the process is affected by dark reactions, which modify the yield appreciably. In the least number of cases our chemical knowledge is sufficient, to predict something in advance."* [Nernst and Noddack (1923): 114]. Also *"the doubt* [cannot] *be suppressed, whether it is practical at all to speak of a 'photochemical law of equivalence',"* however, independent of this *"there remains, of course, the historical fact, that the formulation of the law of equivalence has been extremely fruitful for the photochemistry."* [Nernst and Noddack (1913): 115].

5.7 Organization of Science

5.7.1 Kaiser Wilhelm Institutes

In our discussion of the activities of WALTHER NERNST we have mentioned already, that, in addition to the scientific fields represented by him, he also well understood the aspects of their organization. This does not only concern the smaller parts of the Institutes directed by him, but also Physical Chemistry and other fields in general.

An excellent example of this talent is the foundation of the *Kaiser-Wilhelm-Gesellschaft* (KWG) (Kaiser-Wilhelm-Society), which to a large extent is owed to the engagement of NERNST. The motivation for creating such research facilities in particular for chemistry originated from the strong connection of this science with industry and economy existing since the end of the 19th century. For the challenges arising from this direction the Universities were not adequate any more because of their obligation to combine research with education. There had also developed a need for research institutes, in which the working scientists were freed from the duty to teach. The creation of such separate scientific Institutes existing in addition to Universities and Academies went way back to an idea of WILHELM VON HUMBOLDT. However, at the turn of the 20th century there appeared the need for founding scientific societies within which several Institutes were occupied by following methodically the same general goal.

On December 25, 1903 WILHELM OSTWALD had noted in his diary that he had suggested a *«Chemie-Reichsanstalt»*, which later had been referred to as *«Chemische Reichsanstalt»* (CRA). It followed the model of the *Physikalisch-Technische Reichsanstalt* founded in 1887 in Berlin, where general research goals were pursued by people free of any teaching duties. In 1904, together with his colleague from Leipzig, the historian KARL LAMPRECHT, who also supported the creation of pure research institutes, OSTWALD had visited the USA and had received there ideas about the possibilities for financing such institutions.

During the inauguration ceremonies of the Physical Institute in Leipzig directed by OTTO HEINRICH WIENER on July 8, 1905, the necessity of founding the CRA was mentioned, and OSTWALD gave NERNST a conceptional outline prepared by him for this institution. The latter showed immediately the highest interest, and already on July 19 he reported to OSTWALD: *"In the meantime I have discussed in detail your outline and the whole project with Emil Fischer. We wish to propose that we send your outline, which must be supplemented in some points, to a number of representatives of science and technology."* [Zott (1996): 171]. On August 7 the outline, slightly changed and supplemented partly together with FISCHER, was sent back to OSTWALD. On September 8 NERNST informed his colleagues OSTWALD and FISCHER about a supplementary proposal containing the recommendation, *"that the larger German chemical societies themselves be asked to send also a delegate each into the board."* [Zott (1996): 174].

Also ERNST BECKMANN, who was *Ordinarius* of Applied Chemistry in Leipzig at the time, joined the effort for founding a CRA, which was strongly pursued since October 1905. On October 14 a meeting was organized in Berlin by OSTWALD, FISCHER, and NERNST, at which *"proposals for the foundation of a chemical 'Reichsanstalt'"* were announced. At this meeting, rejecting opinions were expressed, in particular by CARL ALEXANDER VON MARTIUS, the retired Director of the *Aktiengesellschaft* for the fabrication of aniline (Agfa) and by LEO GANS, a chemist and industrialist working in Frankfurt/Main. To their written statement regarding this matter NERNST reacted in terms of an attempt for understanding and suggested to OSTWALD to formulate an opposing statement [Martius (1906); Ostwald (1906a); Gans (1906); Ostwald (1906b)]. However, on January 15, 1907, referring to FISCHER, NERNST admitted: *"About this matter I still want to remark, that for the two of us the proposal by Gans appears quite worthwhile to pursue; for example, in addition to the Reichsanstalt there could exist very well a fund, by* which the board of the new institution supports outsiders, thereby supplementing its working program." [Zott (1996): 181].

In April 1906 in the document «*Die chemische Reichsanstalt*» OSTWALD pointed out the necessity for a speedy establishment of this institution [Ostwald (1906c)]. At the 4th International Conference of Applied Chemistry in Rome (April 26 – May 3, 1906) he had also raised the question of this project. As a result, the chemist and industrialist LUDWIG MOND donated the amount of 205 000 *Mark*. Apparently, in Berlin NERNST and FISCHER had shifted the activities of the inner committee for the preparation of the establishment of the CRA mainly to themselves, presumably, because OSTWALD more and more removed himself from Physical Chemistry, in order to concentrate on philosophy, organization of science, and other matters. Therefore, on January 7, 1908 the colleague from Leipzig wrote to his former collaborator: *"Dear Colleague! Since the position as a member of the inner committee does not indicate the relation, in which I find myself regarding the project of the chemical 'Reichsanstalt', I am honored to announce to you my resignation. Very sincerely yours WOstwald."* [Zott (1996): 186]. Exactly two months later, on March 7, 1908, the «*Verein* (association) *Chemische Reichsanstalt*» was founded. ERNST BECKMANN became the designated President of the CRA.

Since 1909 in the Prussian Ministry of Church, School, and Medical Matters the *"Records concerning: the project of Dr. Althoff concerning the* [Prussian] *demesne Dahlem for state purposes (establishment of a distinguished colony of excellent scientific locations, a German kind of Oxford)"* were kept, which were based on the plans and concepts of FRIEDRICH ALTHOFF, head of the ministry department dealing with the universities and higher education. The ideas of ALTHOFF were incorporated into the secret memorandum of November 21, 1909 of ADOLF HARNACK, theologian and Professor of Church History. This memorandum had the goal to establish research institutes for the natural sciences within a *"Royal Prussian Society for the Advancement of the Sciences"*. The establishment of such a Society had also been the subject of the memoranda of the collaborators of ALTHOFF, namely FRIEDRICH

SCHMIDT-OTT, head of the Ministry Department of the Arts and Sciences, and the physicist working at the time in the Ministry of Education and Cultural Affairs, HUGO ANDRES KRÜSS.

The Emperor WILHELM II approved these ideas. However, regarding the financing one did not rely on the German *Reich*, but instead on private donations and on the State of Prussia. During May 1910 representatives of commerce, economy, industry, and banking met at the Ministry of Education and Cultural Affairs in Berlin, in order to discuss the establishment of a Society envisioned by HARNACK. LEOPOLD KOPPEL also belonged to this group.

On September 17, 1910, this banker already well known as a patron of the natural sciences and having excellent contacts to NERNST, announced, that instead of a department of Physical Chemistry planned by OSTWALD, FISCHER, and NERNST for the CRA, he would be willing to endow and finance an independent Institute of Physical Chemistry, if until the centenary of the University of Berlin the Emperor would guarantee that the amount of 35 000 *Mark* per year for the salary of the director and for fixed expenses would be taken up by the State. Indeed, during this celebration on October 11, 1910 WILHELM II called for the establishment of research institutes and asked for donations. He also agreed that a Society would be founded, which carried his name and which was placed under his protectorate. With that the establishment of the «*Kaiser-Wilhelm-Gesellschaft zur Förderung der Wissenschaft*» (KWG; Kaiser-Wilhelm-Society for the Advancement of Science) was decided. By the KOPPEL endowment alone it had received the amount of one million *Mark*.

When the *Kaiser-Wilhelm-Gesellschaft* was becoming active on January 11, 1911 in a constituting session in the large assembly hall of the Royal Academy of the Arts in Berlin, instead of a single CRA, two Kaiser-Wilhelm-Institutes (KWI) were created: the KWI of Physical Chemistry and Electrochemistry directed by FRITZ HABER and endowed by KOPPEL, and the KWI of Chemistry its first director being ERNST BECKMANN. ADOLF HARNACK became the first President of the Society, and the industrialist GUSTAV KRUPP VON BOHLEN UND HALBACH and the banker LUDWIG DELBRÜCK became the two Vice Presidents. In this way the connection between science, industry, and finance was demonstrated. In a letter from February 28, NERNST, HARNACK, and KOPPEL proposed to the Prussian Ministry of Education and Cultural Affairs, that the two Institutes be placed within separate buildings on a common ground. In this context NERNST represented the Society *Chemische Reichsanstalt*, HARNACK the KWG, and KOPPEL his foundation existing since 1905. Around the middle of May 1911 the area in what today is Berlin-Dahlem was inspected by the architect ERNST EBERHARD VON IHNE, and by HABER, BECKMANN, and KRÜSS. On December 23, 1911 a contract was signed by the President ADOLF HARNACK, by the First Treasurer of the KWG FRANZ VON MENDELSSOHN, and by the Chairman of the Society *Chemische Reichsanstalt* EMIL FISCHER, which fixed the foundation of the KWI of Chemistry.

On October 23, 1912, the two planned KWIs of Chemistry and of Physical Chemistry and Electrochemistry were opened in a ceremonial act. The Emperor attended the opening and delivered a small speech. The actual opening speeches were given by EMIL FISCHER for the Society *Chemische Reichsanstalt*, by ADOLF HARNACK for the KWG, and by AUGUST VON TROTT ZU SOLZ, Minister of Education and Cultural Affairs of Prussia.

The original idea underlying the establishment of a CRA in the sense of WILHELM OSTWALD was definitely abandoned, when during spring of 1913 the Society *Chemische Reichsanstalt* was renamed into "Society for the Advancement of Chemical Research". Furthermore, very soon the KWG aimed at the establishment of additional Research Institutes, which were devoted to Sciences other than Chemistry.

So already prior to the actual foundation of the KWG the idea had come up, to establish also a KWI of Physics. However, because of the existence of the *Physikalisch-Technische Reichsanstalt* this idea was put back initially. However, when at the end of 1913 LEOPOLD KOPPEL indicated, that he would be willing to donate an endowment also for physical research, in 1914 WALTHER NERNST, FRITZ HABER, MAX PLANCK, HEINRICH RUBENS, and EMIL WARBURG as members of the Prussian

Academy of Sciences applied for and justified the establishment of a KWI for Physical Research. For a certain amount of time in each case, this Institute was required to employ suitable physicists, in order to solve important and urgent physical problems by means of theoretical and experimental research. For this purpose a small building was to be erected again in Dahlem. The administration of the Institute was to be controlled by a board, and in addition a scientific committee directed by a "permanent Honorary Secretary" had to be formed. However, immediately prior to the outbreak of the First World War, the finances required for the Institute were rejected by the Ministry of Finance.

ALBERT EINSTEIN had been designated for the position of Honorary Secretary. He had just agreed to come to Berlin. Last but not least this extremely important step for the world center of the natural sciences at the time was the result of the active endeavor of NERNST. It is said that NERNST himself regarded the fact that EINSTEIN had been won for Berlin, as his greatest achievement in the organization of science. In 1913 EINSTEIN had been nominated as a member of the Royal Prussian Academy of Sciences. During the summer of this year NERNST and PLANCK traveled to Zurich, where in 1912 EINSTEIN had just accepted an offer from the *Eidgenössische Technische Hochschule*. In Zurich they invited him to come to Berlin to become a member of the Prussian Academy and Professor without any teaching obligation, in some sense as the successor of VAN'T HOFF. In order to strengthen their request, NERNST had been able to persuade KOPPEL, to contribute a considerable amount to EINSTEIN's salary. On November 12, 1913 the election of EINSTEIN to the Academy has been approved by the Emperor WILHELM II. On December 7 the former accepted the offer of the "Berlin people", and on March 22, 1914 he left Zurich. Early in April he arrived in Berlin, after he had visited Antwerp and Leiden.

Surprisingly, during the War there arose the possibility to move one step closer to the realization of the KWI of Physics, when considering the time after the War, FRANZ STOCK, machine and tool manufacturer in Berlin, donated half a million *Mark*. As a result, on July 6, 1917 the Senate of the KWG decided to operate the Institute starting in October of that year. EINSTEIN was installed as Director as usual in the case of a KWI, and the Board consisted of NERNST and the four other signatories of the memorandum of 1914. A building for the Institute was not erected, and also not after the War. The existing funds would be made available for research in a different way. Therefore, in a circular of the Board of March 1919 it was announced that *"substantial means"* exist, which *"are available to scientific Institutes as well as to individual scientists for allowing or for helping to undertake scientific research"*, i.e., for the purchase of instruments and for the granting of fellowships. Only since 1937 there existed in Dahlem an actual KWI of Physics. It is interesting that in March of 1929 NERNST, EINSTEIN, HABER, PLANCK, MAX VON LAUE, EMIL WARBURG, and FRIEDRICH PASCHEN had advocated the establishment of an Institute of Theoretical Physics as an extension of the KWI of Physics.

Starting in April 1946, OTTO HAHN began to fundamentally change the form of the KWG. On February 26, 1948 these efforts culminated in the foundation of the «*Max-Planck-Gesellschaft zur Förderung der Wissenschaften*» (MPG, Max-Planck-Society for the Advancement of Science).

Concluding we list the Presidents of the KWG and the MPG:

Time of Office	Name	Scientific Field
1911–1930	ADOLF VON HARNACK	theology, church history
1930–1937	MAX PLANCK	theoretical physics (Nobel Prize 1918 Physics)
1937–1940	CARL BOSCH	technical chemistry (Nobel Prize 1931 Chemistry)
1941–1945	ALBERT VÖGLER	steel industry, politics
1945–1946	MAX PLANCK	(see above)
1946–1960	OTTO HAHN	radiochemistry (Nobel Prize 1944 Chemistry)
1960–1972	ADOLF BUTENANDT	biochemistry (Nobel Prize 1939 Chemistry)

Time of Office	Name	Scientific Field
1972–1984	REIMAR LÜST	astrophysics
1984–1990	HEINZ STAAB	organic chemistry
1990–1996	HANS F. ZACHER	jurisprudence
1996–2002	HUBERT MARKL	zoology
since 2002	PETER GRUESS	biology

5.7.2 German Electrochemical Society

The name of NERNST is also connected with a scientific society in Germany since its foundation remains important until today: the German Electrochemical Society, which since 1902 carries the name of ROBERT WILHELM BUNSEN.

The idea to found such a Society goes back to ARTHUR WILKE at the end of the 19th century. He was an electrical engineer, who could win over as the first people supporting this project WILHELM BORCHERS, metal-processing expert and electro-metallurgist, and FRIEDRICH VOGEL, working in Charlottenburg as *Privatdozent* of electrical engineering and electrochemistry. These scientists had planned to publish in 1894 a "Journal of Electrical Engineering and Electrochemistry" (*Zeitschrift für Elektrotechnik und Elektrochemie*) in Halle on the river Saale with the publisher WILHELM KNAPP, who in this way must be counted also as belonging to the *"fathers of the group"* [Jaenicke (1996): 10]. In this journal WILKE was responsible for the electrical engineering and BORCHERS for the electrochemical part. Connected with this was the idea, that this journal should become the organ of a Society still to be founded, following the example of the Society of German Chemists and its *"Journal of Applied Chemistry"*. In order to realize this goal, WILKE succeeded to get support from representatives of industry and of science, for example, from WILHELM OSTWALD.

In the beginning of March 1894 WILKE, BORCHERS, and VOGEL issued an appeal for the foundation of an Electrochemical Society, in which it was pointed out: *"The rapid and excellent development beginning to show nowadays in the case of electrochemistry is highly important for Germany. Since in the case of chemistry as well as of electrical engineering our country is first among all other countries, and it must keep this high rank also in the promising field in-between, in electrochemistry. In fact, also in our case we note a strong drive toward this goal, and we can be proud to point out, that on German soil the electrochemistry, regarding its science as well as its application, finds the strongest support. However, the high rank of Germany in the field of electrochemistry to an increased degree results in an obligation to cultivate this science and technology and, since such a cultivation calls for a common effort, it must appear as an honorable duty of the relevant scientists and technicians to create an organ for cultivating the electrochemistry, its research and its application."* (quoted after [Jaenicke (1996): 14]). We point out, that the high value of the electrochemical research described in this case, had also been emphasized in the application of the Faculty in Göttingen in just the same year 1894 to offer WALTHER NERNST the position of *Ordinarius*, as we have mentioned above.

In fact, NERNST, who at the time had just reached his 30th year of age, was one of the most important representatives of electrochemistry indicated in the appeal and playing a central role in Germany. Therefore, it is not surprising that he belonged to the 16 University teachers, who until March 30, 1894 had expressed their interest in the foundation of an electrochemical Society. On April 21, 1894 65 prominent people participated in the foundation meeting in Kassel, among these 18 from Universities, 14 representatives from industry including WALTHER RATHENAU, 32 chemists and engineers from industry, and ERIC WATSON from South Birkenhead as a foreign member. In addition to OSTWALD and NERNST, the representatives from the academic area included MAX LE BLANC, and CLEMENS WINKLER. As an influential member of the Prussian State Parliament HENRY VON BÖTTINGER had been won, who just at this time played an important role in the fact that in 1894 NERNST could be kept in the Prussian Göttingen. During the foundation meeting WILHELM OSTWALD was elected as the first chairman and HENRY VON BÖTTINGER as the second chairman, and the *"Journal of Electrical Engineering and Electrochemistry"*, the first issue of which had just appeared, was confirmed as the journal of the Society. Already with its second volume in 1895, the journal was renamed *«Zeitschrift für Elektrochemie»* (Journal of Electrochemistry). Subsequently, this title was changed again, and only in 1953 the name *«Berichte der Bunsengesellschaft»* (Reports of the Bunsen Society) was adopted, and still remains valid today.

WALTHER NERNST was also one of those, who acted as First Chairman of the Society. Hence, he belonged to a series of distinguished people, most of whom were physical or industrial chemists. In the following we list the names of these people during the lifetime of NERNST:

Time of Office	Name
1894–1898	WILHELM OSTWALD
1898–1902	JACOBUS HENRICUS VAN'T HOFF
1902–1905	HENRY VON BÖTTINGER
1905–1908	WALTHER NERNST
1908–1911	PAUL MARQUART
1911–1914	MAX LE BLANC
1914–1918	HANS GOLDSCHMIDT
1918–1920	KARL ELBS
1920–1922	FRITZ FOERSTER
1922–1924	AUGUST BERNTHSEN
1925–1926	GUSTAV TAMMANN
1927–1928	ALWIN MITTASCH
1929–1930	MAX BODENSTEIN
1931–1932	HEINRICH SPECKPETER
1933–1934	RUDOLF SCHENCK
1935–1936	HANS GEORG GRIMM
1936–1941	RUDOLF SCHENCK
1942–1945	PETER ADOLF THIESSEN

In 1899 HEINRICH SPECKPETER had obtained his PhD with NERNST, his thesis having the title *"About a Quantitative Electrolytic Separation Method of the Halogens Chlorine, Bromine, and Iodine"*. Also the former students of NERNST, PAUL GÜNTHER, ARNOLD EUCKEN, and KARL

FRIEDRICH BONHOEFFER served in the office of First Chairman after the Second World War during 1947 – 1949, 1950, and 1951 – 1952, respectively. GEORG-MARIA SCHWAB, First Chairman during 1955 – 1956, had carried out his PhD thesis with ERNST HERMANN RIESENFELD, completing it in 1923 about the subject of ozone in the Institute of NERNST in the *Bunsenstraße* in Berlin.

Since the foundation of the Society up until 1922 JULIUS WAGNER, the former colleague of NERNST in Leipzig in the Institute of OSTWALD, was its manager. *"With his legendary love of order later he even withstood the similarly legendary disorder even of Nernst."* wrote MAX BODENSTEIN in 1936 to HANS GEORG GRIMM [Jaenicke (1996): 20].

Also the *Zeitschrift für Elektrochemie* is connected with the activities of NERNST. On June 10, 1894 he was asked by OSTWALD, who was trying very hard to popularize electrochemistry: *"How is it with the popular articles about the modern electrochemistry, which you should and also were willing to write for the journ. of the Soc.? In the interest of the Soc. I would greatly appreciate, if you start with this soon."* [Zott (1996): 71]. As a matter of fact, in the first volume of the new journal edited by BORCHERS and WILKE, NERNST had published two articles [Nernst (1894/95].

For the volumes 2 – 4 BORCHERS and OSTWALD acted as editors of the journal, the title of which did not contain the word *'Elektrotechnik'* (Electrical Engineering) any more since the year 1895/96. Since the year 1898/99 instead of OSTWALD, NERNST acted as the editor of the volumes 5 and 6. Starting with volume 7 of the year 1900/01, there was another change of the editors, since RICHARD ABEGG, the former assistant of NERNST and since 1899 Section Head of the Chemical Institute in Breslau, working now together with his former superior, replaced BORCHERS. The latter had sold his rights of the journal to the Electrochemical Society, such that the Society now had at its disposal its own organ. In addition to ABEGG and NERNST, the students of the latter, HEINRICH DANNEEL and PAUL GÜNTHER, worked as editors of the journal, as can be seen from the following overview [Jaenicke (1996): 195–196] (p. 217).

In 1912 the services of NERNST for the German Bunsen Society were honored with the honorary membership. In 1914 he was presented with the Bunsen Memorial Medal, with which also four of his former students had been honored: HANS VON WARTENBERG 1951, MATTHIAS PIER 1953, KARL FRIEDRICH BONHOEFFER 1955, and PAUL GÜNTHER 1958.

Journal Title	Volumes	Starting from year	Editors
Zeitschrift für Elektrochemie	5–6	1898/99	WILHELM BORCHERS WALTHER NERNST
	7	1900/01	WALTHER NERNST RICHARD ABEGG
	8–10	1902	RICHARD ABEGG
Zeitschrift für Elektrochemie und angewandte physikalische Chemie	11–14	1905	RICHARD ABEGG HEINRICH DANNEEL PAUL ASKENASY
	15–16	1909	RICHARD ABEGG PAUL ASKENASY
	⋮	⋮	⋮
	52–55	1948	PAUL GÜNTHER
Zeitschrift für Elektrochemie, Berichte der Bunsengesellschaft für physikalische Chemie	56–65	1952	PAUL GÜNTHER

Since 1953 the Bunsen Society bestows an award, which in each case should be connected with a name of the three great physical chemists WALTHER NERNST, FRITZ HABER, or MAX BODENSTEIN. In 1955 the first Nernst Award was presented to the electrochemist MARK VON STACKELBERG for his fundamental studies of the electrolysis and its application in analytical chemistry.

5.7.3 Other developments

In addition to his active role in the creation and consolidation of the two scientific organizations or societies mentioned above, there have been several other contributions of NERNST to the organization of science, of which a few should be briefly discussed.

One contribution is connected with ROBERT VON LIEBEN, who during spring of 1899 had come to NERNST in Göttingen, in order to continue his studies. Since that time he had developed a close friendship with his teacher. For example, in Göttingen NERNST and LIEBEN discussed the subject of the just discovered radioactivity and the future prospects of the motorized air travel. When after one year LIEBEN left the *Georgia Augusta* without concluding his studies, within this short time certainly under the influence of NERNST he had invented a gearbox for automobiles and an "electrochemical phonograph". In the case of the latter there exists a joint publication [Nernst and Lieben (1900/01)].

In 1900 LIEBEN returned to his native town of Vienna, and in 1905 he began to occupy himself with the development of telephone relays, since the electromagnetic amplifiers existing at the time did not allow good communication by telephone over large distances. It is very likely, that NERNST pointed out to his former student the research results of ARTHUR WEHNELT demonstrating the generation of cathode rays by means of a glow cathode made from alkali-oxide. By combining this WEHNELT-cathode with the cathode-ray tube of KARL FERDINAND BRAUN, in 1906 LIEBEN was able to obtain a patent for his "cathode-ray relay". Together with his former fellow students in Vienna, EUGEN REISZ and SIEGMUND STRAUSS, LIEBEN improved the tube, being highly impractical so far due to its enormous size and the need for a permanent connection to a vacuum pump. LIEBEN had employed both in the laboratory, which he financed himself. So in 1910 and 1911 additional patents could be obtained. The latter patent deals with the LIEBEN-REISZ-STRAUSS relay, which usually is only referred to as LIEBEN-tube.

During August of 1911 at the initiative of NERNST, a telephone amplifier based on this tube was demonstrated successfully in his Institute in the *Bunsenstraße* in Berlin to leading representatives of science. It is also likely that about half a year later NERNST did play not only a small role in the so-called "LIEBEN-Consortium", founded for the purpose of marketing the patents of LIEBEN. Of course, the Company Siemens & Halske in Mühlheim at the river Ruhr belonged to this Consortium. In 1847 WERNER SIEMENS and the mechanic JOHANN GEORG HALSKE had founded this Company as *«Telegraphenbauanstalt»* (institution for building telegraphs), in order to realize the proposals by SIEMENS for improving the electric telegraphs.

In addition, the Consortium consisted of the *Allgemeine Elektrizitätsgesellschaft* (AEG), to the director EMIL RATHENAU of which NERNST had good relations, as we have mentioned in Section 4.9 in connection with the NERNST lamp. Further members of the Consortium were the *«Gesellschaft für drahtlose Telegraphie, System Telefunken»*, founded in 1903 jointly by the AEG and by Siemens & Halske, and the Company *«Felten & Guillaume Carlswerk Actien-Gesellschaft Cöln»*. The first Director of the former Company was Count GEORG VON ARCO, one of the pioneers of radio-technology and well known because of the *"System Slaby-Arco for wireless telegraphy"* developed by him together with ADOLPH SLABY.

On February 19, 1912, almost exactly one year prior to the early death of ROBERT VON LIEBEN, a contract was signed, providing LIEBEN with the amount of 100 000 *Mark* and his participation in the license fees. The LIEBEN tubes were manufactured in the "Cable Factory Oberspree" of the AEG founded in 1898 and located today within the territory of Berlin. Certainly it is justified, if they are looked at as the origin of all subsequent tube developments.

Apparently, NERNST had immediately recognized and promoted the importance of the invention of LIEBEN. Beyond this recognition, his personal relation with ROBERT VON LIEBEN can be seen from the remark of MAX REINHOFFER in his *"Ceremonial Lecture at the Unveiling of the Lieben Plaque"* during April 1927: *"In his obituary Nernst calls him not only his dear student, but also his devoted friend."*

MAX BODENSTEIN characterized NERNST's position regarding the subject of patents with the following words: *"He also did not think it would be useless to have protection by a patent for technically relevant, but otherwise purely scientific observations. So at one time he emphatically explained to me – as well as also to other people – that because of the fact, that Röntgen immediately made the use of his rays available to the whole world, he did not help the technical development of the x-ray tube and, hence, also its scientific application. In order to technically develop the protected idea of an invention, a Company can spend large amounts, since it can expect its subsequent recovery, many people experiment with unprotected ideas, however, everybody only with modest means, which cannot get the matter moving. ... However, another statement about patents, which is even documented in the literature, can be sure of general consent: the statement, how difficult it is to formulate a patent correctly."* [Bodenstein (1942a): 100].

In connection with the matter of patents NERNST was called quite often to the *Reichsgericht* (Imperial Court) in Leipzig to serve as an expert in a patent court case. For example, on February 3, 1897 he wrote to WILHELM OSTWALD: *"Recently I have been appointed (by the Reichsgericht) as an expert in a case dealing with storage batteries and I will have to come to Leipzig for the hearing, however, probably only during the summer term."* [Zott (1996): 117].

The invalidity action against the so-called "high-pressure patent" (DRP 238450) of FRITZ HABER of September 14, 1909 deserves special interest. This patent dealt with the implementation of the ammonia synthesis at *"very high pressures of about 100 atmospheres, however, preferably of about 150 to 250 atmospheres and more"*. The main argument was, that the results about an increase of the yield of ammonia by working at a higher pressure (see Section 5.5.3), reported by NERNST already in 1907, would have anticipated the content of this patent and also that of the previous one of October 13, 1908 (DRP 235421) about working in circulation under pressure. Hence, WILHELM OSTWALD stated in his expert opinion: *"In summary it must be stated, that on the one hand the behavior of the catalytic ammonia synthesis up to 75 atmospheres were generally and regarding numerical details exactly known, and that on the other hand the extension of the process up to 100 atmospheres did not lead to a situation, which could not be expected or even numerically predicted based on the well known facts and laws. By no means such a small extrapolation as that from 75 to 100 atmospheres can be treated as a product of an inventive activity, which would deserve the protection of a DRP."* (quoted after [Nagel et al. (1991): 38, 40]).

Of course, the plaintiffs hoped that NERNST would provide the crucial evidence, last but not least because of the way in which he had publicly humiliated HABER in 1907 during the 14th General Convention of the Bunsen Society [Nernst (1907b)] (see Section 5.5.3). However, they were completely disappointed.

It is likely that NERNST had two motives for taking the side of his intimate enemy HABER. One motive resulted from his inherent objective respect for the scientific accomplishments of others. The other motive originated more from his practically thinking nature. For example, as we have discussed already, in 1906 he had considered his results in connection with the ammonia production important enough, such that in September of the same year he concluded a contract with the Chemical Factory Griesheim-Electron as a consultant [Hoechst (1966)]. However, in 1908 BERNHARD LEPSIUS, who had submitted the offer from NERNST to the Griesheim factory, had to admit, that the corresponding joint experiments would have produced only negative results. Therefore, in 1910 NERNST denounced the contract with the Company in Griesheim and shifted his contractual relation to the BASF, with which also HABER was associated. Because of the two motives we have mentioned, the expert opinion of NERNST was completely opposite to that of OSTWALD: *"Under the impression of these experimental results varied many times and exactly tested in all directions, I cannot help to express my keen conviction, that in the patent document 238450 one deals with totally new results, and that the information contained in it provides a sound experimental base for a highly valuable novel technical procedure. Therefore, I have no doubt, that the invention described in the presently discussed patent specification to the fullest measure deserves the protection provided by the issue of a patent."* (quoted after [Nagel et al. (1991): 40]).

On March 4, 1912 during the hearing at the *Reichsgericht* in Leipzig,

NERNST stated, that his results would be of scientific value, however, they would not be of technical interest. He did not exclude a technical application. However, since the yield of ammonia achieved by FRITZ JOST appeared too small for him for a technical synthesis, in his publication [Nernst (1907b)] intentionally he had refrained from a discussion of technical questions, in order not to block the way for later inventors. Based on this attitude of NERNST, the action was dismissed.

The peculiarity of NERNST which we have emphasized several times, to look at fundamental and applied research always as a unity with a view on the practical use, can be seen clearly also from the investigation and exploitation of energy sources. For example, this is demonstrated by his patents of June 15, 1912 dealing with fuel cells (*"Procedure for the Operation of Fuel Cells"* (DRP 259231), *"Fuel-cell Element with Resistant Electrodes"* (DRP 264026), and again *"Procedure for the Operation of Fuel Cells"* (DRP 265424)), however, also by his bold idea to utilize the atmospheric electric energy for scientific and practical purposes. For this idea he planned a corresponding facility on top of Mount Generoso at 1701 m altitude, located within the Swiss canton Ticino in a region between Lake Como and Lake Lugano known for its many thunder storms. These preparations came to a stop, when on August 20, 1928 during the mounting of a cable leading to an antenna for special measurements KURT URBAN, a very young collaborator, fell from the slope of Mount Generoso in a fatal accident. In his obituary NERNST acknowledged the engagement and the capability of URBAN and characterized an aspect *"of the scientific exploitation of Franklin's experiments"* pursued with this facility: *"He was inspired by a passionate interest in this work and did not shy away from any trouble and danger connected, of course, with the observation of electric voltages of a magnitude never studied before. His passing away means a loss which is difficult to replace."* [Nernst and Lange (1928)]. Today, on top of Mount Generoso there operates a station focusing on lightning and atomic research.

NERNST also took part in the establishment of a station for the observation of cosmic radiation at the Jungfraujoch, a pass at 3454 m altitude in the Highlands of Bern in Switzerland.

The activity of NERNST as a Full Member of the Prussian Academy of Sciences must also be included in his role in the area of the organization of science. He has formulated the proposals for the election of FRITZ HABER, ABRAM JOFFE, HEIKE KAMERLINGH ONNES, and MARTIN KNUDSEN, effecting their acceptance as a Full or Corresponding Member into the Academy. Among the many proposals cosigned by him we mention in particular those connected with ALBERT EINSTEIN, ERWIN SCHRÖDINGER, the successor of MAX PLANCK at the University of Berlin, and THEODORE RICHARDS, who had wanted to take away from him the primacy of the discovery of the Third Law of Thermodynamics.

5.7.4 Rector of the University and the German Institute for Foreigners

Among the achievements of NERNST in the area of the organization of science we must count also his activity in a high and highest position of the Friedrich-Wilhelm University. On August 1, 1911 at the meeting of the Philosophical Faculty, with 25 of a total of 29 votes he was elected as Dean for the following year. From the records of the faculty meetings, which NERNST had to chair then on each Thursday, we see that extraordinary problems did not come up during this time. The normal, but by no means unimportant tasks included the dissertation and habilitation procedures in the Philosophical Faculty. Last but not least, the work of the Dean was aggravated by the fact, that in addition to the natural sciences and mathematics he had to deal with many further directions of the humanities. On August 1, 1912 in the office of Dean, NERNST was followed by the historian MICHAEL TANGL.

We must remember, that the years 1911 and 1912 just were the years of the culmination of the research dealing with the Third Law (see Fig. 5.17), which had been announced to the scientific community already at the end of 1910. So on January 21, 1911, GEORG BREDIG, at the time Professor of Physical Chemistry in Zurich, wrote to SVANTE ARRHENIUS referring to the award of the Nobel Prize in Chemistry in 1910 to OTTO WALLACH for his work in the field of the alicyclic compounds: *"People* had shaken their head a little because of the coronation of Wallach, in this case also Nernst would have been a more likely candidate." [Zott (1996): 191].

When on October 15, 1921, ten years after his election as Dean, WALTHER NERNST started in his term as Rector of the Friedrich-Wilhelm University, he knew, that about two months later he would receive the Nobel Prize in Chemistry of the year 1920. As the subject of his inaugural speech given in the main hall of the University, NERNST chose the validity of the laws of nature, in order to *"turn to more general questions, perhaps to the most general one in the area of the exact natural sciences."* [Nernst (1921a): 4]. After many interesting remarks he noted: *"If in the case of not a single law of nature the books can be considered as closed, of course, today it was doubly impossible to make a final statement about the general concept of the law of nature. At any rate, science cannot deal with perfectly closed subjects; the struggle of the scientist with the unwieldy matter, which resists its defeat by means of the logical power of the law, will always remain comparable to the fight between Heracles and the serpent of Lerna; from time to time one succeeds to cut off one of its heads, but very soon two new heads are leaping up. However, we must not become discouraged. The hero, who can carry home a head of the serpent as a trophy, at the same time has delivered to the cultured mankind a value which cannot be destroyed; furthermore, new tasks motivate permanently additional work. Science is just not a goldmine, which sooner or later confronts its depletion, but instead it is a rewarding field, always bearing new fruits, however, often only after tiresome work."* [Nernst (1921a): 24–25]. Addressing directly the students and indirectly his colleagues, NERNST explained the basics of his opinion about teaching, study, and research: *"Fellow students, ... Yes I claim even, that he is not a true professor, the lecture of which is always understood in all points by his audience; instead, frequently one must open perspectives to the audience, which become clear to the individual listener in front of his eyes only after a further thorough study. If with every sentence the lecturer tries to adjust himself to the level of his audience, he cannot escape the danger to drop below this level, and that is* the worst which can happen to him. Therefore, don't get discouraged, if during your studies you are confronted sometimes by difficulties which seem insurmountable; the larger is the benefit, if the material is mastered by yourself due to your own work! ... In the case of the lively science one

must always look also at the changing developments, not only at the finished subject; at the far horizon there always appear great questions and high goals: 'Step to the bottom, fly to the top, / Rich Nibelung treasures lie around still undetected!'" [Nernst (1921a): 25–26].

Fig. 5.26   WALTHER NERNST shown in the academic dress of the Rector of the University of Berlin (1921/22).

If we start the sequence of the Rectors of the Friedrich-Wilhelm University with THEODOR SCHMALZ, Professor of Roman and German Laws, then NERNST's term as Rector gets the number 112. Before him many important scholars had occupied this office, some of them even several times. The record with five terms of office was held at the time by the Classical Philologist AUGUST BOECKH. In the following list of predecessors of NERNST, who had been appointed as Rector and came from a field close to him, we find great names:

Nr.	Name		Term of Office	Appointed in the Field
9 23	CHRISTIAN SAMUEL	WEISS	1818/19 1832/33	Mineralogy
44	JOHANN FRANZ	ENCKE	1853/54	Astronomy
45	EILHARD	MITSCHERLICH	1854/55	Chemistry
49 62	HEINRICH WILHELM	DOVE	1858/59 1871/72	Physics

Nr.	Name	Term of Office	Appointed in the Field
52	GUSTAV MAGNUS	1861/62	Physics
59	ERNST EDUARD KUMMER	1868/69	Mathematics
60 73	EMIL DU BOIS-REYMOND	1869/70 1882/83	Physiology
64	KARL WEIERSTRASS	1873/74	Mathematics
68	HERMANN VON HELMHOLTZ	1877/78	Physics
71	AUGUST WILHELM VON HOFMANN	1880/81	Chemistry
74	GUSTAV ROBERT KIRCHHOFF	1883/84	Mathematics and Physics
82	WILHELM FOERSTER	1891/92	Astronomy
104	MAX PLANCK	1913/14	Physics

We add ERHARD SCHMIDT, Professor of Mathematics, who as the 120th Rector headed the University in 1929. Prior to NERNST, EMIL SECKEL, appointed for Roman Law, headed the University in 1920/21, and after him in 1922 the pharmacologist ARTHUR HEFFTER was elected as Rector.

One can report only little about the activity of NERNST as Rector, since the relevant documents of the University archives have been destroyed during the Second World War. NERNST's report for his term of office shows no peculiarities [Nernst and Heffter (1922)]. However, it is remarkable and of importance not to be underestimated, that on July 19, 1922 at the initiative of NERNST the German Institute for Foreigners (*Deutsches Institut für Ausländer*, DIA), existing already independently for some time, has been placed solemnly under the protection of the University, and in the presence of all people interested in international scientific connections could be *"opened within some space of the University building especially made available for this purpose"* [Remme (1926): 51]. KARL REMME was one of the main founders of the Institute and also its first Director. Subsequently, he was in charge of the study of foreigners in Prussia.

The DIA was meant mainly for such foreign students, who had been unable in their home country to prepare themselves sufficiently by learning the German language. It had been exactly this deficiency which NERNST had to deplore in Göttingen when he accepted the American MARGARET ELIZA MALTBY in his laboratory. On October 6, 1893 he had written to WILHELM OSTWALD regarding this matter: *"Officially Miss Maltby is admitted to the laboratory and to the lectures, but unfortunately she still has very great difficulties with her German language and, hence, most of the time she does numerical work."* [Zott (1996): 65]. Possibly it was just the language problem which provided the main motivation for NERNST to engage himself in accepting the DIA into the University of Berlin.

An additional purpose of the DIA consisted of the important task to make sure that the foreign students became familiar with the German culture and the German mentality. Therefore, lecture series and practical sessions were organized, which covered the German history, regional studies, literature, plastic arts, and teaching matters. Furthermore, there were visits to museums, art centers, educational institutions, and schools, and excursions to the environment of Berlin were organized. At the end of the study period an educational trip across Germany was arranged. The Institute had a sizeable library with books on the German language, literature, history, philosophy, art, methodology of education, and other fields, as well as daily newspapers and journals.

The courses lasted eight weeks, and during the vacation period six weeks. Teaching took place within small groups. The language courses differed between the following levels: beginning, lower, medium, and advanced. For the first two levels teaching covered eight hours per week, for the other levels six hours per week. Furthermore, there were continuing education courses with four hours per week. At the end of the language courses one could enter an examination. In the case of success, *"a document will be issued, that the examined person knows the German language sufficiently to be able to follow with understanding the lectures at a German 'Hochschule'. At the registration for study, the document is considered by the 'Hochschulen' to represent a sufficient proof of the knowledge of the German language."* [Remme (1926): 55]. Furthermore, the DIA carried out diploma examinations for the methodology of education and the languages. The former certified the capability to teach the language in a foreign country. The continuing education courses served for the preparation of the diploma examination for the languages. This diploma certified the knowledge of the German language, literature, history, and cultural background.

The reason for accepting the DIA into the University of Berlin just at this time can be seen perhaps in the abruptly increasing number of foreign students. For example, in the *Alma mater* of Berlin during the summer term of 1914, immediately prior to the outbreak of the First World War, this number was 1361. In 1916 it had sunk to 505, in 1920 it had reached again 687, and during the winter term 1921/22 it increased up to 1453. Perhaps, this rapid increase indicated a certain stabilization of the still young Weimar Republic, following the failed right-wing attempted coup, lead by WOLFGANG KAPP during March 1920. During the summer term of 1925 with 1410 foreign students about the same number was reached. The situation of the students for all *Hochschulen* in Berlin can be seen from the following overview:

	Students		
	total	foreign	% foreign
University	9045	1410	15,6
Technical College	4045	710	17,6
Agricultural College	892	88	9,9
Veterinary College	172	47	27,3
Business College	2321	233	10,0
Total	16475	2488	15,1

On January 8, 1922 EMMA NERNST told the Swedish scholar KNUT WILHELM PALMAER, who had taken part in proposing NERNST for the Nobel Prize, that the Ministry of the Interior had asked her husband, *"if he would like to become the successor of Warburg, i.e., President of the 'Physikalisch-Technische Reichsanstalt'. It is likely that my husband will accept and will return to pure physics."* [Zott (1996): 208]. This presumption became true in the year 1922. As we will see in Chapter 6, also in this high office WALTHER NERNST became active in the organization of science.

5.8 Managing a Country Estate, Hunting, and Fish Farming

Since NERNST tended to answer the question of the Secretary of the Interior ADOLF KÖSTER about accepting the Office of the President of the *Physikalisch-Technische Reichsanstalt* positively, he had to start negotiations regarding this matter. In a protocol of the Ministry prepared during their course we note the remarkable passage: *"Nernst strongly insisted that he would be allowed to withdraw frequently for 2 to 3 days to his country estate near Berlin and thereby to remain in good shape. During wintertime the visits of his estate would be about once per month, during summertime more often. He would accept the position of the President only under the condition, that he could spend such holidays at his estate as needed without charging them to his allotted vacation time."* [BAK; Hoffmann (1992): 37]. Here we notice NERNST's lifelong love of life in the country and in free nature, the origin of which can be found in his visits to the demesne Engelsburg of his uncle RUDOLF NERGER during his childhood and youth, as we have discussed already in Section 2.2.

The *"country estate near Berlin"*, mentioned in the protocol, i.e., the estate Rietz very close to the small town of Treuenbrietzen in the Mark Brandenburg (Figs. 4.4, 5.27) has also been discussed already above in connection with the family of NERNST (Section 4.4). To these remarks we can add, that NERNST did use his country estate not only for resting and relaxation, as he had indicated to the Ministry. Corresponding to the nature of the scientist, he complemented his enjoyment of some object or activity by connecting it with his scientific interests whenever it was possible for him. This can be seen particularly clearly from his love of the automobiles in connection with his research dealing with the proc-

 esses within the internal-combustion engines. So on the 50 hectare of land belonging to the estate Rietz he pursued agriculture intensively.

Fig. 5.27 The manor house in Rietz at about 1910.

Probably this occupation and his research dealing with the ammonia synthesis from the elements were the reason, why on October 1, 1913 for his anniversary speech in the German Museum in Munich NERNST had chosen the subject *"The Importance of Nitrogen for Life"* [Nernst (1913b)]. As done already in 1898 by the English naturalist Sir WILLIAM CROOKES, NERNST pointed out the constantly decreasing natural resources of nitrates, which up to then represented the only source of mineralogical nitrogen fertilizer. Since except for a few cases (legumes) the plants utilized in agriculture can take up the nitrogen needed for their growth only in the form of a chemical compound as given by the nitrates but not as N_2, in due time this shortage would lead to terrible famines. Therefore, CROOKES had made an appeal to the scientific community to develop a process which allowed turning the nitrogen from the air into a chemical compound, from which nitrogen fertilizer could be produced. Now NERNST could state with satisfaction that the process developed by FRITZ HABER, and according to which the mass production of ammonia would be realized in the very near future, had solved the problem raised by CROOKES, such that the catastrophe could be avoided.

In addition to agriculture NERNST had turned also to trout farming, in this case again with a certain success, last but not least due to an intensive study of the relevant specialist literature. Still a few years ago the people from Rietz reported that at the time the master of the estate located directly at the entrance to the village in the direction of Treuenbrietzen had made a considerable contribution to the development of the infrastructure of the village by building new roads in the village and nearby. NERNST had also initiated the towerlike addition to the manor house, which at the time had been crowned with merlons (Fig. 5.27).

In order to be able to pursue his passion for hunting, which also had developed from his visits to Engelsburg, NERNST additionally leased a large area covered by forests and meadows. However, we must note that he was too impatient for hunting big game. Similarly as, for example, in the development of scientific instruments, the visible success had to be achieved quickly also during the hunting activities, which was possible only in the case of hunting smaller game animals.

JOHN EGGERT, a student of NERNST in Berlin who belonged to the guests in Rietz, summarized this aspect of his teacher, indicated by his visits to the country on weekends and during the summer vacation, with the words: *"For him this country life was not a pose or a matter of fashion, but a requirement of nature. Proudly he showed off his experimental fields, his carp* [in Rietz: trout] *ponds, and he was devoted passionately to hunting. Occasionally he explained the gigantic fertilizing experiment to burn an appreciable fraction of the supply of coal on the earth."* [Eggert (1943b): 49]. Furthermore, EGGERT reported: *"However,* [in contrast to sport activities] *Nernst loved social life. Twice during each year he gathered his Institute family, during winter in his home, which still in 1912 was illuminated by the Nernst-light and which contained beautiful pastel paintings by the hand of Ostwald, and once during summer at the country estate, where fertilizing experiments and carps could be admired, those protein producers with thermodynamically particularly high economics, as emphasized by Nernst. The whole family of Nernst always made a strong effort to offer a few happy hours to the guests."* [Eggert (1964): 451].

Not only students and collaborators of NERNST visited Rietz, but also his colleagues. The country estate represented an important location for gatherings of all kinds, and it played an important role in connection with the requirements of representative social activities to be taken up by NERNST and his family. However, regarding this matter MAX BODENSTEIN would observe, *"that he and his wife were inclined to take over the social representation of the chemistry of Berlin."* [Bodenstein (1942a): 103]. In fact, the festive events during the summer and the hunting parties during the winter represented high points well known not only within the University but even within the City of Berlin.

Again on August 1, 1914 NERNST had traveled to Rietz together with the members of his Institute. Already for several days the international political situation had deteriorated noticeably, such that – as remembered by JOHN EGGERT – *"the proper spirit did not develop, however, and at the return of the day-trippers to Berlin the war had broken out."* [Eggert (1943b): 48]. KURT MENDELSSOHN, who was only eight years old in 1914, described the situation less convincingly but more effectively: *"The last of these parties took place on Saturday, the 1st August, 1914. The carefree gathering was interrupted by a phone call to Nernst from the local district official who told him that war had been declared."* [Mendelssohn (1973): 79; 111]. The war took away from NERNST his two sons. RUDOLF was killed already during the first days of the war, GUSTAV during the summer of 1917. Since after this loss the memories of the time spent together in Rietz were too painful, during the fall of 1917 NERNST decided to give up the estate. Therefore, on November 20 EMIL FISCHER wrote to his colleague: *"For me it was a surprise to hear, that your estate in Treuenbrietzen will be sold already in the next days. If the administration of such an estate would not be so highly complicated, perhaps I would decide to do it. However, as long as the war lasts my son is tied down as an officer and I myself cannot take up any more any administrative work."* (quoted after [Hoffmann (1992): 38]). Already in 1916 FISCHER's second son WALTER was driven into suicide because of a psychic illness and the situation of the World War. During March of 1917 the third son ALFRED mentioned in the letter as being still alive had succumbed to typhus, with which he had become infected during his service in the medical corps.

How deeply and permanently NERNST had felt about the loss of his sons can be seen from a report of the pastor's wife ELISABETH SEIBT, referring to the decade 1929 – 1939, during which NERNST had been the owner of the estate Oberzibelle already for several years: *"On and off he was a welcome guest with us together with his wife. He had turned to us young pastor people in a friendly way. Annually he attended the service in the church of Zibelle, however, only one time, namely on Sunday before Advent* [on which the dead are commemorated]. *He had lost his two sons in the 1st World War and now he looked for comfort given by the word of God: eternal life in splendor. After the sermon at the last Sunday of the year of the church we were always invited to his manor house, in order to continue with him the discussion about death and eternity not far from the faith."* [Muche (1989): 18].

During this difficult time the need of NERNST for being close to nature satisfied by his life in the country was, however, also – or perhaps more so – sufficiently strong, such that very soon after Rietz had been sold he acquired another estate in Dargersdorf in the Uckermark near the town of Templin. Because of its representative size and beauty, the manor house (Fig. 5.28), popularly called "palace" and burned down unfortunately in 1945, certainly fitted the life style of NERNST. Also his passion for hunting could be satisfied in the best way, since a separate forester's house belonged to the estate and near the estate there were extended forests rich in game.

Fig. 5.28 The manor house in Dargersdorf.

Due to the turbulent times of the last months of the war and of the Revolution of November 1918 as well as of the early years of the Weimar Republic NERNST could not get settled at his new estate and could not take care properly of its management. One event had a particular impact: *"When in Berlin the Spartacus unrest was dominant* [at the end of 1918], ... *it became known by the way, that Nernst would be/was put on the extradition list by the Entente* [because of his participation in war-related research]. *"* [Eggert (1943b): 48; Eggert (1964): 452]. Possibly, in these days of uncertainty, very soon NERNST had sold his real estate in Dargersdorf, thus ensuring if the requests by the Entente would become reality, his family would still be financially secure.

Along with others in 1920 NERNST had been eliminated from the extradition list, and the winner of the Nobel Prize in Chemistry of that year looked for another country estate. Due to the inflation prevalent in Germany at the time, real estate was sold only rarely, NERNST could acquire his third country estate only in 1922. This *Obergut* or feudal estate Oberzibelle was located within the town of Zibelle (Niwica/Poland) in Silesia or more exactly in the *Oberlausitz* (Upper Lusatia) near Bad Muskau, being nine kilometers away. The town of Sorau (Żary/Poland) is also located nearby (22 km). However, in the official address listing of the county of Rothenburg (Upper Lusatia) («*Adreß- und Verkehrsbuch für den Krs. Rothenburg/ OL*») the entry «*Nernst, Walter, Dr. Prof.* *E*[igentümer]. (owner) *Tel. Nr. 1.*» was to be found only in the edition of 1937.

Fig. 5.29 The manor house in Oberzibelle.

If the distances between NERNST's residences in the center of Berlin and Rietz or Dargersdorf were similarly long, in order to reach Zibelle he had to cover a much longer distance, as can be seen from the following comparison:

Location	Direction	Distance [km]	
		by road (today)	as the crow flies
		from the center of Berlin	
Rietz	southwest	90	64
Dargersdorf	north	96	62
Zibelle/Niwica	southeast	> 175	144

The manor house in Oberzibelle was a bit smaller than the previous ones: *"There were nine rooms altogether and a flat for the administrator."* [Mendelssohn (1973): 158; 213]. On a postcard with a view of the building written on October 8, 1926 to WILHELM OSTWALD, in the imprinted text «*Rittergut Oberzibelle, Schlossansicht*» (feudal estate Oberzibelle, view of the palace) the notation '*Schloss*' (palace) has been changed by NERNST into 'house'. On the other hand, the *"huge library"* of NERNST was *"particularly admired"* [Muche (1991): 17] by the neighbors in Zibelle JOHANNA ZIEGLER and HILDEGARD HÖGEL, temporarily employed at the estate.

The amount of land belonging to the estate was quite large. In 1922 from the feudal estate Dubrau (Dąbrowa Łus) located nearby a total of 286.94 ha of farmland and forests were split up in favor of NERNST. At the end of the 2nd World War the amount of land owned by the NERNST family in Zibelle was 222 ha. 107 ha of this area were farmland, 73 ha were covered by forests and 42 ha by fish ponds.

As in Rietz, also on this property NERNST could pursue his beloved hunting activities as well as a sizeable and productive agriculture and carp farming.

During the years until his official retirement from the University of Berlin in 1933, NERNST made use of his property in the Upper Lusatia predominantly on weekends and during the summer vacation. After that, it became his proper residence. However, different from his life on the estate in the Mark Brandenburg, the time in Zibelle was much quieter.

Here family festivities and anniversaries still were celebrated, such as NERNST's 70th birthday in 1934, when his entire family came together for the last time. However, the big social events that were arranged in Rietz prior to the First World War did not take place any more. Towards the end of his life WALTHER NERNST lead a withdrawn life in Oberzibelle and – due to the dark period in Germany – being increasingly lonely, until the night of 17th to 18th November 1941 this *"life attached to the soil in the country as he had become used to since his earliest youth due to his grandparents"* [Eggert (1943b): 49] came to its end.

In October 1990 the fate of NERNST's last residence and its condition is described by the following report of an eye witness: *"Everything had disappeared from the earth. A little wall one had left standing. The former manor house had not been destroyed by bombs, no, but by the dismantling by Poles for the rebuilding of Warsaw, it is said."* [Muche (1991): 17].

5.9 The First World War

The behavior of WALTHER NERNST during the First World War, which initially was conducted on a European scale, but finally expanded into a global war involving 32 nations, may be described as ambivalent. On the one hand, he actively participated in the war and in the war-related research, and on the other hand, he looked for steps leading to end the war. In this case, both of these activities to a large extent temporally were not separated from each other.

The activities of NERNST during these terrible times on various occasions have been reported as a more decorative, but sometimes also questionable way, for example, in the treatment by KURT MENDELSSOHN [Mendelssohn (1973): 80–82; 112–116] or in certain passages by RICHARD LEPSIUS [Lepsius (1964): 605–606]. In contrast, a study prepared by students of the University of Leipzig [Großkreuz and Heitzsch (1985): 17–20] is quite instructive, aside from a few tendentious remarks. Here we shall include mainly only statements by NERNST himself or by people very close to him.

5.9.1 War-related research: gas warfare, explosives, ballistics

Already ten days after the start of the war, on August 11, 1914, the *Geheimrat* together with his beloved motor-car entered the *«Kaiserlich Freiwillige Automobilkorps»* (Imperial Volunteer Automobile Corps). This corps belonged to the military forces of the First Army, the commander of which was General ALEXANDER VON KLUCK [UAHUB: II, 3]. The First Army was operating on the extreme right wing of the German Army at the Western Front. Based on the strategy developed already in 1905 by ALFRED Graf VON SCHLIEFFEN, it had the task to perform a quick turn passing through Belgium and Northern France and reaching the region of Paris. In this way it was supposed to beat down the left wing of the French Army, to circumvent Paris, and to cause a quick end of the war in the West. Later NERNST told about himself: *"In the years 1914 and 1915 he participated in several engagements"* [UAHUB: II, 44]. One of these *"engagements"* was the First Battle of the Marne, lasting from the 5th until the 9th of September 1914, and stopping the German advance only a short distance from Paris, an advance, which up to then had been hardly held up. The planned *"Blitzkrieg"* (lightning war) turned into a wearing positional battle. Already during the first year of the war NERNST was decorated twice due to *«Tapferkeit vor dem Feinde»* (bravery in front of the enemy) – as it was termed by MAX BODENSTEIN [Bodenstein (1942b): 142]: on October 1, 1914 with the Iron Cross of 2nd Class (E.K. II) and on June 21, 1915 with the Iron Cross of 1st Class (E.K. I). This war decoration had been founded during March 1813 by the Prussian King FRIEDRICH WILHELM III for the duration of the war of liberation and for all ranks. In 1914 the Emperor WILHELM II renewed this decoration for the First World War.

According to his own statements NERNST had *"participated in numerous gas attacks in a leading position"* [UAHUB: II, 44]. However, the actual establishment of the gas war was soon charged to FRITZ

HABER regarding the science and to CARL DUISBERG regarding the industry. As for the Army Command, MAX BAUER was held responsible, since he had been trained in natural science and technology, and at the beginning of the war as a Major in the Operation Section of the Army High Command, he was in charge of Section II dealing with heavy artillery, trench-mortars, fortresses, and ammunition. Already in 1914 he proposed to the Minister of War and the actual Chief of the General Staff ERICH VON FALKENHAYN to look into the use of chemical weapons in the current situation of positional warfare, when initially one was only thinking of irritant substances that make the enemy temporarily incapable of fighting. Following the personal directive by VON FALKENHAYN, at the beginning of October 1914 BAUER formed a commission aiming at this goal, consisting of scientists, representatives of industry, and military officers. DUISBERG promised his cooperation only reluctantly, NERNST, however, *"with immense eagerness"* – as stated by EMIL FISCHER. Most of the demonstrations of the weapons developed on the basis of the experimental results of the commission took place on the firing range in Wahn near Cologne. So in October 1914, the so-called *Ni*-bullet was tested, the name resulting from the word *'Niespulver'* (sneezing powder) irritating the eyes and the respiratory passages and being a salt of dianisidine. However, the use of 3000 such bullets near Neuve-Chapelle at the end of the month was not even noticed by the enemy. Now additional chemists, among them EMIL FISCHER and FRITZ HABER, were called up by BAUER to join the commission. ERICH VON FALKENHAYN informed Major GERHARD TAPPEN, Chief of the Operations Section of the General Staff, about the failure of the *Ni*-bullet, which he attributed to the incapability of NERNST and DUISBERG. The brother of GERHARD TAPPEN, HANS TAPPEN, was a chemist and in the beginning of November he proposed to the High Command of the Army to fill shells with the eye-irritating liquid xylene bromide, subsequently called *'T-Stoff'* after him. NERNST and DUISBERG were hurt when in the middle of December and in the beginning of January 1915 highly visible demonstrations were carried out with this chemical of their competitor. However, neither at the Eastern nor at the Western Front the use of the *T-Stoff* could accomplish any success, which was also the case for further tear-gas shells being tested. In the middle of December 1914 VON FALKENHAYN demanded, that instead of such substances, which only temporarily incapacitate fighting, substances should be used which permanently put the enemy out of action. HABER followed this demand by switching from the testing of only irritating substances to warfare agents leading to fatal results. However, NERNST still insisted on the use of shells filled with irritating gas. At the end of 1914 HABER proposed to the High Command of the Army the use of chlorine, available in large amounts and compressed into steel cylinders, at a time when there were difficulties in the procurement of gun equipment and gun powder. When on April 21, 1915 near Ypres for the first time about 150 tons of this deathly gas in a cloud of six kilometer width reached the enemy lines, the gas war in the sense of the Army Command had started.

Fig. 5.30 Walther NERNST during the First World War: as a scientific consultant at the Front (left), with his wife in Cologne (right).

In the obituary by NERNST for MAX BAUER in 1929 some of these events are touched upon in a reflective way, and at the same time his personal relation to his former *Kriegskameraden* (comrade in arms) and information about his activities during the war can be read off: *"Shortly after the first Battle of the Marne I was officially – in my capacity as 'Benzinleutnant' with the First Army – in the Great Headquarter, which at the time was located in Charleville close to Sedan; Bauer, being a Major in the Operations Section of the High Command of the Army, had heard of my presence. He looked me up and we discussed in detail certain war-technical questions, as an immediate result of which it turned out, that still in the same evening, accompanied by the Major of the Artillery Michelis – now a retired Major General, in order to perform experimental tests on the firing range Wahn located near the large chemical works Leverkusen, I drove in my car to Cologne. I hardly exaggerate, if I say that the further application of the aspects generated together with Bauer lead to an overall change of the conduct of the war. ... Subsequently, our meeting resulted in a steady working team during the war and later in an always growing friendship, which also was expressed continuously in terms of a strong mutual support. During the war we often drove together along the front and here I could observe how high was the reputation of the relatively young officer everywhere with the Generals and the Army Commanders."* [Nernst (1929): 1; 3].

Since 1915 NERNST acted *"as a scientific consultant of the trench-mortar battalion I"* [UAHUB: II, 2], which he helped to found. Regarding this matter, JOHN EGGERT, who during the war was an assistant in NERNST's Institute in the *Bunsenstraße* in Berlin, reported: *"After a few months being decorated with the E.K. II he returned again, in order to work on a number of war tasks with the few remaining members of the Institute. However, soon he went again to the Front, this time to the East together with the trench-mortar battalion I, in the novel equipment of which he had been actively involved, so actively, that in the Institute during an experimental test a few windows were blown out."* [Eggert (1943b): 48; Eggert (1964): 452].

As we had pointed out already, very soon NERNST's achievements in the area of war-related research were highly recognized. So at the end of August 1915 the *Berliner Illustrirte Zeitung* (Illustrated Newspaper of Berlin) observed: *"Apparently the rough war promotes everything rather than the quiet peaceful research. And still in this war the German scientific work celebrates its triumph. ... And with the Iron Cross of 1st Class decorating the breast of the Geh. Reg.-Rat Prof. Dr. Nernst, director of the Chemical Institute of the University of Berlin, at the same time the chemical research looks upon this as an honor for itself.*" [BIZ (1915)]. In fact, NERNST had turned to problems of explosives. The background for this was described by the chemist RICHARD LEPSIUS, whom NERNST had met as a student in Göttingen and Berlin, in the PhD examination of whom he had taken part in 1911, and with the father of whom, BERNHARD LEPSIUS, he had cooperated on the occasion of his contract with the chemical factory Griesheim-Elektron. RICHARD LEPSIUS remembered: *"At the beginning of the positional war one had developed mortars fired with explosives the power of which caused the fillings of the shells to explode, and less explosive* (fillings) *correspondingly had a smaller effect in the trenches of the enemy. Nernst developed the idea to use compressed gases, say, carbonic acid, as a much softer energy of firing."* [Lepsius (1964): 605]. In 1917 RICHARD LEPSIUS himself had issued a document *"Regulations for Blowing Up Using Liquid Air"*, which was translated into Bulgarian.

However, NERNST had also occupied himself with explosives in the usual sense. JOHN EGGERT mentioned that *"following his direct suggestion, Willy Marckwald had discovered the explosive guanidine perchlorate being highly important at the time."* [Eggert (1964): 452]. Connected with the study of explosives is a theoretical paper, on which on December 9, 1915 NERNST reported in the Physical-Mathematical Class of the Academy of Berlin [Nernst (1915)]. During the evaluation of gas explosions using the membrane manometer developed in 1909 by MATTHIAS PIER, in the Institute in Berlin in the presence of excess oxygen strong oscillations of the membrane were observed, the theoretical discussion of which NERNST had intended with this paper. At the end he remarked: *"Of course, the arguments given above also apply in the case, in which one does not deal with the course of gas explosions, but with the recording of*

$$\left[HN\!=\!C\!\begin{array}{c}\nearrow NH_2 \\ \searrow NH_3\end{array} \right]^+ ClO_4^-$$

Guanidine perchlorate

the pressure caused by explosives or gun powder. Here at the same time we recognize the difference between very abruptly and relatively slowly acting explosives." [Nernst (1915): 900].

Already before the war Nernst had developed ideas about the nature of gas explosions in homogeneous mixtures. Starting from these ideas, in 1916 his student HANS CASSEL presented a paper dealing with the inflammation of mixtures of hydrogen and oxygen in different ratios [Cassel (1916)]. An exact definition of the inflammation point could be given. The inflammation was effected by an adiabatic compression of the mixture contained in a steel cylinder by means of falling pistons. A minimum of the inflammation temperature was found for the mixture $H_2 : O_2 = 2 : 3$. CASSEL could confirm the behavior of the inflammation temperature depending on the starting pressure, concluded from NERNST's ideas. According to these ideas the inflammation temperature increases for decreasing starting pressure and vice versa.

After 1916 NERNST's coworkers JOHN EGGERT and HANS SCHIMANK occupied themselves with explosives. In January 1917 they presented lecture demonstrations of their theory to the *Bunsen-Gesellschaft* [Eggert and Schimank (1917)]. In the introduction they explained their motivation: *"In spite of the strong interest met and always will be met by the field of explosives, the number of lecture demonstrations, which explain the most essential properties of this group of substances in a simple, vivid, and harmless way, is quite small. This was the motivation for us to think about a few arrangements, which may satisfy these requirements."* [Eggert and Schimank (1917): 189]. In this way several properties of explosives could be illustrated: their gas development during slowly running reactions, the connection between reaction rate and pressure, and *"the difference between the explosion of a powder, the detonation of an explosive, and the sharp report of a detonator."* [Eggert and Schimank (1917): 192]. MAX LE BLANC thanked *"the lecturing gentlemen very much for these pretty lecture demonstrations, which surely will be extremely valuable for some other lecture."* [Eggert and Schimank (1917): 192]. Continuing the subject, on December 4, 1917 EGGERT and SCHIMANK submitted a paper dealing with lecture demonstrations using acetylene-silver (Ag–C≡C–Ag) [Eggert and Schimank (1918)]. During abrupt heating this explosive substance decayed into silver and carbon with a sharp crack without generating a gaseous component. Therefore, the phenomenon of the sharp crack was explained in terms of a rapid expansion of the air, which is caused by the amount of heat set free during the decay. The presented experiments were meant to demonstrate this fact and its conclusion, that in contrast to silver azide (Ag – N $\overset{(+)}{=}$ N $\overset{(-)}{=}$ N) or mercury fulminate (Hg(O– $\overset{(+)}{N}$ ≡ $\overset{(-)}{C}$ l)$_2$), acetylene-silver is hardly useful as a detonator. Very soon these assumptions about the acetylene-silver had to be corrected, since it appeared that gases continue to develop during its explosion, however, there appeared differences depending on the procedure for producing the acetylene-silver (precipitation from ammoniacal or nitric solution of silver nitrate). The corresponding experiments were performed by EGGERT alone, and he published the results [Eggert (1918)], since *"Mr. Dr. Schimank did not participate in this study, because essentially its subject was a chemical one."* [Eggert (1918): 150]. EGGERT suggested that in the case of the substance obtained from the ammoniacal solution one deals with a mixture of acetylene-silver and silver hydroxide, in the case of the precipitate from the nitric solution with acetylene-silver and silver nitrate.

The mentioned results published by NERNST and his coworkers were certainly not important enough to the conduct of the war in order to keep them secret. Therefore, they are not in contradiction to the statement by MAX BODENSTEIN in connection with this matter that NERNST *"occasionally acted as an inventor ... Of course, this happened during the World War, where his ability along this direction was taken advantage of by the Army Command, without publishing much on this matter."* [Bodenstein (1942a): 99]. Certainly this statement also applies to NERNST's research on ballistics which was important for the development of the mortars similarly as the studies dealing with the explosives. There exist a few letters of NERNST to DAVID HILBERT, with whom he had developed a friendship since his time in Göttingen, when the latter was asked for advice and help in the case of mathematical problems, which are probably associated with these ballistic studies. However, in 1917 in a popular note NERNST discussed the ballistics of the mortars [Nernst (1917)].

In its «*Hochschul- und Personalnachrichten u.ä.*» (College and Personal News and the like) dated under March 1, 1917 the German Bunsen Society published the following news: *"A Kaiser-Wilhelm Foundation for Military and Technical Science has been established. It is its purpose, in cooperation between the most outstanding scientific experts of the country and the experts in the Army and the Navy, to advance the development of the means for conducting the war arising from natural science and technology."* [ZfE (1917): 104]. NERNST acted as chairman of one of the six subcommittees, that were financed by the Foundation and whose research was coordinated by it. In the following compilation we list the different subcommittees:

Subcommittee	Chairman
Chemicals and gases	FRITZ HABER
Metals and non-ferrous metals	FRITZ WÜST
Synthetic oils and lubricants	EMIL FISCHER
Physics, ballistics, telegraphs and telephone	WALTHER NERNST
Engineering	ALOIS RIEDLER
Aircraft and aeronautics	HEINRICH MÜLLER-BRESLAU

FRITZ WÜST was Professor of Research on Iron, ALOIS RIEDLER Professor of Mechanical Engineering, and HEINRICH MÜLLER-BRESLAU Professor of Statics of the Construction of Buildings and Bridges. Since 1895 the latter had occupied himself also with aircrafts, and in this context he had advised Graf (Count) ZEPPELIN in the construction of the wing unit of his airships.

In his role as Chairman on February 22, 1917 NERNST had approached ARNOLD SOMMERFELD with the request to cooperate in the subcommittee in the area of wireless telegraphy. On March 2, 1917 NERNST informed his colleague in Munich about the admission into the subcommittee and he expressed the desire to discuss with him the *"flight of the drawn mortar bullets"*.

In 1917 NERNST was given the *Ordre pour le Mérite* (peace class) succeeding the late Graf ZEPPELIN. This *Ordre* had been founded in 1842 by King FRIEDRICH WILHELM IV of Prussia for the sciences and the arts. As "peace class" (*Friedensklasse*) it was intended to supplement the «*Militärorden*» established one hundred years earlier by FRIEDRICH II. Finally, during Spring of 1918 NERNST received the highest Bulgarian war decoration [UAHUB: II, 2].

During the war NERNST was active also for propaganda purposes, as shown by his appearance at an event of the «*Bund deutscher Gelehrter und Künstler*» (Association of German Scientists and Artists), in short denoted as «*Kulturbund*». The event took place in the evening of June 3, 1916 in the Main Assembly Hall of the German *Reichstag* (parliament building) in Berlin. At the beginning of the war the *Kulturbund* was founded, in order to motivate representatives of the sciences and the arts to act in the interest of Germany. Among others, the members of the Executive Committee included MAX PLANCK, MAX LIEBERMANN, WALTHER RATHENAU, the Court Architect ERNST VON IHNE, the poet and writer HERMANN SUDERMANN, as well as WILHELM VON WAL-DAYER, Professor of Anatomy at the University of Berlin and 1st Chairman of the *Kulturbund*. The motivation for this event was, that after two years of a war, the end of which – as was emphasized again and again – was not to be expected soon, by means of speeches by four well-known people to demonstrate to the German people and to the befriended and the enemy countries abroad, that scientifically, technically, and morally as well as far as regarding supplies Germany was well prepared for an arbitrarily long continuation of the war. The four speakers and their subjects were selected accordingly. So the first speech was given by MAX RUBNER, Professor of Hygiene at the University of Berlin, its subject being *"Our Feeding"*. NERNST followed him with the contribution *"The War and the German Industry"* [Nernst (1916a)]. The last two speeches were devoted to the mood and the moral: *"The Spirit in the Army"* by the writer and dramatist WALTER BLOEM, who now only wanted to be a soldier, and *"The Spirit in the Country"* by RUDOLF EUCKEN, Professor of Philosophy in Jena. The latter was the father of NERNST's student ARNOLD EUCKEN, and in 1908 he had received the Nobel Prize in Literature *"because of his serious search after the truth, the penetrating power of his thinking, and the far-sightedness, the warmth and power of the presentation, with which he has represented and developed his ideal world-view in many works"*.

The speech of NERNST centered on the supply of raw-materials which were indispensable particularly during a war. Because of the blockade, Great Britain had set up against Germany after it had entered the war (August 1914), there appeared problems regarding this matter. *"So it is a thankful task to describe the accomplishments of our industry during the war; perhaps the war can last still very long, but the fight against the consequences of the English blockade can essentially be looked at as finished. Now which were the proper obstacles which had to be taken?"* A few important raw-materials *"partly can be found in Germany, and in this case we want to mention iron and coal as examples. However, a multitude of other materials are imported from abroad, as such we mention copper, rubber, sulfur, cotton, Chile saltpeter."* [Nernst (1916a): 13].

NERNST emphasized approvingly that these difficulties had been recognized immediately. His friend WALTHER RATHENAU, at the time Chairman of the Board of the AEG, together with WICHARD VON MOELLENDORFF, working as an engineer with the AEG and later becoming a politician and theorist of economics, had approached the War Minister ERICH VON FALKENHAYN and had proposed to him the establishment of a Commission helping to keep the expected problems with the raw-materials under control. Then already on August 13, 1914 VON FALKENHAYN ordered the establishment of a War Department of Raw-Materials (*Kriegsrohstoffabteilung*, KRA) attached to his Ministry and appointed WALTHER RATHENAU as the director. GEORG KLINKENBERG, who is well known as a pioneer of the construction of power stations with more than 70 power stations built by him, became the deputy director. On May 1, 1915 Major JOSEPH KOETH took over the directorship of the KRA.

(NERNST wrote to *"Major Kroth"* [Nernst (1916a): 16], perhaps a case of confusion with EWALD KROTH, who was favoring the automobile world.) Between spring of 1915 and summer of 1916 the KRA registered all raw-materials and controlled their processing. Toward the end of the war, the importance of the *"widely branching organization, operating with quiet confidence"*, as emphasized by NERNST [Nernst (1916a): 16], and its merits became smaller and smaller, although the network of the KRA always increased, subsidiary companies and war committees became involved, and until 1918 the number of its members increased up to 2500.

In his speech NERNST explained how the problem of the supply of the raw-materials, mentioned above and obtained previously only by means of imports, was solved. In particular detail without being afraid of *"indiscretions"* he treated the Chile saltpeter, *"a salt of the nitric acid, a nitrogen-oxygen compound, the value of which is only given by the nitrogen content, and so during the discussion of this raw-material we are led to the 'nitrogen problem', ... likely the most important one which occupied the German industry at the beginning of the war."* [Nernst (1916a): 19]. Nitrogen compounds are contained in an important class of mineral fertilizers indispensable for the plant production and feeding as well as in the gun powders. In the latter case they are nearly exclusively compounds of nitrogen with oxygen. Under war conditions the development of explosives based on different materials was too expensive and time-consuming. *"So at present one must say: nitrogen-oxygen compounds are part of the conduct of war, and the side, which experiences a shortage of these compounds, must surrender completely at the mercy of the other side."* [Nernst (1916a): 21]. The unexpectedly long duration and the dimensions of the war made it necessary to produce gun powder and explosive materials in large quantities from the own raw materials. *"In this case our highly developed chemical industry could jump in providing help, radical help, and at that within an extremely short time."* [Nernst (1916a): 21]. NERNST mentioned that it had become possible to turn nitrogen-hydrogen compounds such as ammonia into nitric acid. It is interesting that due to reasons of secrecy he did not mention the possibility for the production of ammonia by means of the large-scale process developed by HABER and BOSCH. On the other hand, regarding the technical conversion of ammonia into nitric acid, he emphasized: *"For the evaluation of the latter process it is important, that also during the processing of hard coal considerable amounts of ammonia are generated, which can be increased nearly without limits."* [Nernst (1916a): 21]. It is interesting to note that the article in the *Berliner Illustrirten Zeitung* (Illustrated Newspaper of Berlin) one year ago, mentioned above, was clearer regarding this matter: *"Geheimrat Haber, the Director of the Kaiser-Wilhelm Institute of Physical Chemistry and Electrochemistry, has given us valuable directions how we can obtain the needed ammunition also without the import of the Chile saltpeter."* [BIZ (1915)]. As expected of him, NERNST concluded his speech – however, not without critical remarks – showing an external optimism, which indicated traits of his true way of thinking to those who are familiar with the nature of NERNST: *"So we can look calmly into the future. It appears that in the area I have discussed the dangers are definitely banished. ... Also in the future within the German land an energetic continuation of work in the industrial world will not be missing. ... And we are also certain that if peace will be won, the German industry will appear on the scene unbent and with undiminished creative power, in order to heal the wounds affected to the well-being of the fatherland as much as possible."* [Nernst (1916a): 22–23].

5.9.2 The effort on peace negotiations

It is conspicuous that in this last sentence of his speech NERNST talked about peace but not about the victory of Germany and its Allies. Almost as a reply to this, immediately following the words of NERNST the *"speech of the Captain Walter Bloem"* begins with the verse composed by himself: *"Are you dreaming of the day of peace? / Who wishes so may dream! / War! is the key word, / Victory! and so it rings on!"* [Bloem (1916): 24]. Also MAX RUBNER used quite different words for ending his speech, which preceded that of NERNST: *"I see before me, like in the sea of blood, the enemies presenting their arms, that is the victory. And powerful rises the German eagle proud and rejuvenated upward to the sun. Per aspera ad astra."* [Rubner (1916): 12].

Even if the participation in the war and in the war-related research described in the preceding Section as well as its recognition by the Army Command appear to contradict this, WALTHER NERNST must be counted yet among the people who strongly pushed for ending the war. This can be demonstrated already starting from an early point in time, and not only after July 19, 1917, on which day three Parties together holding the majority in the German Parliament adopted a resolution of peace. The resolution requested a negotiated peace, since in the meantime the objective of the war formulated in 1914 by the Emperor WILHELM II had turned out to be a lie. In its introduction the resolution said: *"As on August 4, 1914, also at the beginning of the fourth year of the war for the German people the word of the Emperor's speech is valid: 'We are not driven by the addiction to conquer[, we are animated by the unbending will to hold on to the place given us by God.]'. Germany has taken up the arms in order to defend its freedom and independence and for the safety of its territorial possessions. The Parliament strives for a peace of understanding and of the permanent reconciliation of the peoples."*

NERNST's daughter EDITH told the science historian FRIEDRICH HERNECK about the opinion of her father which had developed already during the first months of the war: *"He had only the thought, that the war must be finished as fast as possible, after it had been approved by the whole Parliament and no return was possible any more."* [Zanthier (1977)].

The activities which NERNST had developed in connection with the military use of gas and with the equipment of the mortar battalion should be considered more under the aspect of his effort on peace negotiations. In this case we must note the statement about their highly esteemed teacher expressed by FREDERICK ALEXANDER LINDEMANN (Viscount CHERWELL) and FRANZ (Sir FRANCIS) SIMON: *"A man of his enterprise and initiative could not stay at home when the war broke out in 1914. ... He had some hand in the introduction of gas warfare, which he always maintained was the most humane way of using shells, and for this reason he was placed on the list of those to be extradited in 1919."* [Cherwell and Simon (1942): 102]. Certainly, in its first explorations the Entente had decided correctly, since the use of poisons and, hence, of poisonous gases during the conduct of war by international law is rated as a war crime due to the "Laws and Customs of War on Land (Hague IV)", adopted in the year 1907 at the Second Peace Conference in The Hague. However, the Entente had not considered the fact, that NERNST had occupied himself only with irritating and not with poisonous substances.

At this point also the judgment of EINSTEIN, the declared and widely audible opponent of the First World War, is important to note, which he wrote in the obituary for WALTHER NERNST: *"He was neither a nationalist nor a militarist."* [Einstein (1942): 196].

Certainly, also the award of the *Ordre pour le Mérite* had no military background. Within the *«Hochschul- und Personalnachrichten»* (College and Personal News) of the issue of October 1, 1917 of the *Zeitschrift für Elektrochemie* (Journal of Electrochemistry) one finds the note: *"Berlin (University). Geh.-Rat Prof. Dr. W. Nernst was elected as Ritter* [knight] *of the Ordre pour le Mérite for the sciences and the arts with the right to vote."* [ZfE (1917): 323]. This formulation and the kind of the graphical presentation support the observation we have mentioned. Since the beginning of the war, military decorations and promotions were emphasized by being printed in italics and were placed ahead of the news ordered according to the towns, as it happened, for example, in the issue of August 1, 1915: *"Geh.-Rat Prof. Dr. W. Nernst – Berlin received the Iron Cross of 1st Class."* [ZfE (1915): 404]. Also we must not forget that during the World War NERNST published a few important papers which are not connected with war-related research. In this case we mention his *"Attempt to Return to the Assumption of Continuous Energy Changes Starting from Quantum-theoretical Considerations"* [Nernst (1916b)], his papers dealing with the determination of chemical constants [Nernst (1916c)], and with the photochemical law of equivalence of EINSTEIN [Nernst (1918b)]. Further, we should remember that already during the summer of 1915 MAX PLANCK had received the *Ordre pour le Mérite* [ZfE (1915): 491].

On different occasions NERNST had been reproached because of the fact that he belonged to the 93 German intellectual people from many parts of the sciences and the arts, the names of whom were placed under the dated appeal «*An die Kulturwelt*» (*"Manifesto of the Ninety-Three German Intellectuals to the Civilized World"*). It was the purpose of this document to justify the invasion of Belgium by German troops on August 3/4, 1914 against international law. On August 2 Belgium had turned down the ultimatum demanding the right for the German troops to march through, and only on August 4 war had been declared against it by Germany. The forerunner of the appeal had been a *"Declaration [Kundgebung] of the German Universities to the Universities Abroad"*, which during September 1914 had been initiated by Scientists of the University of Tubingen. Although it was welcomed by the German Government and its support had been promised by all 22 German Universities, it was rated too weak and long-winded by GEORG REICKE, 2nd Mayor of Berlin at the time. Being a poet himself, with two *"colleagues"*, the dramatist HERMANN SUDERMANN and the comedy author LUDWIG FULDA, he composed the appeal «*An die Kulturwelt*», which on October 4, 1914 was published by all major newspapers of Germany, and which was mailed to many scientists abroad. For example, in this document, also referred to as *"Manifesto of the 93"*, distorting the facts it says: *"As representatives of the German Science and Art to the whole world of culture we raise protest against the lies and slander, with which our enemies want to dirty the pure case of Germany in this difficult fight for survival forced upon it. ... It is not true, that Germany has caused this war ... , that we have wickedly violated the neutrality of Belgium ... , that our Military Command has disregarded the international law ... , that the fight against our so-called militarism is not a fight against our culture, ... You, who know us well, ... Believe us! Believe us, that we will carry this fight to the end as a people of culture, to whom the legacy of a Goethe, a Beethoven, a Kant is similarly sacred as its home and its native soil. For this we vouch for you with our name and with our honor!"* [Böhme (1975)]. As a selection of the people known as the undersigned we mention the Nobel-Prize winners (indicating the field and the year of the award) and a few other scientists, who were also close to NERNST: ADOLF VON BAEYER (chemistry 1905), EMIL VON BEHRING (medicine 1901), PAUL EHRLICH (medicine 1908), RUDOLF EUCKEN (literature 1908), EMIL FISCHER (chemistry 1902), WILHELM FOERSTER, GERHART HAUPTMANN (literature 1912), FRITZ HABER (chemistry 1918), ADOLF VON HARNACK, FELIX KLEIN, PHILIPP LENARD (physics 1905), MAX LIEBERMANN, WILHELM OSTWALD (chemistry 1909), MAX PLANCK (physics 1918), WILHELM CONRAD RÖNTGEN (physics 1901), WILHELM WIEN (physics 1911), RICHARD WILLSTÄTTER (chemistry 1915). The biochemist ALBRECHT KOSSEL (physiology 1910), OTTO WALLACH (chemistry 1910), and EDUARD BUCHNER (chemistry 1907) belonged to the German intellectual people who received the Nobel Prize before August 1914 and at this point in time would have been in a position to agree to the mentioning of their name under the appeal. At the beginning of the war the latter voluntarily entered the military service, and in 1917 in Romania he suffered heavy injuries to which he finally succumbed.

Regarding the signature of her father under the appeal, NERNST's daughter EDITH remembered: *"During his absence from Berlin it was quasi enforced by my mother in a great hurry. On the telephone she was told: also Planck, Fischer, etc. have signed it."* [ZANTHIER (1973b)].

After the war, a few months before his suicide, on April 17, 1919 regarding the appeal EMIL FISCHER wrote to SVANTE ARRHENIUS: *"The assertions on Belgium contained in it, which we as the undersigned believed at the time, now cannot be maintained any more, after we know that during the first years of the war our Government has completely lied to us on many points."* [Zott (1996): 201]. Whether his overall statement regarding the *"undersigned"* is actually also completely valid in the case of NERNST, may be questioned. Then the retraction by all of those still living, the names of whom appear under the appeal, as proposed by ARRHENIUS, was approved by NERNST, in addition to FISCHER, HABER, PLANCK, RUBNER, WALDEYER, and HEINRICH RUBENS within the Academy of Berlin. However, RUBENS had not signed the appeal to begin with.

Of course, the appeal was a complete failure abroad. Instead, the reality of the German conduct in war was viewed negatively. Dated October 10, 1914, ROMAIN ROLLAND wrote: *"These days I have received your peculiar appeal to the world of culture, with which the Imperial Army Corps of the German intellectual people has bombarded Europe. ... This mobilization of the regiments of the pen ... has, I believe, supplied a few more reasons to fear the organization of the 'Reich', but not a single reason to value it higher. Not without consternation the cultural world has read what the most famous names of Germany's science, art, and thinking ... have certified by the document: it is not true, ... , that day is day and night is night!"* (from the German translation in [Rolland (1966): 94–95]). On December 11, 1914 SVANTE ARRHENIUS wrote: *"Dear friend Wilhelm* [OSTWALD]. *... Now if 100 German well-meaning Professors get together and write an appeal 'It is not true' about this territory, still nobody will believe it."* and he continued: *"Yesterday I quietly celebrated the Nobel day. ... I was thinking about all the noble and great people, which visited us during these days. Now they are trying to prove that all members of other countries are wrong; they themselves are the only ones thinking correctly."* [Zott (1996): 193, 195].

However, this way of thinking cannot be associated with WALTHER NERNST. Also during the war he stuck to his high regard for the international character of science. In his speech in 1916 in the building of the *Reichstag* he confessed: *"I regard it as a matter of honor to leave it still not ignored, that from the circle of the English and French scholars close to myself some indications have reached us, that over there also during these times one operates in the spirit of the word by Helmholtz, according to which science winds an untearable bond around all nations of culture."* [Nernst (1916a): 19]. ALBERT EINSTEIN generalized this nature of NERNST, which can be seen from this special example, with the words *"He judged things and people almost exclusively by their direct success, not by a social or ethical ideal. This was a consequence of his freedom from prejudices."* [Einstein (1942): 196].

It was a peculiarity of NERNST, that he knew how to judge correctly not only scientific and technical developments, but also political and military matters. Certainly in this case the first capability was a prerequisite of the second; however, possibly both supplemented each other favorably. NERNST also attributed an analogous interactive relationship to his friend WALTHER RATHENAU, *"since he put his hopes on the revolutions of our time in philosophy and in natural science and technology, which all he knew exactly, and strongly believed in a logical development of the state out of the modern conceptions."* [Nernst (1922a): 7–8]. His talent for accurate judgment on military matters can be demonstrated with the problem of unrestricted submarine warfare, i.e., the sinking of all merchant ships of the enemy without warning. This kind of warfare, that was against the international law similarly as the British food-blockade, was ordered in January 1916 by the Chief of General Staff ERICH VON FALKENHAYN after a military failure in 1915, so as to impose a blockade against Great Britain and force a victorious end of the war or at least an honorable peace. Due to the increasingly precarious situation of the theatres of war on the ground, in the beginning of 1917 the Imperial German Navy started unrestricted submarine warfare, to facilitate a turn of the war. However, this step did not accomplish a decisive weakening of Great Britain. NERNST remembered: *"When I discussed with Dr. Rathenau the chances of the planned declaration of the unrestricted submarine war at the time, both of us, of course, were perfectly clear about the technical shortcoming of our instruments of power in this area."* [Nernst (1922a): 7].

The efforts of NERNST directed at the goal of ending the war or accomplishing peace negotiations by no means were only of a verbal or theoretical nature. On the contrary, in a practical peace mission he met in Brussels several times the Belgian business man and banker FRANZ PHILIPPSON of German descent. The first meeting on May 16, 1915 was followed by meetings on June 16 and 17, on November 13, 1916, as well as finally on December 29, 1917. Probably these meetings were arranged by ROBERT GOLDSCHMIDT, a son-in-law of PHILIPPSON [Haag (1984): 330], GOLDSCHMIDT being the physical chemist, who in the spring of 1910 had drawn the attention of ERNEST SOLVAY to NERNST, and who subsequently had been one of the three secretaries of the 1st SOLVAY-

Congress in Brussels in 1911 (see Section 5.5.5). By attending this congress and also the second one in 1913 NERNST had already become acquainted with the Belgian capital.

The subject discussed during these meetings becomes evident from the protocol written by hand by NERNST on June 17, 1916: *"Herr Geheimer Regierungsrath Professor Dr. Nernst from Berlin and Herr Franz Philippson, Generalconsul, in Brussels have held discussions in the house of the latter, 18 rue Guimard, with the purpose of a preliminary discussion to take place among those interested (Belgium, Germany, England, France) in order to arrange a peace conference. Such a preliminary discussion must take place in Rotterdam between Belgian, German, English, French representatives, which are meeting with the knowledge of the corresponding Governments, but with no obligation whatsoever of the latter, and to be precise one of the Belgian representatives will occupy the chair during the negotiations. The possible result of the preliminary discussion would be transmitted to the corresponding Governments, which would have to decide subsequently, if an official conference would have to take place. Herr Nernst will inform the German Government about the above matter and as soon as he has obtained approval of the latter he will instruct Herrn Philippson about this; upon the receipt of the necessary passports the latter will immediately proceed to France, in order that this matter can be submitted to the other concerned Governments under the initiative of His Majesty the King of Belgium [ALBERT I.]."* [Haag (1984): 349]. Already nine days later NERNST could report to PHILIPPSON from Berlin: *"I am happy to be able to tell you, that, as long as private representatives from the three other concerned countries meet at a neutral location, correspondingly also German representatives would be prepared to come for the purpose of a most confidential and non-committal discussion."* [Haag (1984): 352].

In the years 1915 and 1916 NERNST traveled to the meetings under the instruction of the Chancellor (*Reichskanzler*) THEOBALD VON BETHMANN HOLLWEG, to whom he had also suggested this idea. During the year 1917 there happened many and lasting changes. So during July BETHMANN HOLLWEG was toppled and at the end of October also his successor GEORG MICHAELIS. On November 1, 1917 GEORG VON HERTLING was appointed as Chancellor and Prussian Prime Minister. Under him the Supreme Army Command under PAUL VON HINDENBURG and ERICH LUDENDORFF could expand its power undiminished against the national Government. After the October Revolution in Russia, on December 25 an armistice with this enemy in the east was arranged resulting in peace negotiations. These and other circumstances were the reason, why on December 29, 1917 NERNST admitted to PHILIPPSON, that now only he himself had felt motivated for this meeting.

We must not forget that the terrible war had inflicted great harm on WALTHER NERNST by taking both sons from him and his family, as we have mentioned already several times (Sections 4.4, 5.5.8, and 5.8). Here honor and painful loss were always close to each other. So in the «*Hochschul- und Personalnachrichten*» (College and Personal News) of the *Zeitschrift für Elektrochemie* of November 1, 1914 we can read in immediate succession: *"The gentlemen Geh.-Rat Prof. Dr. W. Nernst, Prof. Dr. W.* [perhaps correct: E. = Eduard] *Jordis and Dr. M. Rohmer* [perhaps correct: W(ILHELM) ROHN] *received the Iron Cross* [2nd Class]. *Berlin. In the battle for the fatherland was killed Rudolf Nernst, son of the Geh.-Rat Prof. Dr. W. Nernst, the same, who at the 21st General Assembly of the German Bunsen Society in Leipzig* [end of May 1914] *accepted the Bunsen Memorial Medal in the name of his father."* [ZfE (1914): 585]. Without directly referring to the fate of *"friend Nernst"*, who *"as a voluntary automobile driver behind and at the front gained so much merit, that he received the Iron Cross"*, on December 20, 1914 GUSTAV TAMMANN wrote to SVANTE ARRHENIUS: *"The celebration of peace, Christmas, is around the corner. A sad celebration for many people in the country, who mourn the loss of sons."* [Zott (1996): 199]. In fact, NERNST's daughter EDITH remembered, that the Christmas celebration in 1914 in the Berlin home, to which her father had invited ALBERT EINSTEIN who had been won as a colleague for Berlin, happened in a depressed atmosphere, of course, in particular because of the loss of her brother RUDOLF [Zanthier (1977)]. The events of the year 1917 occurred similarly as in 1914, however, within a larger temporal distance and in reversed order, as one can read in the Section *"Mixed News"* of the *Chemiker-Zeitung* (Newspaper for Chemists): *"In the battle for the fatherland were killed in action: ... on April 21 in an air battle Gustav Nernst, Lieutenant, recipient of the Iron Cross 1st and 2nd Class, 21 years old, son of Prof. Walter Nernst, who has lost already his other, older son in the war."* ([CZ (1917): 533] of June 30, 1917) and *"Geh. Reg.-Rat Prof. Dr. W. Nernst in Berlin was made a Ritter* [knight] *entitled to vote of the Ordre pour le Mérite for science and the arts."* ([CZ (1917): 669] of August 28, 1917). Before the end of the year NERNST traveled to PHILIPPSON in Brussels, in *"December 1917"* he had started his monograph on the Third Law with the words *"During times of sorrow and distress ..."* [Nernst (1918a): III] (see Section 5.5.8). Later, on February 14, 1921, he admitted to his friend ARRHENIUS: *"Of course, my wife and myself are still always filled with deep sorrow; both of our splendid sons have neither experienced my appointment as a member entitled to vote of the Ordre pour le Mérite nor presently the Nobel Prize!"* [Zott (1996): 205]. However, corresponding to his nature WALTHER NERNST never publicly complained about the deeply felt harm of the *"times"*.

5.10 Political Activities

When on November 11, 1918 the German delegation lead by the Permanent Secretary without portfolio at the time, MATTHIAS ERZBERGER, in the Forêt de Compiègne (France) signed an armistice, this corresponded to an unconditional surrender. *"In these eventful days"* on November 14, 1918 ALBERT EINSTEIN answered a letter of SVANTE ARRHENIUS and he concluded in it: *"It is quite peculiar with which elasticity already now most people have gotten used to the completely new situation; this is the most surprising result of all surprises."* [Zott (1996): 202]. With this he could hardly have meant WALTHER NERNST.

Under, in many respects, difficult conditions of the immediate postwar period the latter made a great effort for the return to normality. Of course, to this belonged the functioning of research, a goal he pursued by means of his activity within the organization of science. So in the year 1919 NERNST accepted his election into the Senate of the *Kaiser-Wilhelm-Gesellschaft* as successor of the late WILHELM VON SIEMENS. Initially this was a temporary arrangement until the official appointment on October 28, 1919. In the *Kaiser-Wilhelm-Institut für Metallforschung* (for metal research) in 1920 he served as a member of the Administrative Committee and in 1921 in its Board of Trustees.

The activity of NERNST as Rector of the University of Berlin during the period of October 15, 1921 until October 15, 1922, was reported already above (Section 5.7.4), and must also be included in his accomplishments in the organization of science. From the way in which NERNST performed in this office we can see that he did look upon science and its organization as non-separable from general politics. In particular, some events of the year 1922 provide the possibility to illustrate this statement in an exemplary fashion. These events which also affected NERNST personally were caused by right-wing extreme, national, and anti-Semitic opponents of the Weimar Republic. On June 24, 1922 the Foreign Secretary WALTHER RATHENAU was murdered, since in this community he was treated as a conceding politician (*Erfüllungspolitiker*) in the matter of the reparation payment and as a symbolic figure of the "Jewish Republic". Closely connected with this case one must look at the murder threat expressed against the convinced democrat and Jew ALBERT EINSTEIN. In 1922 at the centennial of the Society of German Natural Scientists and Doctors, the oldest scientific organization of Germany, he was supposed to give the ceremonial lecture about the relativity theory. On July 6, 1922 EINSTEIN wrote to MAX PLANCK, the Chairman of the Society: *"It is said that I belong to the group ..., against which assassination attempts are planned by national groups ... Now nothing helps except for patience and leaving town."* (quoted after [Wazek (2005): 222]).

NERNST took the occasion of his speech at the annual memorial celebration of the founder of the University of Berlin on August 3, 1922 to condemn the murder directly and the murder thread indirectly in front of the public and to present in an appeal his opinion regarding this matter and the connection between science and politics: *"Only a few days before his* [RATHENAU's] *death one evening in a small group in my home we discussed certain questions of the international scientific exchange; then he expressed the beautiful word that for the recovery of Europe it would mean a progress, if at least a few areas of the human culture could be withdrawn from the battle of the daily politics. This word, which initially referred to the external politics, is no less valid for the inner life of our nation. Let us hope that the abhorrence of the wicked crime and at the same time the condemnation of raw violence, even if it concerns only the violent disregard of the opinion of dissidents, unifies the wellmeaning people for a joint effort, and that in Germany the weapons of prudent critique, led by the honest conviction and, where suitable, with academic calmness, again receive more recognition."* [Nernst (1922a): 8].

These events in Germany are obviously connected with the fact that EINSTEIN as well as NERNST felt induced to respond in writing to the German translation of the educational poem «*De rerum natura*» (On the Nature of Things) of the Roman poet LUKREZ, which had been carried out by the classical scholar HERMANN DIELS from Berlin and which appeared posthumously [Einstein (1924); Nernst (1924b); Rösler (1999): 284–287] – and this not only because of its content of philosophy of nature concerning the doctrine of the Greek philosophers DEMOCRITUS and EPICURE. So one reads in the *"preamble"* by EINSTEIN: *"Upon everybody who is not completely absorbed by the spirit of our time, but instead occasionally feels as a spectator in front of his fellow men and especially of the mental attitude of them, the work of Lukrez will exercise its spell. ... As the main goal of his work he presents the liberation of man from the slavish fear, caused by religion and superstition and nurtured and exploited by priests for their purposes. ... His admiration of Epicure, Greek culture, and language in general, which he places highly above Latin, is touching. ... Where is the modern nation which harbors and expresses such a noble attitude in front of a fellow woman?"* [Einstein (1924); Rösler (1999): 284–285]. In NERNST's book-review one reads: *"The work of Lukrez is dominated by the following three aspects, which have lost nothing of their moral and logical power: (1) Freedom of the scientific research, in particular liberation from the enslavement by the priests guided by idolatry; ..."* [Nernst (1924b): 1741–1742; Rösler (1999): 286].

Except for his activity as Dean and Rector of the University of Berlin, NERNST was also directly involved in local politics. MAX BODENSTEIN remembered, *"that occasionally he voluntarily had acted within the administration, ... for some time as a city councilor, about which I do not know anything except for this fact."* [Bodenstein (1942a): 102].

In 1921 even the opportunity was given to WALTHER NERNST to enter high politics. During this year the Republican WARREN G. HARDING became the 29th President of the USA, and on August 25, 1921 concluded the separate "Peace Treaty of Berlin" with Germany independent of the Peace Treaty of Versailles. As a potential candidate for the position of the Ambassador in the USA, the extremely influential Personnel Manager of the Foreign Ministry, EDMUND SCHÜLER, proposed to his superior, the Permanent Secretary EDGAR VON HANIEL, to recommend NERNST for this position. Both the Foreign Secretary FRIEDRICH ROSEN and the Chancellor JOSEPH WIRTH agreed to offer the well-known scholar an opportunity to go to Washington as Ambassador. However, it is said that NERNST declined this offer, supplying the quite implausible reason, that his knowledge of the English language would be insufficient for this position. Probably, more likely he did not want to leave the field of science, in particular, since he felt responsible and competent for the development of science in Germany during the post-war period to a higher degree than for politics, and since he knew that he had received the Nobel Prize in Chemistry of 1920, *"when the telegrams arrived on Friday* [February 11, 1921] *morning."* (Letter to ARRHENIUS of February 14, 1921 [Zott (1996): 205]).

5.11 Visits to the USA and to South America

EDMUND SCHÜLER, who became known because of the extensive "Schüler Reform" of the entire Foreign Service named after him, belonged to the personalities in the Foreign Ministry, who after the defeat of 1918 supported the idea that the accusations of representatives of the German commerce and industry evident since the turn of the century and expressed more strongly during the World War must be now taken seriously. These accusations targeted the fact that the diplomats were not sufficiently familiar with the international economy and the global trade as well as with their conditions. On the other hand, NERNST had demonstrated that he was quite competent to handle such questions. The fact that he had also developed a certain familiarity with the situation in the USA might have motivated SCHÜLER to propose the worldwide well-known scholar for the position of the Ambassador in Washington.

This fame was the reason why among his students who came to him from all parts of the world there were also some from the United States as, for example, his PhD students in Göttingen MARGARET E. MALTBY (dissertation 1895: *"Method for Determining Large Electrolytic Resistances"*), JOHN K. CLEMENT (dissertation 1904: *"About the Formation of Ozone at High Temperatures"*), EDMUND S. MERRIAM (dissertation 1906: *"On the Theory of the Residual Current"*), and IRVING LANGMUIR (dissertation 1906: *"About the Partial Reunification of Dissociated Gases during Cooling"*). Already in the group around WILHELM OSTWALD in Leipzig he had developed a lasting acquaintance with the American physical chemists ARTHUR A. NOYES and MORRIS LOEB. In connection with the Third Law of Thermodynamics he experienced a controversy with THEODORE W. RICHARDS. NERNST had always admired the great American physicist JOSIAH WILLARD GIBBS, as can be seen also from his apparently last publication [Nernst (1939)].

However, NERNST also knew about life in the USA from his own experience from different occasions and visits. His first trip across the Atlantic concerned his attempt to obtain a US-patent for his light bulb. So on April 7, 1898 he boarded the steamer *«Lahn»* in Bremen in order to travel to New York. From aboard the ship he told WILHELM OSTWALD (later erroneously dated with *"1897"* by another hand): *"Unfortunately, I cannot attend the certainly very interesting meeting* [of the German Electrochemical Society], *since I have to travel to New York urgently. ... I am fine, in a short time the difficult, i.e., the financial and the business part of the light-bulb issue should be dealt with."* [Zott (1996): 128]. After 1898 when NERNST was able to sell the patent of his light bulb to the AEG, in the United States he could win GEORGE WESTINGHOUSE to take up the commercial development and the marketing of the NERNST lamp. In 1901 the "Nernst Lamp Company" was founded, which had its production site in Pittsburgh. The yttrium oxide Y_2O_3 needed for the NERNST pin was fabricated from the mineral gadolinite $Y_2FeBe_2[O|SiO_4]_2$, which was obtained from a mine at the Barringer Hill in Texas belonging to the Company.

On March 14, 1903 GUSTAV TAMMANN wrote to SVANTE ARRHENIUS: *"Nernst looks quite well, he shows an immense mobility, not infrequently in Berlin, next week he goes to America."* [Zott (1996): 157]. This time it was the purpose of his trip to the USA to present lectures on physical chemistry. In the same year also WILHELM OSTWALD had received *"a letter from the little university town of Berkeley, California, near San Francisco from the Professor of Physiology over there, Jaques Loeb"*, in which the latter invited him to present a lecture on the relations between physical chemistry and biology at the opening of his new laboratory. JAQUES LOEB had been born and educated in Germany and was *"a glowing admirer of the new physical chemistry, to which he must owe the major part of his successes."* [Ostwald (1927): 320].

The following year brought OSTWALD again to the USA to attend an international congress on the arts and sciences, which was organized in connection with the World Exhibition in St. Louis. However, in the invited lectures physical chemistry was represented by JACOBUS HENRICUS VAN'T HOFF and ARTHUR A. NOYES, whereas OSTWALD had been invited by the Department of Philosophy for a contribution on the methodology of the sciences. Also NERNST made an effort to represent properly the German physical chemistry at the World Exhibition. During July, 1903 he approached OSTWALD regarding this matter: *"Assuming that the circular of the Prussian School Administration regarding the chemical exhibition in St. Louis, also signed by myself, is already in your hands, with these lines I would like to ask you most kindly in particular for your kind cooperation in a suitable representation of physical chemistry. Especially, I would like to suggest if you would not like to consider among others the exhibition of an almost complete collection of your instruments used in the physical-chemical laboratory course."* To this inquiry he received the answer: *"The typical instruments of my laboratory will be exhibited by the mechanic Köhler of the Institute."* [Zott (1996): 157–158]. After NERNST's light bulb had been awarded the *Grand Prix* at the World Exhibition in 1900 in Paris it was also presented in 1904 in St. Louis.

In our discussion of the Third Law of Thermodynamics we had mentioned already, that in 1906 NERNST had talked about his discovery within his presentation of the "Mrs. Hepsa Ely Silliman Memorial Lecture" at the Yale University in New Haven (Section 5.5.2). This lecture series had been endowed by the children of HEPSA ELY SILLIMAN after her death. In particular her son AUGUSTUS ELY, a banker and author, *"presented eighty thousand dollars to Yale University"* to establish this series of Memorial Lectures. HEPSA ELY SILLIMAN had published a theory of the origin of meteorites based on FARADAY's work on chemical affinity and electromagnetism. The lectures should emphasize astronomy, chemistry, geology, and anatomy. The first lecture was given in 1901 with the intention to illustrate *"the presence and wisdom of God as manifested in the natural and moral world."* In the meantime for the 6th Silliman Lectures from October 22 until November 2, 1906 NERNST had traveled from Liverpool to New York on board the steamer "Oceanic" between September 26 and October 3.

During April 1914 WALTHER NERNST traveled to Argentina in order to lecture for six weeks at the National University in La Plata founded in 1905. In the case of NERNST for the first time a German scientist had been appointed at an Argentine University, to represent over there the European science following French, Italian, and Spanish scholars. At this time the German Bunsen Society informed NERNST that on May 22, 1914 following FRIEDRICH KOHLRAUSCH (1908) and the industrial chemist IGNAZ STROOF from Griesheim (1911) he would receive the Bunsen Memorial Medal. In a letter of April 23 from La Plata to MAX LE

BLANC, the First Chairman of the Society, NERNST expressed thanks, regretted that he could not personally accept the Medal, and added: *"Just during these days in my electrochemical lectures at this University I demonstrate experiments dealing with the dissociation voltage and with the influence of the concentration of the metal ions upon the e.m.f.; on this occasion I had to think also quite vividly of you, and not without satisfaction I could note that these experiments work out quite well also over here, i.e., that our theories partly devised under the Polar Star appear to be valid also under the Southern Cross – certainly another encouraging sign of their generality! The attached little paper is written in the zone of the antipodes; hopefully on the northern hemisphere it does not find too many antipodes!"* [Nernst (1914b)]. This paper *"About the Application of the New Thermal Law to Gases"* [Nernst (1914c)], dated *"La Plata, April 1914, Physical Institute of the University"*, was NERNST's *"modest little gift in return"* [Nernst (1914b)] for the Memorial Medal. On May 22, 1914 at the General Meeting of the Bunsen Society at the request of LE BLANC the paper was read by HANS VON WARTENBERG, and the Memorial Medal was accepted by *«Studiosus Rudolf Nernst»* in the name of his father.

In Argentina NERNST received also several honors: *"The University and the Chemical Society of Buenos Aires made him an Honorary Member, the Directors and Professors of the Argentine Universities, the German Scientific Association, and the German Club organized for him special festivities."* [IMWKT (1914): 1285]. At the festive session of the German Scientific Association in his speech NERNST raised the question whether a world-view could be found with the help of the natural sciences. For an answer he started with the verse of GOETHE from the first part of *«Faust»*: *"Mysterious on a bright day, / Nature does not allow lifting its veil, / And what it does not reveal to your mind / You cannot take from it using levers and screws."* However, contrary to these words the measuring instruments developed recently in physics and chemistry have allowed a deep insight into nature. On the other hand, in the sense of GOETHE the reason for a world-view must be left to the poets, since for them a formula derived from natural science would be much too complicated and, hence, impractical. Before NERNST left Argentina on May 15, 1914, at a farewell banquet his scientific accomplishments were emphasized by the Argentine Minister of Education. The visit of the German scholar may have been the motivation for the fact, that the Argentine Ministry and the German Embassy considered the speedy realization of an exchange of professors. WILHELM JOST summarized NERNST's visit to South America with the words: *"After all we know about it, it must have been a truly triumphant journey,* [not only through Argentina] *also through Uruguay and Brazil."* [Jost, W (1964): 528].

Apparently, during his visits to the USA, NERNST had gained substantial knowledge about the industry, the people, and the different general conditions in this country. It is said that after the United States had entered the First World War on April 6, 1917, he had asked the Emperor WILHELM II for an audience in order to explain to him the seriousness of the situation due to this development. *"Immediately afterwards he reported to Carl Bosch: He had pointed out to the Emperor how highly superior America would be technically compared to Germany. The Americans would build the needed ships, equip the troops, and bring them over here. According to his opinion, now peace would have to be achieved at any price. The situation would be hopeless. The Emperor had turned quite serious and had looked at his Generals, whereupon Ludendorff took the word and answered briefly, he would have detailed reports from America, and Nernst's fear would be totally unjustified. Then he, Nernst, had been dismissed with thanks."* [Jost, W (1964): 528].

During the year 1928, WALTHER NERNST took another trip to the United States in order to accept one of the oldest and most distinguished awards, which was and is still given for outstanding accomplishments in the natural sciences and technology: the Franklin Medal of the Franklin Institute in Philadelphia, Pennsylvania. This institution was established in 1824 by the manufacturer SAMUEL V. MERRICK and the geologist WILLIAM H. KEATING for continuing the legacy in memory of BENJAMIN FRANKLIN. Since 1824, the award ceremony has taken place at the end of a week-long celebration in April. In 1928 NERNST was honored for his "Applications of Thermodynamics to Electro and Thermo-Chemistry".

Other recipients of this Medal in addition to many other important personalities were, for example, 1920 SVANTE ARRHENIUS (Physical Chemistry), 1926 NIELS BOHR (Distinguished Work on Dynamic Atom), 1927 MAX PLANCK (Law of Radiation), 1934 IRVING LANGMUIR (Fundamental Research in Chemistry and Physics), and 1935 ALBERT EINSTEIN (Contributions to Theoretical Physics). On January 17, 1933 NERNST congratulated his former student LANGMUIR on a postcard for winning the Nobel Prize in Chemistry in the year 1932, and he added: *"P.S. I hope to come to the U.S.A. next summer!"* Obviously, this wish was not realized.

At the beginning of July 1955 an honorable invitation from Philadelphia was sent to *"Dr. Walther Nernst / Professor of Physical Chemistry / University of Berlin / Berlin, Germany : The President and Board of Managers of the Franklin Institute request the pleasure of your company at Medal Day ceremonies at the Franklin Institute on Wednesday, October 19, 1955."* At the time the city of Berlin was politically divided, and the letter first reached the Free University (FU) founded in 1948 in the Western Section of Berlin. Regarding the recipient of the letter, its *"Academic Information Office"* noted: *"FU unknown – 18/8/55"*.

Chapter 6

President of the Physikalisch-Technische Reichsanstalt (PTR) (1922 – 1924)

On January 6, 1922 NERNST wrote to ARRHENIUS, with whom he had reconciled again after a longer period of disagreements: *"For the two of us the festive week in Stockholm will always represent the most glamorous and most beautiful life memories of all my many journeys! By the way, this time the delay of this letter has a good reason, I think. Having returned home I found an inquiry regarding the Presidency of the Reichsanstalt, since Warburg resigns effective April 1. ... A few days ago on very careful consideration I have accepted."* [Zott (1996): 208]. Thereby NERNST announced that he would exchange the position of an *Ordinarius* and of the Rector of the University of Berlin with that of the President of the Physikalisch-Technische Reichsanstalt (PTR, equivalent of a German Bureau of Standards).

6.1 Brief History of the PTR

When this happened formally on October 15, 1922, NERNST became the fourth President of an important scientific institution, which exactly 35 years ago, during October of 1887, had started its operation at the time in a rented space or in rooms made available by the *Hochschule* in Charlottenburg.

Prior to this, KARL HEINRICH SCHELLBACH, Professor of Mathematics and Physics at the *Friedrich-Wilhelm-Gymnasium* and at the War

Academy in Berlin and teacher of the Crown Prince FRIEDRICH WILHELM, who was to become Emperor FRIEDRICH III, in July 1872 submitted the proposal to the Government to establish a State Institute for the promotion of the exact sciences and the precision technique. The Prussian Ministry of Education and Cultural Affairs forwarded the memorandum of SCHELLBACH, which was signed also by HERMANN VON HELMHOLTZ and by EMIL DU BOIS-REYMOND and which was supported by the Crown Prince, to the Academy of Sciences for its evaluation. However, the latter rejected the proposal for establishing such an Institute.

In spite of this, the idea was pursued further by the people interested, and its realization was supported by the Crown Prince. In addition, the Field Marshal HELMUTH VON MOLTKE could be won for the plan, because as Chairman of the Prussian Central Board of Directors for surveying he was interested in a promotion of the precision technique. After about nine years of practically unsuccessful endeavors at the end of 1882 VON MOLTKE convened a meeting at which the subject of the State Institute was discussed in detail, and which subsequent to further negotiations in June 1883 led to a new memorandum. This was signed by two military officers, two mechanics, and eight scientists. To the latter belonged WERNER VON SIEMENS, HERMANN VON HELMHOLTZ, WILHELM FOERSTER, HANS LANDOLT, and CARL ADOLF PAALZOW.

However, in the end a document became decisive, with which on March 20, 1884 WERNER VON SIEMENS approached the German Government. After prior to this he had offered to the Prussian Minister of Education to donate to the State of Prussia a property of $19\,800$ m^2 area, now he declared to be ready to donate to the State also the amount of $500\,000$ *Mark* in real estate or in capital for building a technically oriented research institute for the natural sciences. As a result in 1886, the Government submitted the plan for building a *Physikalisch-Technische Reichsanstalt* to the *Bundesrat* (Federal Council) and the Parliament. The former approved it, whereas the budget commission of the latter and all Parties except for one rejected it. Only after an additional effort by the Crown Prince, finally did the budget commission and the majority of the

Parliament give their approval, such that for the fiscal year 1887/88 the funds for the *"establishment of a 'physikalisch-technische Reichsanstalt' for the experimental advancement of the exact natural sciences and the precision technique"* were inserted into the budget of the *Reichsamt des Inneren*. So on April 1, 1887 the planning work could begin. The construction began on the property in Charlottenburg donated by WERNER VON SIEMENS. Until 1897 ten buildings were erected on an area of 2.5 ha, such that in a memorandum of that year for the Parliament it could be stated: *"So now the most important condition for the pursuit of its tasks, namely the sufficient amount of space for its practical needs, has been well satisfied for the 'Physikalisch-Technische Reichsanstalt', and it is justified to say that the completed impressive facility is not reached by a similar one in the world."*

In 1888 HERMANN VON HELMHOLTZ was appointed as the first President of the PTR. In addition to his outstanding scientific accomplishments in physics and physiology, and his personal interest to escape the teaching requirements of the University, another factor in his appointment the fact may have played a role – a certain family relation between the engineer and industrialist and the physicist existed because of the marriage in 1882 of SIEMENS' son ARNOLD with HELMHOLTZ' daughter ELLEN.

Until May 1945, together with the German *Reich*, the operation of the *Physikalisch-Technische Reichsanstalt* also ended, six physicists had followed HELMHOLTZ to the office of the President of the PTR, and they are listed in this table.

President of the PTR	Time of Office
HERMANN VON HELMHOLTZ	1888–1894
FRIEDRICH KOHLRAUSCH	1894–1905
EMIL WARBURG	1905–1922
WALTHER NERNST	1922–1924
FRIEDRICH PASCHEN	1924–1933
JOHANNES STARK	1933–1939
ABRAHAM ESAU	1939–1945

In order to better evaluate the accomplishments of NERNST for the PTR during his brief period of office, we present an overview of the structure of this institution before and after the activity of NERNST. At the foundation of the PTR there were two Sections, where Section I dealt with physical research and the technical Section II with inspection and testing. A test document indicating the activity of Section II is shown in Figure 6.1.

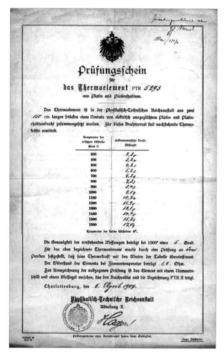

Fig. 6.1 Test document of the PTR for a thermocouple of April 6, 1907 (The handwritten remark at the right upper edge indicates: *"Private property of Geh. Nernst. May/1907."*).

Under the presidency of EMIL WARBURG in 1912 the PTR acquired a Laboratory of Radioactivity, in which HANS GEIGER performed important experiments with his counter tube and where he developed it further into a precision instrument, and in 1913 a High-Current Laboratory as well as a branch laboratory on the *Telegraphenberg* (Telegraph Mountain) near Potsdam, in which contrary to Charlottenburg, magnetic measurements could be performed without external interference. In the winter of 1914/15, WARBURG was able to attract ALBERT EINSTEIN and WANDER JOHANNES DE HAAS as guest scientists. During this time they discovered the gyromagnetic effect named after them, by proving the existence of AMPÈRE's molecular currents.

In 1914, WARBURG effected a change in the organization of the PTR, the new structure of which can be seen from the following scheme:

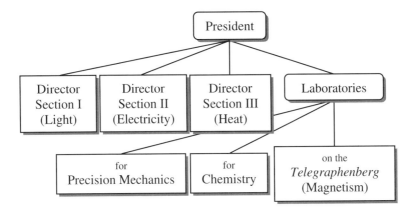

Each of the three sections was divided into a purely scientific and a technical-scientific operating subsection.

During the term of FRIEDRICH PASCHEN, WALTHER MEISSNER built a low-temperature laboratory. MEISSNER had been employed by the PTR right after he got his PhD in 1907. Under his direction in 1925 in this laboratory the liquefaction of helium was achieved, the PTR being the third location worldwide with this distinction following Leiden and Toronto. Subsequently, a series of new superconductors were discovered in this laboratory. Together with his coworker ROBERT OCHSENFELD in 1933, MEISSNER discovered the effect which represents the most fundamental property of a superconductor and which subsequently was named after them. According to this effect, an external magnetic field is expelled from the interior of a superconductor, if its temperature is lowered below its critical temperature. The economic crisis becoming more and more serious after 1928 prevented PASCHEN from realizing additional planned laboratories.

During his presidency, JOHANNES STARK had expanded the basis of a few working areas of the PTR or he had started some new ones. Among others, we mention the acoustics, the X-ray physics and technology, the frequency measurement, the test of instruments related to traffic with optical methods, and the lubrication technology. On the 50th anniversary in 1937 the total number of employees had increased more than fivefold compared to 1898, from 80 in the year 1898 up to 443. In 1937 nearly 80 laboratories belonged to the PTR, which were accommodated in addition to the old territory within two factory buildings and in a high-rise building. Due to the start of the War in 1939, plans for a new building complex for the PTR could not be realized. In 1938 the position of a permanent deputy of the president had been created.

During the presidency of ABRAHAM ESAU the laboratories reporting directly to the President, having reached a large number in the meantime, were combined into two Sections each of which was managed by a Director: Section V for Atomic Physics and Physical Chemistry, and Section VI for Mechanics and Acoustics. Because of the increasing number of air raids on Berlin, the main part of the PTR was evacuated to Thuringia, mainly to the town of Weida located 11 km south of Gera, and to Silesia. Only a small part remained in Berlin-Charlottenburg.

In addition to the Presidents we have mentioned, many well-known and often also famous scientists were working at the PTR, which is indicated by the following selected list: In addition to GEIGER, MEISSNER and OCHSENFELD, Scientific Members were LEO LOEWENHERZ (1887 – 1892, Director, precision mechanics), OTTO LUMMER (1887 – 1904, Member, optics), WILHELM WIEN (1890 – 1896, Assistant, optics), FERDINAND KURLBAUM (1891 – 1904, Member, optical electronics), OTTO SCHÖNROCK (1894 – 1935, Member, Laboratory Head, polarimetry), EDUARD GRÜNEISEN (1899 – 1927, Director, electric measuring technique), FRIEDRICH DOLEZALEK (1900 – 1901, Assistant, electric measuring technique), ERNST GEHRCKE (1901 – 1946, Director, optics), WALTHER BOTHE (1913 – 1930, Member, radioactivity), WALTER NODDACK (1922 – 1935, Member, chemistry), and MAX VON LAUE (1925 – 1934, Consultant, theoretical physics). In addition to others, the following voluntary collaborators were working at the PTR: ERNST PRINGSHEIM (1893 – 1904, optics), JAMES CHADWICK (1913 – 1914, radioactivity), and WERNER KOHLHÖRSTER (1922 – 1930, ultraviolet radiation). In addition to EINSTEIN and DE HAAS, as a guest at the PTR there worked, for example: MARGARET E. MALTBY (1899 – 1900, electric conductivity), HANS LANDOLT (1910, chemistry), ROBERT POHL (1925, physics), IDA NODDACK (1929 – 1935, chemistry), ERIKA CREMER (1932 – 1937, low temperatures), and EDUARD GRÜNEISEN (1932 – 1933, elastic constants).

The precision measurements of the spectral energy distribution of the black-body radiation performed at the PTR by OTTO LUMMER and his coworkers provided for MAX PLANCK the essential basis for the discovery of his radiation law and, hence, for the discovery of the quantum theory. Also WIEN's radiation law originated from precision measurements performed at this important research laboratory.

6.2   Activities of NERNST at the PTR

In the letter of January 6, 1922 to ARRHENIUS mentioned at the beginning of this Chapter NERNST commented on the start of the following semester and the change in the office of the Rector: *"So it appears that until April 1 a successor for myself must be found."* [Zott (1996): 208], however, his wife and he himself *"at any rate will continue until the end (Oct. 15) in the role of the Rector with simple, but very many dinners!"* [Nernst (1922b)]. As his successor NERNST could have imagined physical chemists such as FRITZ FOERSTER, GEORG BREDIG, HANS VON WARTENBERG, MAX BODENSTEIN, MAX VOLMER, or RUDOLF SCHENK. Since *"Haber's Institute now also nearly became a Univ. Institute, i.e., the true physical chemistry is already represented"*, one could also think of *a "radioactive gentleman"* such as OTTO HAHN or KASIMIR FAJANS, among whom ERNST HERMANN RIESENFELD would further represent the physical chemistry as a section head in the *Bunsenstraße*. It cannot be understood why in this connection NERNST did not include also WILLY MARCKWALD, who had occupied himself successfully in the *Bunsenstraße* with radiochemistry.

Later ERIKA CREMER, who at the time was a student at the Institute of Physical Chemistry, described the situation caused by the departure of NERNST from the University: *"At the time none of the German physical chemists could have declined this offer. The* [appointment] *committee was confronted with a difficult task. ... In this case only God himself* could have the proper distance to make the correct decision. In this matter one went even as far as Sweden contacting Svante Arrhenius at the age of 63 at the time. He pointed the choice toward Bodenstein." [Cremer (1971): 966]. During spring of 1923 MAX BODENSTEIN took over the Chair and the Institute of NERNST at the University of Berlin.

Of course, already in 1922 NERNST had obtained his discharge document, with which the appointment as Honorary Professor at the University of Berlin was connected at the same time [UAHUB PhF: 353–354]. A ceremonial farewell colloquium was organized in the lecture hall in the Bunsenstraße (Fig. 6.2).

Fig. 6.2 Farewell colloquium in 1922 for WALTHER NERNST in the lecture hall in the Bunsenstraße (starting from the right, respectively: 1st row: (1) RIESENFELD, (2) NERNST, (3) MARCKWALD, (4) WALTER NODDACK, (5) PAUL GÜNTHER, 2nd row: (2) KURT BENNEWITZ, (3) JOHN EGGERT, (5) FRITZ BORN, 3rd row: (2) CLARA VON SIMSON, (4) KARL FRIEDRICH BONHOEFFER, (5) FRANZ SIMON).

When NERNST left the University his connections with the *Reichsanstalt* were not new any more. Since 1905 he had been an active member of the Advisory Board and, hence, he was well familiar with the operation of the PTR. Already in 1894 and then again in 1905 he belonged to the inner group of the candidates to succeed HERMANN VON HELMHOLTZ and of FRIEDRICH KOHLRAUSCH, respectively [Hoffmann (1990): 40]. The first case was likely connected with the effort to keep NERNST in Prussia after he had been offered the position of *Ordinarius* in Munich (see Section 4.6). At the time, on October 25, 1894 following the inquiry by FRIEDRICH ALTHOFF, EMIL FISCHER had given him the answer: *"Since primarily Nernst is a physical chemist, further since his strength is not the performance of measurements, and finally since according to my opinion it would be a disaster for the young man to occupy already now an administrative position like the Presidency of the 'Physikalische Anstalt', I answer your question with No."* [ZSAM: 7; Hoffmann (1990): 40].

In December of 1921 the question to offer NERNST the position of the President of the PTR as the successor of EMIL WARBURG, resigning because of his age, was pursued persistently by the Ministry of the Interior for the third time. At the first confidential discussion in the Ministry NERNST showed *"an inclination to accept the position of the President of the Physikalisch-Technische Reichsanstalt"*, he asked however, *"to keep the matter strictly confidential, until he will give an answer in the middle of January."* [ZSAP-N: 6]. Apparently, NERNST regarded the offered position very highly. This can be seen from a remark which, according to JOHN EGGERT, it is said he made *"in a small internal ceremony"* prior to his departure from the University: *"Up to now I have worked with older students; from now on, when the curtain rises for the final act of my life, I will have to deal with young 'Geheimräten'."* [Eggert (1943b): 49]. Therefore, after consulting with WARBURG, already on January 4, 1922 NERNST announced: *"With a deep feeling of thanks I am ready for the so highly honorable offer to accept the Presidency of the Physikalisch-Technische Reichsanstalt."* [ZSAP-N: 12]. Because of the not yet completed term as Rector, effective April 1, 1922 NERNST at first took over the new responsibilities only in an acting capacity. Then the official date of the beginning of his Presidency of the PTR was October 15, 1922, the day of the change of the Rector of the University of Berlin.

Soon it should become apparent, that this was not the beginning of the *"final act"* of the scientific career of the 58 year old *Geheimrat*. The primary reason for this was the fact, that the expectation to be able to perform and to supervise physical research at a high level at the *Reichsanstalt* was strongly disappointed. In Germany just in 1922/23 the strong inflation had caused abruptly an almost total devaluation of the German

Mark. The resulting severe economic crisis had the consequence that the President of the PTR to a high degree and nearly exclusively had to deal with organizational tasks and administrative issues. It became necessary to overcome the pressing financial problems which threatened the further operation of the PTR. Already in March 1922 the Advisory Board had noted that the budget of 9.7 million *Mark* for 1922 corresponding to that of 1921 would not be sufficient. Because of his exceptional talents and engagements in the organization of science NERNST was able *"to steer the 'Anstalt', suffering particularly because of the inflation, through the difficult time without getting jeopardized."* [ZSAP-P: 32], as it was noted on November 2, 1924, half a year after NERNST had resigned from the Presidency.

Immediately after assuming his new office the financial restrictions of the Government enforced by the inflation and by the endeavors of industry and economic associations confronted NERNST with the task to incorporate the *Reichsanstalt für Maß und Gewicht* (Imperial Institution for Measure and Weight), the former Imperial Normal-Calibration Commission, into the PTR [Stenzel (1976)]. In December of 1922 he submitted to the Parliament a statement of the PTR regarding this matter, after the Parliament had drawn up a memorandum on the question of the incorporation. There followed repeatedly hearings by the relevant committees of the Parliament and complicated negotiations about the details of the merging of both Imperial Institutions. During the summer of 1923 the Imperial Institution for Measure and Weight was incorporated into the PTR as Section I. In this process it lost about 40% of its previous staff. NERNST ensured that the Ministry recognized this loss as the rigorous reduction in staff requested from the PTR because of the inflation, and last but not least by means of this success, he spared the PTR additional essential reductions during these difficult times.

By taking over about 50 people from the Imperial Institution of Measure and Weight the number of employees of the PTR, which before amounted to about 160 (1920: 164), was even increased by more than 30%. NERNST had formulated the problem of activities at the PTR, connected with this increase of high importance to him, in a statement submitted to the Parliament already in December of 1922: Because of this enormous increase in the number of staff compared to University Institutes, it would be nearly impossible for the President to keep the overall view and to cooperate himself effectively. Immediately after assuming his new office in spite of the difficulties generated by the inflation, NERNST was able to succeed in getting the new position of a Head of the Administration approved. However, the resulting relief regarding the administrative work load did not mean that NERNST could devote himself to research activities more than in only a limited way.

When on January 4, 1922 NERNST expressed his agreement to direct the PTR he indicated his intentions connected with this decision: Since the *Reichsanstalt* more and more is looked at as an administrative bureaucracy, it would become important for him *"to carefully cultivate the academic spirit of the institution"* and to achieve *"a return to the tradition created by von Helmholtz"* [ZSAP-N: 8; Hoffmann (1990): 42].

This endeavor to strengthen again basic research should be served among others by the establishment of a photochemical laboratory and by the restructuring of the PTR according to the following scheme:

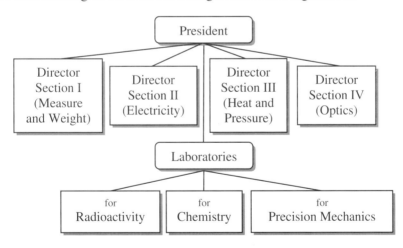

The subdivision of the Sections into scientific and technical Subsections was given up in favor of laboratories which were operating scientifically primarily according to thematically oriented aspects. In addition to the three laboratories listed above, the main library and the main machine shop were directly under the President.

Scientifically the Institution benefited from the fact that in 1922 NERNST brought his former student WALTER NODDACK to the PTR and put him in charge of the laboratory for chemistry. NERNST suggested to him to search for the still unknown chemical elements of the group 7 of the Periodic Table. In 1924 NODDACK's future wife, Ms. IDA TACKE, transferred from the AEG to the PTR, where she worked as a guest scientist. Already in 1925 both accomplished the discovery of the new element named rhenium by themselves in enriched products of gadolinite $Y_2FeBe_2[O|SiO_4]_2$ by means of X-ray spectrum analysis. They were supported by OTTO BERG employed as a physicist by the Siemens & Halske Company. Since 1928 in collaboration with WILHELM FEIT they were able to isolate the element rhenium also chemically. Furthermore, in 1925 NODDACK and TACKE claimed that by means of X-ray spectroscopy in enriched fractions of columbite (a series of mixed crystals: niobite $(Fe,Mn)(Nb,Ta)_2O_6$ – tantalite $(Fe,Mn)(Ta,Nb)_2O_5$) they had observed also the so-called ekamanganese, which they called masurium. However, since the observation could not be reproduced, the discovery was not recognized (however, see [Assche (1980): 214]: *"Noddack, Tacke, and Berg very clearly described the chemical and physical experiments leading to the observation of elements 43 and 75. The elements 43 and 75 were identified through L and K X-ray spectra, the analysing power of which does not leave much doubt on the very existence of these in their samples. ... These authors eventually discovered the last missing stable element (75) and more than 14 years before the discovery of fission itself, the first fission product ($^{99}43$)."* Only in 1937 in Palermo EMILIO SEGRÈ, winner of the Nobel Prize in Physics of 1959, and the mineralogist CARLO PERRIER in a molybdenum sample irradiated in Berkeley with deuterons detected the missing element between manga-

Group 7		
25	manganese	Mn
	ekamanganese	
43	masurium	Ma
	technetium	Tc
75	rhenium	Re
107	bohrium	Bh

nese and rhenium because of its radioactivity. They gave it the name technetium.

During the Presidency of NERNST at the PTR WALTHER BOTHE performed his experiments demonstrating the validity of the energy-conservation law in the case of elementary radiation processes, for which he received the Nobel Prize in Physics of 1954.

During spring of 1923 in his report to the Advisory Board NERNST had recommended to establish at the *Reichsanstalt* the position of a consultant for theoretical questions according to the requirements of modern physics. Subsequently, at the relevant Secretary of the Interior, KARL JARRES, he asked for providing the additional finances needed in this case. In fact in the same year MAX VON LAUE, Professor at the University of Berlin since 1919, could be won over as a part-time consultant in the area of theoretical physics. In 1933 VON LAUE played an important role in clarifying some issues in the field of superconductivity, which then led to the discovery of the MEISSNER effect by WALTHER MEISSNER and ROBERT OCHSENFELD.

Among the accomplishments of NERNST at the PTR in the area of the organization of science we may also include his effort to lead the institution out of international isolation due to the First World War and to pave the way in particular for a cooperation with the corresponding establishments in England and the USA. With this goal in mind, in January 1923 he had invited politicians and diplomats to visit the PTR. Therefore, on January 22 the *Deutsche Allgemeine Zeitung* (German General Newspaper) could report that *"among the guests the President Ebert, the American Ambassador"* had been seen, one had obtained some information *"about the relation between the Institution and the two large, similar Institutions in America and England"*, and had heard *"much welcome praise of the state of the German science expressed by all guests"* [Hoffmann (1990): 43]. With this effort NERNST had achieved an important step for the further development of the PTR.

At the end of February 1924 NERNST submitted his resignation to the Secretary of the Interior KARL JARRES. Being nearly 60 years old at the time, among other reasons he justified this step by the fact, *"that during these in particular for our Institution difficult times it appears more appropriate, if another suitable, perhaps younger person will take my position. Because these circumstances have meant that presently a large number of organizational measures must be carried out by the President of the Institution, on which, at least for the next years, he must absolutely devote his full attention, if proper justice should be given to the matter. I myself must fear that most likely to a large extent this task would absorb the last force I can muster, if I personally do not want to despair completely on coping with it. However, I believe that generally one will understand, and in this case I could appeal to numerous testimonies of outstanding colleagues, if I prefer to use the time for work remaining available to me more for scientific tasks including the completion of new editions of my books."* [ZSAP-N: 55–56]. Even the urgent plea by JARRES and the hardly realistic promise to alleviate the difficulties, which had led to hindrances affecting the work of the President, could not change NERNST in his decision such that effective May 1, 1924 he resigned from the office of the President of the PTR in order to accept again a position as *Ordinarius* at the University of Berlin. Still in June in a Parliament debate the Ministry of the Interior was reprimanded, because one had been unable to keep an outstanding scientist such as NERNST at the PTR due to bureaucratic reasons, which further restricted its financial options, and led to an unbearable burden with administrative work.

Only after more than half a year, during which LUDWIG HOLBORN served as acting head of the Institution, a successor of NERNST had been found in FRIEDRICH PASCHEN, who took up this office on November 1, 1924. HOLBORN had worked at the PTR since 1889 and since 1914 as Director of the Section dealing with heat and pressure. As possible successors of NERNST initially JONATHAN ZENNECK, WILHELM WIEN, and MAX WIEN were discussed, however, they either declined the offer or they were not accepted by the Advisory Board of the PTR or by the Secretary of the Interior. One had also considered offering the position to a person from the ranks within the PTR, in which case HOLBORN and ERNST GEHRCKE were eligible. Also MAX PLANCK had indicated that he would accept the office of the President, however, finally he remained at the University of Berlin.

NERNST had strongly supported the PTR not only during its search for a new President, but also by means of his activity in its Advisory Board to which he belonged. In this case, in particular, he pushed the idea that the scientific research at the PTR would be expanded and strengthened. So in 1927 he suggested investigations of ferromagnetic materials at low temperatures, which should be beneficial for the generation of extremely high voltages for the splitting of atoms. 1930 in the Advisory Board he discussed the necessity to create a temporal standard better than the existing astronomical one. This could be accomplished by the use of quartz resonators. Two years later he recommended the establishment of a laboratory of optoacoustics and electroacoustics with the justification: *"If the Reichsanstalt would want to be complete, then it must arrange for such a laboratory."* [ZSAP: 23b].

Regarding more general problems of the PTR NERNST tried to strengthen the relations with the Universities and with their alumni, in order to reduce in this way the bureaucratic trends of the *Reichsanstalt*. By the end of the 1920s in the so-called New-Building Committee, he supported an extended new-building and renovation program. Regarding the question of the patent rights of the members of the PTR, he was of the opinion that these belonged to the jurisdiction of the Institution only in the case if the content of the patent was connected with the official business of the concerned employee. NERNST has also been an active member of the Personnel Committee which exercised a consulting function in all important decisions regarding the personnel.

As a member of this committee, in 1933 at the meeting of the Advisory Board NERNST argued against the procedures of the Government in connection with the appointment of a new President and against the fact, that *"the former President F. Paschen was practically expelled from his office"* [Hoffmann (1990): 44]. In fact, on May 1, 1933 PASCHEN was prematurely dismissed from this office. On March 8, 1933 he had seen to it that a swastika flag hoisted on the roof of a building of the PTR was taken down again, and on the following day he had protested against the flag being run up again. As a result the operating Nazi unit at the PTR wrote a letter to WILHELM FRICK, the new Secretary of the Interior, in which they indicated, that *"already since many years a selection in democratic direction is carried out. The operating unit is convinced that in connection with the new appointment at the position of the President you, dear Mr. Secretary, will take care of this matter."* [ZSAP: 1]. On April 21, 1933 at the order of the Ministry of the Interior ALBERT EINSTEIN had been removed from the Advisory Board.

For the new President, JOHANNES STARK, well known for his nationalistic and anti-Semitic attitude, the behavior of NERNST at the meeting of the Advisory Board in 1933 must have provided a fresh reason for the highest displeasure against the celebrated scientist, who had not only brought EINSTEIN to Berlin, but who also accepted and defended his and other "Jewish" theories. STARK indeed was determined to "take care of this matter" in the demanded sense. Already in June 1922 in a letter to the Ministry of the Interior STARK had expressed his displeasure about the appointment of NERNST as President of the PTR, which in an immensely exaggerated way he also implied for the whole community of the German physicists. Still in the same year in an insulting manner he informed the public about his criticism in his document *"dedicated to the German physics during the present crisis"*. The fact that exactly in 1924 NERNST obtained the Chair of Experimental Physics at the University of Berlin, STARK certainly considered to be an affront against himself, since he had also applied for this position. At any rate, subsequently STARK referred to NERNST who was ten years older as his *"personal enemy"* [Stark (1987): 83]. Of course, President STARK could not allow that his Institution would officially congratulate such a predecessor at his 70th birthday. Soon after STARK assumed office, NERNST's influence upon the matters of the PTR vanished, since in establishing the Nazi principle of the leader (*Führerprinzip*), in addition to the Full Assembly the former also abolished the Advisory Board. The same happened also in 1934 to the accomplishment initiated by NERNST and realized under PASCHEN: the position of a consultant in the area of theoretical physics, occupied by MAX VON LAUE, was cancelled.

Chapter 7

Professor of Experimental Physics at the University of Berlin (1924 – 1933)

In order to be active again as a researcher and teacher in the area of the natural science, WALTHER NERNST was much interested in the Chair of Experimental Physics, which already 30 years before had been a possibility for him as the successor of the late AUGUST KUNDT. At the time in Göttingen one did not want to lose the *Extraordinarius* at the age of thirty years to Prussia. Now since the death of HEINRICH RUBENS in the year 1922 this Chair had remained unoccupied, and ARTHUR WEHNELT was the acting Director of the Institute. The application of the disappointed PTR President NERNST must have been quite welcome to the Philosophical Faculty, because as successor of RUBENS certain nationalistic groups represented by ERNST GEHRCKE had favored PHILIPP LENARD and JOHANNES STARK, of course with their approval. In the files of the archive of the Humboldt University in Berlin one can find the statement of the appointment committee: *"After lengthy consultations the Faculty refrained from putting the Nobel laureates P. Lenard and J. Stark on its list, because due to their passionate and not always objective opposition against the modern theoretical physics these important scholars would jeopardize the fruitful cooperation among the physicists in Berlin. If the submitted list would not lead to a result, the Faculty asks to be heard again."* (quoted after [Haberditzl (1960): 412]). WALTHER NERNST had been placed at the top of the list. This rejection of LENARD and STARK and the proposal to appoint NERNST was signed by the physicists MAX PLANCK, MAX VON LAUE, and ARTHUR WEHNELT, by the chemists MAX BODENSTEIN, WILHELM SCHLENK, and ALFRED STOCK, by the mathematicians ERHARD SCHMIDT, and RICHARD VON MISES, as well as by FRITZ HABER and HEINRICH JOHANNSEN.

The appointment of NERNST took place already on May 3, 1924. This now really meant the start of the *"final act"* in his academic career. At the age of nearly sixty the scholar had joined the series of the Directors of the Physical Institute of the University of Berlin boasting the names of outstanding physicists, as is shown by the list on the left.

Director of the Physical Institute	Term as Director
PAUL ERMAN	1810–1851
GUSTAV MAGNUS	1851–1870
HERMANN VON HELMHOLTZ	1871–1888
AUGUST KUNDT	1888–1894
EMIL WARBURG	1895–1905
PAUL DRUDE	1905–1906
HEINRICH RUBENS	1906–1922
WALTHER NERNST	1924–1933

Fig. 7.1 WALTHER NERNST during a physics lecture in the Physical Institute of the University of Berlin.

The temporal sequence of the directorships of NERNST at University Institutes in Berlin was opposite to that of their establishments. In Section 5.3 we have mentioned already, that the building in the *Bunsenstraße*, containing the Physical-Chemical Institute directed by NERNST from 1905 until 1922, had been planned as an extension of the ensemble consisting of the Physical Institute and the attached Director's villa (Fig. 5.2). Since 1924 for nine years this building became the location of the professional activities and of the private home of NERNST.

At this point we briefly cover the history of this building which turned out to be so important for the history of physics. Already at the end of January 1867, at the relevant Prussian Ministry of Education and Cultural Affairs, GUSTAV MAGNUS had called for a newly to be created Institute building, in which the instruments and the laboratory of the Physical Institute, placed so far in separate locations, could be accommodated in the same building. In addition, the building was to have two lecture halls for physics lectures including rooms for preparations, several work rooms, as well as the space for the official residence of the Director. In spite of the principal agreement by the Ministry, the negotiations took a long time and had not reached a result when MAGNUS died on April 4, 1870.

In connection with the offer from the University of Berlin, HERMANN VON HELMHOLTZ had expressed the condition that a dedicated building for the Physical Institute as requested by his predecessor MAGNUS must be erected. Since this request and the appointment of HELMHOLTZ on February 13, 1871 took place during the period of the German-French War (July 1870 until May 1871), the command of which on the German side had been taken up by Prussia, the realization of the building had to be postponed for the time being due to financial reasons. After the War, EMIL DU BOIS-REYMOND, who was interested in a new building for the Physiological Institute immediately next door to the new Physical Institute, was able to persuade the Minister of War ALBRECHT VON ROON to transfer an area of 7760 m^2 to the Ministry of Education and Cultural Affairs. Within this area between the *Neue Wilhelmstraße*, the *Reichstagufer*, the *Schlachtgasse*, and the *Dorotheenstraße* there were the workshops for the artillery. In the northern half located at the *Reichstagufer* the Physical Institute including the Director's villa (1873 – 1878) and subsequently the 2nd Chemical Laboratory and the Technological Institute (1879 – 1883) could be built for the Philosophical Faculty (Fig. 5.2). On the other hand, in the southern half located at the *Dorotheenstraße* the Physiological Institute including the official residence of the Director (1873 – 1877) and the Pharmacological Institute (1879 – 1883) were built for the Medical Faculty (Fig. 7.2).

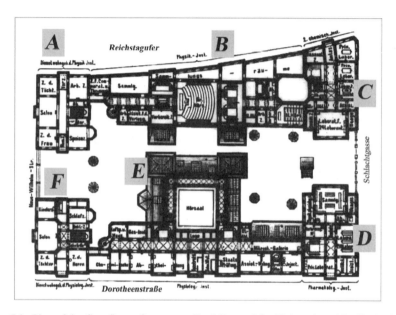

Fig. 7.2 Plan of the first floor of a group of buildings of the University of Berlin (1896): (A) Director's residence of the Physical Institute, (B) Physical Institute, (C) 2nd Chemical Laboratory (and the Technological Institute in the basement), (D) Pharmacological Institute, (E) Physiological Institute, (F) Director's residence of the Physiological Institute.

In November 1872 the first plans for the building of the Physical Institute discussed with HELMHOLTZ were submitted by PAUL EMMANUEL SPIEKER to the Ministry. At the beginning of April 1873 after many changes the complete project was finalized. Under the supervision of SPIEKER and after 1874 of OTTO FERDINAND LORENZ the construction was carried out by FRITZ ZASTRAU and MORITZ HELLWIG. *"During spring of 1878 it was completed far enough, that the move out of the old space in the University building could take place. ... After its completion among the German teaching institutes of similar kind the Physical Insti-*

tute in Berlin was the largest and the most beautiful." had been stated by HEINRICH RUBENS 31 years later [Rubens (1910): 284–285].

Compared to the Physical-Chemical Institute, now NERNST had to direct a large facility for research and for teaching. On July 28, 1926 in this context he wrote to ARRHENIUS: *"The day after tomorrow in this S.S.* [summer term] *I have my last lecture Experimental Physics I (five hours per week). Certainly, this lecture for beginners is somewhat craftsman-like, however, a 'craft not without a golden base' and above all for me not uninteresting. Though admittedly, since I have about 40 PhD students in the laboratory, it is high time for me to take a vacation."* [Zott (1996): 215].

On the other hand, a large part of the work within the Institute had to be carried out by coworkers. To these belonged ARTHUR WEHNELT, PETER PRINGSHEIM, WILHELM ORTHMANN, as well as the students of RUBENS, MARIANUS CZERNY and GERHARD HETTNER, a son of the mathematician GEORG HETTNER, the whose lectures NERNST had attended during his winter-term 1883/84 in Berlin. The support by these people and other capable scientists made it possible for NERNST to pursue his own research interests, as will be discussed in the following.

7.1 Solutions of Strong Electrolytes

Except for the two years of his activity as President of the PTR, since his time as an assistant in Leipzig NERNST had always occupied himself with problems in the field of electrochemistry. His former student JAMES R. PARTINGTON listed the research subjects which followed after the derivation of the NERNST equation: *"Other of Nernst's researches on electrochemistry are on decomposition potentials, the residual current, a theory of contact potentials, the theory of the lead accumulator, the determination of the ionic product of water by the acid-alkali cell, the electromotoric activity and deposition of alloys, and the diffusion of electrolytically deposited hydrogen through platinum and palladium, and the proof that a palladium electrode becomes charged with hydrogen in a solution of reducing agent, so behaving as a hydrogen electrode. He devised a sim-*

ple apparatus for demonstrating the migration of a colored ion in electrolysis by the moving-boundary method, and an apparatus for measuring transport numbers by the Hittorf method. He proposed a method for measuring the hydration of ions from transport measurements in solutions containing an indifferent solute. A Wheatstone bridge with condensers in the ratio arms was used to measure the internal resistance of a cell. Nernst's theory of the dropping mercury electrode was based on that of a concentration cell, and Palmaer proved experimentally that the mercury drops cause changes of concentration of mercury ions in solution. ... " [Partington (1953): 2858].

Also as *Ordinarius* of Experimental Physics NERNST returned once again to electrochemistry and to the theory of electrolytic solutions. In this case it concerned again a problem of the strong electrolytes. Already about a quarter of a century earlier, in 1901, in connection with concentrated solutions of this class of substances he had entered a polemic with PHILIPP ABRAHAM KOHNSTAMM, ERNST COHEN, ARTHUR A. NOYES, and above all with SVANTE ARRHENIUS. For a longer period this even led to a serious break from his Swedish friend. It started with a paper by ARRHENIUS dealing with the dissociation of strong electrolytes [Arrhenius (1901a)]. In a letter of November 19, 1900 to WILHELM OSTWALD, ARRHENIUS had foreseen a possible annoyance: *"I informed Jahn about the matter in advance, he answered quite superciliously, as a true 'Berliner' gradually became used to. ... If there would exist the smallest error in my calculations, the three interested people (Jahn, Planck, Nernst) would attack me and would beat me up quite severely."* [Körber (1969): 160].

However, NERNST reacted surprisingly strongly, which is demonstrated – according to OSTWALD's opinion – by his *"very unpleasant reply"* [Nernst (1901a)] and by his remarks in a letter to OSTWALD of February 19, 1901: *"I would not have expected that A*[rrhenius]. *could get so excited to produce such outrageous nonsense; apparently, he has forgotten the most elementary principles of the theory of solutions. I regret this in his own interest; ... However, what really has annoyed me, that is his method to assign the theory of the electromotoric action of the ions now to Helmholtz as much as possible. I look at this as an act of the most pronounced hostility from his side and I have already formulated my answer correspondingly."* [Zott (1996): 142]. The fact that in addition to the technical discrepancies there also was another reason for a more profound animosity, was indicated by NERNST in another letter (of February 24, 1901) to OSTWALD: *"The reason of the animosity of A. against myself is, ..., as far as I know, the following. Because of much evidence, I am strongly convinced, that in Sweden regarding the estate of Nobel dirty cheating is exercised by having the money allocated to quite different purposes (institutes, etc.) instead to its true destination, and that by now the prizes are held back 5 – 6 years."* [Zott (1996): 145].

Also HANS JAHN reacted upon the controversial paper in a reply [Jahn (1901)]. Therefore, ARRHENIUS felt for himself the need to compose [Arrhenius (1901b)] *"an answer to the peculiar attacks by N. and J."*, as he remarked on April 25, 1901 [Körber (1969): 165]. He believed that this could end the dispute. NERNST published another paper on the theory of concentrated solutions [Nernst (1901b)]. When on July 24, 1901 he sent this manuscript to OSTWALD, he remarked: *"That a fellow such as Arrhenius (apparently due to complete laziness) must degenerate to such silly prattle as indicated by his two notes!"* [Zott (1996): 149].

When in the 1920s NERNST turned again and for the last time to the solutions of strong electrolytes, the intention was to look into an emerging problem, but not to enter a polemic with other scientists. The starting point was the theory of the solutions of strong electrolytes and of the interaction between their ions created since 1923 by PETER DEBYE and ERICH HÜCKEL.

The DEBYE-HÜCKEL theory explains the deviation of dilute solutions of strong electrolytes from the behavior of ideal solutions, which is manifested, say, in the statement *"The strong electrolytes do not obey the law of mass action."* [Eggert (1941): 498]. The key element is the fact that the electrostatic interaction between the ions is taken into account, which is missing by definition in ideal solutions similar to any other interaction. This concept was justified by weighty arguments, which had been found already by different scientists, but in particular by NIELS BJERRUM, since strong electrolytes disintegrate much stronger into ions than expected from the classical theory of ARRHENIUS and in many cases these are practically completely dissociated as has been noted already in 1904 by ARTHUR A. NOYES.

In addition to this assumption of the complete dissociation, the theory of DEBYE and HÜCKEL is based on two additional prerequisites. The ions are treated as non-polarized point charges with a spherically symmetric field, and in addition to the COULOMB forces all other intermolecular forces are neglected. These assumptions restrict the applicability of the theory to highly dilute solutions.

In this theory each ion is surrounded by a cloud or atmosphere of ions, the total charge of which in magnitude is equal to and by its sign it is opposite to that of the central ion. The consideration of the charge distribution within this cloud allows to calculate the electrostatic interaction potential and to derive an expression for the inter-ionic interaction energy. Because of this energy the ions cannot move perfectly freely in the solution. One finds that all properties depending on the ion concentration, like the electrolytic conductivity and the osmotic pressure, have a smaller value than that corresponding to the chemical-analytical, i.e., to the true concentration. Hence, at least in the limit of very dilute solutions the theory of DEBYE and HÜCKEL could explain how the COULOMB fields of the free charges qualitatively and quantitatively influence these properties.

However, in spite of many experimental confirmations of the theory, there were also a few shortcomings. One of them showed up in the consideration of the energetic behavior of the electrolytic solutions. For example, for the electric work W_e needed in order to dilute one mole of a 1-1-electrolyte (as, for example, NaCl or KNO_3) starting from the concentration c mol·l^{-1} by adding an infinite amount of solvent, one can derive the relation

$$W_e = -N_L^3 e_0^3 \varepsilon^{-1} \sqrt{(1000\varepsilon kT)^{-1}\pi \cdot (\Sigma_i n_i z_i^2)^3} \cdot c = -B\sqrt{c}$$

where e_0 denotes the electric elementary charge, k BOLTZMANN's constant, N_L the LOSCHMIDT number, ε the dielectric constant of the solvent,

T the absolute temperature, n_i the amount of the ion i, and z_i the number of charges of the ion i. By means of thermodynamic calculations one finds the heat of dilution Q_d

$$Q_d = W_e \cdot \left(1 + \frac{T}{\varepsilon}\frac{d\varepsilon}{dT}\right) = -B \cdot \left(1 + \frac{T}{\varepsilon}\frac{d\varepsilon}{dT}\right) \cdot \sqrt{c},$$

which can be measured by calorimetry. Since $W_e < 0$, we must have also $Q_d < 0$, if ε is independent of the temperature (dε/dT = 0). However, in the case of water at 291 K one finds $-(d\varepsilon/dT)|_{291} > (T/\varepsilon)|_{291}$ and, hence $Q_d(291\,\text{K}) \propto +\sqrt{c} > 0$. This result *"contradicts a superficial consideration"* [Eggert (1941): 520].

Such deviations from the experience could originate from the assumptions of the DEBYE-HÜCKEL theory. In particular, several scientists considered it possible, that also in the case of strong electrolytes neutral or other complex composites of ions certainly could exist in the solution in very small amounts and perhaps also in higher concentrations. The only point was to determine them experimentally beyond any doubt. In 1926 NERNST took up this problem by suggesting to his collaborator WILHELM ORTHMANN to carry out precision measurements of the heat of dilution of strong electrolytes. As we have just explained and as noted also by BJERRUM in 1926, these were well suited for a comparison with experience.

For the measurements a differential calorimeter was used. This consists of two calorimeters ideally being equal, such that no temperature difference develops between them if the same amount of heat is introduced to both. Therefore, with this instrument one can measure very accurately the values of the reaction heat, if the reaction takes place in one calorimeter and if a well known amount of heat, say, by means of electric heating, is introduced to the other, exactly corresponding to the reaction heat in the case of equality between the temperatures. Using this technique, NERNST and ORTHMANN could improve significantly the calorimetric method. 1-1-electrolytes were investigated, the heat of dilution of which was determined as a function of the concentration. It was found that the relation $Q_d(291\,\text{K}) = \text{const.} \cdot \sqrt{c}$ shows the theoretically expected positive sign starting with concentrations below 0.1 mol·l^{-1} and displays also the theoretically calculated value only in the case of higher dilution.

In 1927 the measurements were extended to different temperatures. By using thermodynamics NERNST concluded from the obtained results, that a term must be contained in the quantity Q_d corresponding to the energy of the splitting of the molecules during the dilution of the solution, the molecules having been non-dissociated before. Hence, this part is based on a chemical process, whereas the other is caused by the removal of charges in a dielectric during the dilution and is connected with a physical process. In 1927, instead of the still non-dissociated molecules BJERRUM had also assumed associated ions, with a larger percentage fraction in the solution of a given concentration, the smaller was the distance between their centers. Similar to LARS ONSAGER in 1927 and to other scientists also NERNST could specify the "true" degree of dissociation of strong electrolytes.

NERNST and ORTHMANN reported the results of their studies at the Academy of Berlin [Nernst and Orthmann (1927); Nernst (1928a)] and at the Bunsen-Society [Nernst (1927)], and, of course, had published them in the *Zeitschrift* [Nernst and Orthmann (1928); Nernst (1928b)]. Later LUDWIG EBERT said about the last investigations carried out by NERNST in the field of electrochemistry: *"Probably the fraction of the free charges was underestimated by Nernst. However, for each attempt to solve the question, how the effect of the forces of the ions and the influence of the equilibria in solutions of strong electrolytes superimpose themselves, the gained experimental data will remain of a lasting value, in particular if both factors are to be treated as a function of the temperature."* [Ebert (1943): 265].

7.2 Vibrating Strings and the Neo-Bechstein Grand Piano

Toward the end of his term as Director of the Physical Institute and also beyond NERNST occupied himself once again as design engineer and inventor. However, in this case this new invention was a *"leisure-time occupation, a 'physique amusante' as he called it."* [Bodenstein (1942a): 101]. We refer to the Bechstein-Siemens-Nernst- or the Neo-Bechstein-grand-piano, an electrophone, which again can be considered as a pioneer achievement within the history of the development of this musical instrument. *"Still today it is mentioned even in very brief summaries of the history of the musical instruments, because it can be considered as a typical example of a historical period of this development."* [Herrmann (1972): 41]. For example, one can read in an encyclopedia of music: *"Also the Neo-Bechstein-grand-piano constructed by W. Nernst in 1930 belongs to the group of electric musical instruments with electric sound pick-up, in which case the sound not only continues to ring arbitrarily long with the selected intensity, but can also be varied within modest limits regarding its dynamics and timbre. If nevertheless the Neo-Bechstein-grand-piano did not gain general acceptance, this indicates that the function of the electric musical instruments lies in another area."* [Seeger (1966): 259–260].

Due to his occupation with musical instruments and with the question of their construction, WALTHER NERNST belonged to a series of natural scientists, who had already made remarkable achievements in this area. At first, one would think of ERNST FRIEDRICH FLORENZ CHLADNI, who in 1790 created his *Euphon* based on rod vibrations and in 1800 the *Chlavicylinder*, but also of BENJAMIN FRANKLIN, who in 1761 developed the glass harmonica, a friction instrument, which became known in Europe at least because of the *"Adagio & Rondo for glass harmonica, flute, oboe, viola, and cello, in C major"* (KV 617) by WOLFGANG AMADEUS MOZART.

The invention of NERNST concerns the modern class of the electronic musical instruments or electrophones, which can be divided into two groups. In the case of the instruments of the first group (A) vibrations are generated mechanically by means of strings, tongs, plates or else, which are transformed into electric oscillations by a microphone, a mechanical reproducer, or a photocell. The electric oscillations are processed further electronically. The other group (B) contains instruments in which the vibrations are generated electronically and which do not have any mechanically vibrating parts except for the key- and the switch-mechanism and the membrane of the loudspeaker (Fig. 7.3).

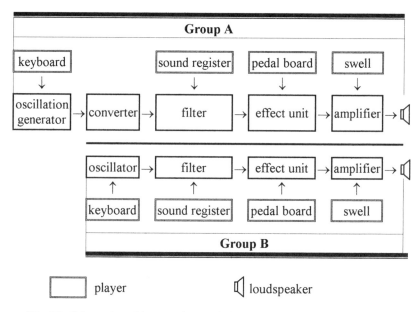

Fig. 7.3 Schematics of the two classes of the electronic musical instruments.

Sometimes previously the electric and the electronic musical instruments were distinguished from each other (for example [Seeger (1966): 259–260]), in which case NERNST's grand piano must be counted among the former. Today the definition can be used *"All instruments are denoted as electrophones, in which electric processes play a role."* [Enders (1988): 158], in contrast to the mechanical or the acoustical musical instruments (aero-, chorda-, idio-, membranophone). In this case the notation 'electrophones' does not apply exactly to the electromechanical instruments (group A). They are a mixture between both groups. Hence, the Neo-Bechstein-grand-piano is a chorda- (string instrument) and an electrophone at the same time.

At the beginning of the 20th century a few composers had certain visions about the creation of music which could only be generated electronically. To these musicians belong the Russian ALEKSANDR N. SCRIABIN and the American HENRY D. COWELL. In 1913 the Italian painter and futurist LUIGI RUSSOLO designed an instrument for the reproduction of environmental noise generated from noise and electronic "musical boxes". At this time, i.e., before NERNST was working on his invention, physicists and engineers developed the first electronic instruments commercially available, which belonged, however, to group B.

The keyboard-sphaerophone of JÖRG MAGER was one of the early instruments of this kind. In this case an electron tube in a low-frequency circuit was used for generating the electric oscillations. The sound frequencies were regulated by means of a keyboard, which activated different capacitors attached in parallel to the coil of the tank circuit. In 1920 also the Russian physicist LEV S. TERMEN fabricated his electronic musical instrument called *theremin* (*therminovox, termenvoska*) or initially *aetherophone* because of the use of radio waves. It was played without coming into contact with it by moving the hands in a weak electric field between two rod antennas. Although in 1934 EDGAR VARÈSE in his *"Equatorial for two aetherophones"* and in 1944 BOHUSLAV MARTINŮ in his *"Fantasia for aetherophone, oboe, string quartet, and piano"* have composed works for this instrument, and although it has been used often in film music, at the appearance of the first keyboard synthesizer this early form of music synthesizer was forgotten. This situation changed when, for example, ROBERT MOOG created modern *theremins*, in which the capacitors were replaced by transistors and which can be attached to a computer by means of a musical instrument digital interface.

Still before 1930 FRIEDRICH TRAUTWEIN constructed the *trautonium*, which enjoyed a wide distribution. In this instrument, glow lamps together with capacitors attached in parallel generate a relaxation oscillation rich in harmonics, the fundamental frequency and intensity of which can be varied by means of a band manual acting like a potentiometer. For charging the capacitor TRAUTWEIN had used the anode current of an electron tube and in this way had made it possible to control the sound frequency via the grid bias voltage of the tube. Subsequently, the instrument was improved more and more, since 1930 by the composer and physicist OSKAR SALA at the Music Academy in Berlin. He and many other well-known composers, including PAUL HINDEMITH, ARTHUR HONEGGER, and RICHARD STRAUSS, have created original works for the *trautonium* of SALA.

One should mention also the *Ondes Martenot* by the music teacher and radio engineer MAURICE MARTENOT, who created this electronic instrument in 1928 after he had been stimulated by TERMEN in 1923. The sound generation of this keyboard instrument is based on the principle of the heterodyne oscillator similar to the *theremin*. It was used, for example, by ELMER BERNSTEIN in the USA for film music, and also by EDGAR VARÈSE and ARTHUR HONEGGER.

Also some of the so-called organs belong to the electronic musical instruments of the group B such as that of LAURENS HAMMOND named after him and for which he applied for a patent in the USA in 1934. Within this instrument 91 metal gearwheels of different sizes, having a sinusoidal cog profile and being driven simultaneously by an electromotor, generate an induction current in little coils with a pin-shaped magnet core, the coils being permanently assigned to each gearwheel. The frequency of the induction current depends on the number of cogs and on the number of revolutions. Pressing a key acting as a switch results in an electronic amplification of the corresponding induced current, which is made audible via loudspeakers. Harmonics can also be added. Previously, these organs were used in music entertainment as well as in movies and the radio for the accompaniment with sound and for the generation of noise.

The instruments of group B are only poorly suited for imitating the tone of traditional instruments, since they only generate the harmonics, but cannot include in the imitation any non-stationary relaxation processes which are characteristic of the tone of the mechanical musical instruments. Therefore, it is the purpose of these electronic instruments to obtain new tone qualities and thereby to extend the possibilities of the musical expression. However, the early electronic instruments developed in the 1920s and the 1930s we had mentioned, primarily followed the example of the traditional musical instruments regarding the tone quality.

So the *trautonium* can imitate string and wind instruments as well as drums and castanets. However, the imitation did not justify the existence of these products as we have seen, such that in the end they did not find a wider distribution.

At this time when people experimented with high intensity on the construction of electronic musical instruments in many countries and especially in Germany, NERNST, of course, could not simply step aside from this modern trend, being a natural researcher. With the electromechanical string instrument (Fig. 7.4) created by him in 1930, in some sense he also started the group A of this class of instruments. In this case in terms of the tone qualities nothing had to be imitated. Instead, one could build on a tradition of sound generation which was several hundred years old. Therefore, the research had to concentrate only on the electronic means to which the sound generation was to be adjusted in order to achieve a continuing improvement of the traditional instruments. In this sense MAX BODENSTEIN made the following assessment on NERNST's invention: *"I feel that the Neo-Bechstein relates to the grand piano of today like the latter to the cembalo."* [Bodenstein (1942a): 100].

It had been the original goal of NERNST to build a radio piano, by which the music should be transmitted directly from a radio station, i.e., without an acoustic transmission by a microphone. In this case he wanted to utilize the experience and knowledge he had gained before together with his collaborator HANS DRIESCHER during studies for improving the loudspeakers. In 1929 during the work dealing with this piano, the young engineer OSKAR VIERLING changed from the Imperial Institute of Telegraph Technique (*Telegraphentechnisches Reichsamt*) in Berlin to the University in order to study physics. At the *Reichsamt* as an assistant of MAGER, VIERLING had worked on the electric amplification of pianos and violins together with HARALD BODE. In 1928 he had already formed the idea to record sound generated by the strings in the piano using microphones and, following its amplification, to make it audible in a loudspeaker. Therefore, he turned to NERNST at the University. In this way the latter heard also of developments carried out in Hungary by the physicist FRANKÓ. In this case, a traditional piano had been equipped with microphones in order to reproduce the sound in a loudspeaker. Therefore, during the construction of his musical instrument NERNST could not only utilize his own and DRIESCHER's experiences and patents, but also those of VIERLING and FRANKÓ. In 1930 at the Radio Station Hamburg one had also tried to feed the alternating currents, generated during the recording of the string vibrations of a piano using electric sound pick-up, directly to the transmitter, i.e., to realize NERNST's original idea. The acoustics expert FRITZ WILHELM WINCKEL, at the time still a student at the *Technische Hochschule* in Berlin-Charlottenburg, in 1931 emphasized the achievement of NERNST and DRIESCHER in connection with the creation of the Neo-Bechstein-grand-piano in a comparison with the earlier developments we have mentioned: *"However, then the problem was solved only half-way because the resonance filtering of the harmonics was missing. Only Geheimrat Nernst together with his collaborator Hans Driescher was able to create an instrument, which in every respect took into account the physical principles of the theory of vibrations."* [Winkel (1931): 841].

Regarding the acoustic experiments to be performed, in particular concerning the mechanics of the touch of the keyboard, NERNST had approached the piano factory C. Bechstein, because the latter had much experience with the fabrication of grand pianos. The factory had been founded by CARL BECHSTEIN senior in Berlin about 75 years earlier. The great success of the company was partly due to the fact that one had been able to create an instrument which well satisfied the requirements of the trend of the pianists regarding the touch of the keyboard which appeared together with the romantic piano music. In particular, this concerned the fact that the material did not experience fatigue and breakage, as it happened not infrequently in the case of the traditional instruments. In January 1887 HANS VON BÜLOW, who was a consultant of the company, presented the concert-grand piano built by the company to the public with the *"Piano Sonata in B minor"* by FRANZ LISZT, and the piano was considered a sensation. Subsequently, pianists such as BÜLOW, LISZT and others have brought BECHSTEIN's instruments to world fame. At the

World Exhibition of 1896 in Berlin CARL BECHSTEIN received a gold medal for his instrument.

Later NERNST's daughter EDITH VON ZANTHIER reported, that on the other hand her father was unhappy about the collaboration with the Bechstein Company because of its relations with the rising ADOLF HITLER [Zanthier (1973a)]. In fact, the son EDWIN of CARL BECHSTEIN senior and mainly his wife HELENE belonged to the earliest patrons of the later dictator. After the death of his father, EDWIN BECHSTEIN together with his brothers CARL and JOHANNES as a businessman had taken on the management of the company. Later he let himself being paid off; however in 1923 he bought himself into the company again when it became a joint-stock company. Since around 1924 the couple had supported HITLER with large amounts of money. HELENE BECHSTEIN, who had taught HITLER good manners and correct behavior, is said to have expressed: *"I would wish he would be my son."*

For the electrical part of his grand piano NERNST was able to interest the Siemens & Halske Company, which then became responsible for the complete electrical equipment from the microphones to the loudspeaker, and which after some negotiations in the end secured for itself the right of first refusal. Hence, one could observe: *"By means of contractual agreements with the companies Siemens & Halske A.G. and Telefunken G.m.b.H. as well as with Herrn Geheimrat Professor Dr. Nernst, the Bechstein Company has the exclusive rights for the whole world to exploit the patent rights applied for and obtained by Herrn Prof. Dr. Nernst and his collaborators for the fabrication of electric pianos."* [Bechstein (1931)]. In the newspaper *«Der Mittag – Düsseldorfer Stadtanzeiger»* (The Noon – City Advertiser of Düsseldorf) it was observed: *"The cooperation of three names with world-wide reputation: Nernst-Siemens-Bechstein in this case has created an instrument of high perfection."* [Bechstein (1931)].

Since the sound pick-up was done by means of microphones, in the case of their grand piano NERNST and DRIESCHER could do without the soundboard, i.e., without *"the soul of the instrument"*, which *"in its method of construction is based on the old experiences and the mysteri-*

ous art of the piano-maker without being accessible to a theoretical calculation." [Winkel (1931): 840–841]. In the case of the grand piano it was essential to touch the strings only very lightly, since vibrations become purer when the force of the touch was smaller. Furthermore, DRIESCHER had found out that in the case of a light touch the formation of the tone is particularly beautiful, since otherwise a faint crackling noise can appear due to the generation of disturbing induction currents in the microphone magnets. The use of electron tubes made it possible to amplify with exact proportionality the vibration amplitudes electrically converted within the microphones. As demonstrated by the experiments, under these conditions it was sufficient to use thinner and shorter strings than in the conventional grand piano, such that a cheap iron frame could serve for the stringing and with only 1.4 m the length of the instrument corresponded to that of a baby grand piano.

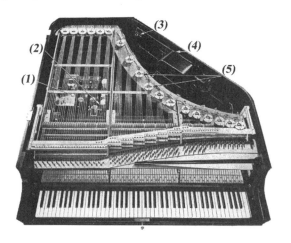

Fig. 7.4 The Neo-Bechstein grand piano in the opened state: (1) amplifier, (2) volume control, (3) record connection, (4) radio receiver, (5) microphones.

The light touch of the strings was achieved by means of so-called micro-hammers. Although DRIESCHER had patented several configurations for the touch of the strings, in the fabricated grand pianos only that shown schematically in Figure 7.5 was always used. Upon the touch of the key, the large hammer hits the touch strip. The micro-hammer is attached to the large hammer using a small leather band in such a way that it remains mobile and that its mass beats against the string fabricated from steel. This one and the other configurations of DRIESCHER allow the pianist to keep the technique of the touch he is used to. At the same time the knocking noise is suppressed which appears at the touch in the case of the conventional pianos.

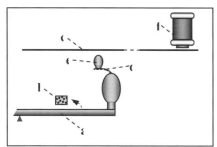 

Fig. 7.5 (left) Scheme of the key transmission in the case of the Neo-Bechstein-grand-piano: (a) lever arm of the key, (b) touch strip, (c) steel string, (d) leather band, (e) micro-hammer, (f) microphone; (right) WALTHER NERNST working on a single-string model of the Neo-Bechstein-grand-piano.

The microphones serving for the direct conversion of the mechanical vibrations generated in this way into electrical oscillations did not require a membrane. Only after it had been amplified, the alternating current generated directly in the magnet system of the microphones was converted into audible sound in the loudspeaker. It had turned out, that one did not need a microphone for each of the 87 strings, and that instead 18 were enough, under the magnet systems of which usually five strings were arranged together, respectively (Fig. 7.4). After its generation the alternating current passes through a system of capacitors and filters. Its adjustment allows – following an expression by NERNST – to deliver a special timbre by order of the customer by enhancing certain harmonics. In this way it was possible to improve significantly the quality of the high notes and of the bass, which had shown a lower quality in the grand pianos before. Finally the alternating current reaches the loudspeaker via a three-tube amplifier, the volume of which can be controlled to adjust the playing of the instrument to the specifics of the surrounding space.

Also the left pedal and a lever pull on the right-hand side of the keyboard carry new functions. Due to the absence of the sound-board, since the fading time of a tone is about five times longer than in the case of a regular grand piano, by utilizing the left pedal one achieves the effect of a harmonium. If the pedal is pushed gradually, one can cause a uniform rise of the tone, an effect which CHLADNI already tried to apply. Pulling the lever effects the application of a mechanical felt damping and, hence, it provides the possibility to play the instrument as a spinet. Because of the existing amplifier and loudspeaker, also a record player and a radio receiver could be connected to the grand piano (Fig. 7.4). Hence, one could play a part of an orchestra or a singing voice of a record and play oneself the solo part or the piano accompaniment on the instrument.

The final version of the grand piano we have described had a brief prehistory. An initial instrument for testing had been built by the E. Werner Company, before during the first months of the year 1931 the first Neo-Bechstein-grand-piano was fabricated according to the gained knowledge. Then NERNST suggested building a *"second instrument for testing"*, because he had noticed some flaws in particular in connection with the amplification. *"Concerning the amplifiers I feel that two output valves are not enough. ... It appears that for all further plans the possible suppression of the difference tones* [generated by the overdriving of the electron tubes] *represents the main subject regarding the construction of the amplifier as well as the selection of the best loudspeakers. We are not even sure if the difference tones really originate only in the output part, or not at least partly in the smaller electron tubes,"* NERNST had written to DRIESCHER on May 24, 1931 (quoted after [Herrmann (1972): 44–45]).

The development of the new instrument was followed with great interest by the amateur musicians MAX PLANCK and ALBERT EINSTEIN as well as by the philosopher and psychologist CARL STUMPF known among other things because of his *"tone psychology"*. Also NERNST's wife was asked several times to judge the tone qualities of the grand piano under development.

As DIETER B. HERRMANN had found out [Herrmann (1972): 46], on August 25, 1931 the Bechstein Company called a press conference, at which the completed Neo-Bechstein-grand-piano was presented and was introduced likely by MAX NAHRATH with a sonata of BEETHOVEN. As can be seen from a text prepared by the Company, the acceptance was overwhelming [Bechstein (1931)]. For example, one could read in the *Königsberger Hartungsche Zeitung* (newspaper): *"Since the invention of the Hammerklavier* [pianoforte] *two hundred years ago at any rate in the area of the piano fabrication nothing has been constructed, which is similarly suited to cause a complete change and to expand the expressionist range."* [Bechstein (1931)]. The important musicologist ALFRED EINSTEIN wrote: *"If yesterday E.T.A. Hoffmann, the 'Ghost-Hoffmann', could have descended again from the heaven of music upon the Earth as a guest and during his walking tour across the native Berlin in the house of C. Bechstein at the Zoo* [Budapester Straße 9a] *could have attended the first presentation of a new instrument, he would have found a dream fulfilled: an instrument joining the vox humana with the vox mundana, an instrument, which at the same time is a carrier of art and is also physical, alive, and mechanical, an instrument uniting in itself piano, radio, and voice apparatus. ... The surprising, important, lasting: that primarily it is a piano, and at that a good and cheap instrument. ... Nobody who was listening yesterday could have avoided the impression that an event of perhaps immeasurable importance has happened for the piano fabrication."* [Einstein, Alf (1931)].

In fact, because of the savings mentioned (absence of the soundboard, iron frame for the stringing, etc.) at the price of 2800 *Reichsmark* the instrument was cheaper than the smallest conventional baby grand piano. With his invention NERNST had wished to create an instrument for music at home at a reasonable price, as told by his daughter EDITH: *"With this he hoped to be able to provide pleasure to many families."* [Zanthier (1973a)]. However, he applied his brilliant professional and organizational capabilities with strong engagement for constructing this instrument. Expressions such as *"leisure-time occupation"* and *"physique amusante"* [Bodenstein (1942a): 101] are due to the specific custom of NERNST to indicate often exactly the opposite of that which is really meant. However, certainly it had not been his declared goal to create an instrument which would be rated as a milestone in the history of the musical instruments, as it was described, for example, by MAX BODEN-

STEIN, ALFRED EINSTEIN, or by news items such as *"The new miraculous instrument should start a renaissance of the piano."* (*8 Uhr-Abendblatt* (8 O'clock Evening Paper), Berlin), *"Because of the Nernst grand piano we are advanced not only a few steps, but at once several kilometers. Therefore, the Nernst grand piano will cause not only a revolution in the area of the construction of instruments, but also in the area of the composition of piano music."* (*Kölnische Zeitung* (Cologne Newspaper)) [Bechstein (1931)], and others.

Also in the following years the Neo-Bechstein-grand-piano generated some interest of noted musicians. To these belonged the German composers EUGEN D'ALBERT and MAX VON SCHILLINGS, the Czech composer ALOIS HÁBA, and the American conductor BRUNO WALTER. As a conductor MAX VON SCHILLINGS was known as a patron of modern music, and D'ALBERT even predicted a future within music for NERNST's instrument.

Among the electronic musical instruments of the early period NERNST's grand piano was the most successful economically because it was produced only as a series with 150 copies, of which immediately after the announcement of the instrument ten copies were bought by the American entertainment industry. Also the radio stations used the grand piano for some time. Still in the 1970s the Danish-German tenor HELGE ROSVÆNGE used the instrument during singing instruction because of the possibility to control the volume. However, compositions for the Neo-Bechstein-grand-piano have not become known.

Today functioning Neo-Bechstein-grand-pianos from the 1930s can be found only in the Museum of Musical Instruments in Berlin and in the Collection of Musical Instruments of the Museum of Cultural History in Vienna. However, the Japanese company Yamaha with its headquarters in Hamamatsu, which was founded in 1887 by TORAKUSU YAMAHA for the fabrication of musical instruments, and which is famous today as a producer of motor-cycles, engines, and electrophones, is the only company manufacturing three electronic instruments, the forerunners of which were developed in the 1930s. To these belong the Neo-Bechstein-grand-piano, which, hence, has not been pushed out by the electric organs and the synthesizers like almost all other electrophones of that time.

As in the case of his other great invention, the light bulb, NERNST took his occupation with the construction of the grand piano as an opportunity for research in the natural sciences. This is demonstrated, for example, by the research of SIEGFRIED SAWADE, which was suggested and supported by NERNST and was carried out at the Physical Institute from 1930 until 1932, and which was concluded in 1933 by the dissertation [Sawade (1933)]. In this case one had started from the point that there did not exist a perfectly satisfactory theory of the vibrating string, since all theories developed before were based on assumptions, which were violated in practice, such as the assumptions about the hammers touching the string. In his treatment of the associated problems SAWADE primarily investigated the behavior of the micro-hammers, which were used in the Neo-Bechstein-grand-piano. In order to observe the behavior during the touch he used spark photography. During his experiments SAWADE focused on the influence of the mass of the hammer, the mass of the string, the location of the touch, and the surface of the hammer upon the time during which the hammer is in contact with the string. He also studied the energy transfer during the touch with a soft hammer. Furthermore, by means of the recording of the decay process of the sound using an oscillograph he could clarify the main differences in the tone quality between conventional and electronic pianos. Finally, SAWADE treated the question, how within NERNST's grand piano the missing damping effect of the soundboard can be substituted by artificial additional damping.

We should mention that also OSKAR VIERLING, who had carried out important preliminary work regarding NERNST's invention and in 1931 had created a separate instrument with his *Elektrochord*, obtained his PhD in 1935 with a thesis on *"The Electroacoustic Piano"* [Vierling (1936)]. Later he himself had indicated that he got his PhD with NERNST, although at that time the latter had already retired. Subsequently, OSKAR VIERLING represented his field in several positions as Professor. In 1941 he founded an electrotechnical Company that continues too being renowned today.

7.3 Studies in Cosmology and Astrophysics

Still up to the last years of his life WALTHER NERNST devoted himself to an additional research area: cosmology and astrophysics. Already a long time before him his friend SVANTE ARRHENIUS had worked in this field. In 1903 the latter had written the first textbook on Cosmic Physics, at the outset in the German language in order to achieve a larger distribution [Arrhenius (1903)]. ERNST HERMANN RIESENFELD, the brother-in-law of ARRHENIUS, expressed the opinion: *"In this way cosmic physics was established as an independent science. The large upturn, which this science had experienced in our century, not in the smallest degree is a result of the investigations by Arrhenius and of the writing of the first textbook for this discipline."* [Riesenfeld (1930): 27].

With his research in the field of cosmology NERNST followed a number of important scientists who had occupied themselves before him with the science of the origin, the development, and the overall structure of the universe. NERNST himself counted among these predecessors in addition to IMMANUEL KANT and PIERRE SIMON DE LAPLACE the physicists HERMANN VON HELMHOLTZ, Lord KELVIN, and LUDWIG BOLTZMANN. NERNST motivated his turn to the cosmology in the following way: *"Progress in this area can only occur because of new facts and the new ideas resulting from them. ... [Since] in the meantime one has gained empirical knowledge or can use logical means, which were not available to the excellent predecessors, ... it is probably not presumptuous, if presently one expects success from a renewed treatment of the problem."* [Nernst (1921b): 17].

His first activity in this field happened during the years of his position as *Ordinarius* of Physical Chemistry at the University of Berlin. Already in September 1912, however, *"more on the sideline"* [Nernst (1921b): 2], he presented his initial ideas at the 84th Convention of the Society of German Natural Scientists and Physicians in Münster [Nernst (1913c)]. However, at this stage his considerations were dealing *"less with the attempt to introduce a new cosmic conception, but instead with an illustration of the thermodynamic approach."* [Nernst (1912b): 9]. In 1921 he pursued these ideas again in his corresponding lectures at the Prussian

Academy of Sciences in Berlin, at the Association of Engineers in Vienna, and at the Urania in Prague. During the same year he summarized these ideas in the publication *"The World Structure in the Light of Research"* [Nernst (1921b)]. Already in 1923 its translation into Russian arranged by ABRAM JOFFE was published in the Soviet Union, being only less than a year old at the time. In his address in 1922 at the memorial ceremony at the University of Berlin, dealing with the "new stars" (novae, *stellae novae*) or according to the modern terminology created by FRITZ ZWICKY and WALTER BAADE with supernovae [Nernst (1922a)], NERNST referred to this publication. Also as President of the PTR he discussed in lectures his ideas on cosmology. For example, in September 1923 at the *Technische Hochschule* in Berlin he talked about the *"Energy Balance of the Universe"*.

It was a statement of his teacher LUDWIG BOLTZMANN in his inaugural speech at the Academy of Sciences of Vienna in the year 1886, which for the first time provided for NERNST the motivation to occupy himself with the cosmos. This was before the time when NERNST studied in Graz. BOLTZMANN had noted that up to then all attempts to prove the nonreality of the thermal death of the universe, concluded for the first time in 1867 by RUDOLF CLAUSIUS from the Second Law of Thermodynamics, were unsuccessful. Hence, also BOLTZMANN did not want to undertake an attempt in this direction. NERNST indicated later: *"This point which I read as a student made the largest impression on myself, and since then it remained always in my mind if there would not show up a way out anywhere."* [Nernst (1921b): 1]. Regarding the question of his competence for a scientific occupation with the cosmology, the scholar, who in the meantime had become highly prominent in the field of Physical Chemistry, emphasized, *"that in particular the physical chemist would be prepared not in the poorest way, if one has to develop a judgment on questions of cosmology, which are dealing both with the physical as well as with the chemical part."* [Nernst (1921b): III]. In 1912 in a lecture [Nernst (1913c): 116] PAUL GÜNTHER extended this statement, likely referring to the concluding remarks of his teacher, by pointing out that one of the fundamental problems of physical chemistry, namely the physical and chemical transformation of the character of a system due to very slow processes changing the existence of the chemical elements, could be investigated only with objects representing systems being independent of each other and in different stages of development, and that a large number of such objects is offered by the astrophysics. More generally we know: *"The physical chemistry is in direct contact with the astrophysics in the question of the expansion of the material state space. ... Because of these relations between the two sciences it can be understood that in particular most recently Nernst and Arrhenius have turned to cosmological questions."* [Günther (1924): 454–455].

However, regarding questions of astrophysics NERNST liked to consult with astronomers, or he went back to their results, in order *"to avoid in this way the danger which always appears if a scientist enters a field further away from his own area."* [Nernst (1921b): III]. PAUL GUTHNICK, Professor of Astronomy at the University of Berlin and Director of the Observatory in Potsdam-Babelsberg, and his coworkers were among these experts.

In his cosmological considerations NERNST started from an assumption, which ARRHENIUS had adopted already in 1906 in his work «*Das Werden der Welten*» (The development of the Worlds): the universe resides in a stationary state, i.e., on the temporal average as many stars become extinct as new ones are generated. Hence, this concept represents the opposite of the thermodynamic one, the consequence of which is the thermal death. In fact for its derivation it was assumed that the universe is a closed thermodynamic system. Since we have $dQ = 0$ due to the closed feature, one obtains for the change of the total entropy $dS = dQ/T + d_iS$ (Q heat energy; T absolute temperature), that the entropy can change only because of the entropy production d_iS, which can never be negative due to the Second Law of Thermodynamics. Since $d_iS > 0$ in the case of irreversibility, i.e., for all natural processes, and $d_iS = 0$ in the case of reversibility or the equilibrium state, the entropy will always increase until it reaches a maximum in equilibrium. Then the temperature will be the same everywhere, and macroscopic processes could occur only in the case of an external input, which, however, is impossible under the given conditions. The total energy of the universe has been transformed irreversibly into heat. In this sense the universe is dead having everywhere the same temperature.

In his thoughts NERNST started from the point that the total energy in the universe, irreversibly transformed into heat by means of radiation, *"moves into the sea of the ether"*, in order to be stored there in form of the zero-point energy introduced by MAX PLANCK and to represent the energy content of the ether [Nernst (1921b): 1]. Already in 1912 he had explained the basic features of this concept at the Convention of the Natural Scientists and Physicians we have mentioned [Nernst (1913c): 116], where he attributed particular importance to the radioactive decay processes for getting over the hypothesis of the thermal death. He quoted verbally the corresponding passages in his publication of 1921 [Nernst (1921b): 2–5] (see also [Huber and Jaakkola (1995): 55]). He thought that all chemical elements can decay radioactively. However, in the case of the elements which are stable in the context of today this process would be so slow that it could not be observed by a measurement. Therefore, within all atoms he saw the supply of huge amounts of energy, compared to which the heat energy due to the kinetic and potential energy of the atoms and the chemical component is negligibly small. Furthermore, the radioactive processes would be absolutely irreversible in the sense, that apparently the decay products in no way can be transformed back again into the starting material. This had been concluded by NERNST using the Third Law of Thermodynamics, because he had calculated that the temperature needed for the reversal of the radioactive decay is much higher than the maximum temperature calculated by Sir ARTHUR EDDINGTON for the center of a star. Since it is also associated with a degradation of the energy and even with a degradation of matter, *"the theory of radioactive decay thus doubled the likelihood of a Götterdämmerung of the Universe."* [Huber and Jaakkola (1995): 55]. In spite of this NERNST could imagine a process which counteracts the radioactive decay processes. This would mean that in the end the atoms of all elements in the universe would change into an original substance (*Ursubstanz*), which should be identified with the ether. Then analogous to the kinetic theory of gases one can imagine a multitude of configurations, to which belong also some highly improbable ones. In this way also once in a while an atom can become transformed back again, which even could be a heavy one according to the probability. Although such processes cannot be observed experimentally, because of the immensely long lifetime of most elements, the extremely small density of matter within the cosmos – estimated too high by NERNST, and the rarity of such events, in 1912 NERNST felt that *"at any rate the indication"* would be *"not quite without interest, that presently a not too improbable opinion is possible, according to which the matter existing in the universe and its energy content would reside in a certain state of persistence, and that, hence, an end of all activities at least must not be stated any more as an absolute consequence of our concept of nature."* [Nernst (1913c): 116; Nernst (1921b): 5].

We see that NERNST justified the existence of a stationary cosmos with his concept of an ether, within which matter could be regenerated again from fluctuations of the contained zero-point energy. When nine years later he noted during his further treatment, that *"in the meantime some new aspects had come up"* [Nernst (1921b): 5], one had to think in particular of concepts regarding the ether. The opinion expressed by NERNST in 1916, that the ether contains large amounts of zero-point energy in the form of vibrational energy [Nernst (1916c)], independent of him had been stated also in 1921 in a paper by EMIL WIECHERT [Wiechert (1921)]. In some sense NERNST was supported even by ALBERT EINSTEIN, since in 1920 in his publication *"Ether and the Theory of Relativity"* [Einstein (1920)] the latter introduced the ether again, which had been eliminated in connection with the Special Theory of Relativity. However, in contrast to NERNST and WIECHERT he did not attribute material properties to it. EINSTEIN felt the need for this step when he applied his Theory of Gravitation to the universe as a whole which he assumed to be stationary similar to NERNST. The resulting dissatisfactory consequence of a collapse of the cosmos under its own gravitation was overcome by EINSTEIN by introducing the cosmological constant, which effected the stability of the universe by generating the necessary pressure and was understood in terms of the energy content of the empty space. This is exactly the subject of NERNST's statement: *"A hypothesis, which should cover for us the loss of applicability postulated by the Second Thermal Law of Irreversible Processes, cannot avoid to draw support from the energy content of the light-ether (or, if you like, of the 'empty space')."* [Nernst (1921b): 1–2].

However, the constancy of the vacuum light velocity postulated by EINSTEIN could not be reconciled with the concept of the ether proposed by NERNST. If in principle NERNST accepted EINSTEIN's Theories of Relativity and defended them against all attacks, he attributed to the laws of these and of at least all deterministic theories a certain approximative character. However, in this case the statement of the Special Theory of Relativity regarding the light velocity within the empty space did not have to be absolutely correct. In his inaugural speech as Rector of the University of Berlin in 1921 he clearly expressed this opinion: *"The works of Galiäi[sic] and Newton are marvelous as on the first day, however, they did not bring to us the definite laws of the motion of the celestial bodies. Nobody will want to claim that, say, the Theory of Relativity will bring this completion; pretty soon the absolute constancy of the light velocity, with which it operates, will turn out to be an approximation."* [Nernst (1921a): 13].

The ideas about the stars, from which in the beginning of the 1920s NERNST started on his explanations on cosmology, were about the following: All fixed stars have a mass varying only relatively little, as taught to us by *"a, however, not very rich experience"* [Nernst (1922a): 16]. For NERNST this mass spans the order of magnitude of about $0.5 - 5$ $m_\odot$; however, in fact it can take up also values around 50 $m_\odot$ ($m_\odot$ Sun mass $\approx 2 \cdot 10^{30}$ kg), the density (ca. 10^{-5} up to 3 kg·m^{-3}) and, hence, the radii (ca. $4.5 \cdot 10^8 - 5.5 \cdot 10^{10}$ m) of which fluctuate strongly. (In the case of the average values within the brackets the A0-stars (white dwarfs), the average density of which is about 10^5 kg·m^{-3} and the average radius about $1.5 \cdot 10^7$ m, were ignored). Except for more rare objects, surface temperatures of about $3000 - 12000$ K were assumed. One had found that in the case of the hottest stars there exists only a single type regarding their spectral character and their density. On the other hand, in the case of those radiating with a red color, i.e., being relatively cold, there are two types strongly differing from each other in their density, namely the giant and the dwarf stars. From this one had concluded that in this way different steps of the evolution of stars manifest themselves. NERNST described this situation in the following way: *"The present dominating concept about the sky of the fixed stars consists of the fact that one assumes, a star is formed out of compression of world dust, as formulated already by Kant; in this case under always progressing compression the mass attains at first red heat, then yellow heat, and finally the brightest white heat, in order then to cool down again gradually because of the immense heat output due to radiation, and in this way to transform itself into a dark star along the path of the yellow and the red star. Since during its evolution the star becomes more and more dense as we have said, and therefore shrinks more and more, during the initial stage of the evolution a red or a yellow star is much more extended than during the later stage, therefore, during the early stage of course* [it is] *also much brighter than during the later stage."* [Nernst (1922a): 16].

The British physicist and astronomer Sir ARTHUR EDDINGTON had developed a theory, which tried to explain the approximate equality of the masses of the stars we have mentioned. He started from the point that the equilibrium of a very large and hot mass is due to the balance between the gravitational force directed toward the interior and the sum of the expansion force and the radiation pressure, both of which are acting in the opposite direction. Under this assumption EDDINGTON was able to calculate a mass m_{max} comparable to the observed values, above which the radiation pressure at the surface of the star completely compensates the effect of gravitation. Therefore, a star with a mass $m > m_{max}$ would be unstable, since all mass portions $m - m_{max}$ would be weightless and, hence, could be removed from the star with only a small amount of force.

With the assumption that the radioactive decay of the atomic nuclei within the masses of the stars would represent their actual energy source, NERNST could derive some quantitative results on the process of the star evolution. According to PAUL GÜNTHER this represented *"the essential progress of Nernst's concept compared to the previous ideas, which have also assumed already radioactive processes within the stars."* [Günther (1924): 455]. However, it was quite clear that the radioactive elements known at the time, of which uranium has the largest atomic number, cannot supply the necessary amount of energy. Therefore, NERNST assumed the existence of so-called transuranic elements having a decay energy higher than that of the known radioactive elements. Since a superposition of the decay laws of different such elements approximately results again in an exponential law, one could base the calculation on a single transuranic element with the decay rate ν or the half-life $\tau_{1/2} = \ln 2/\nu$ acting for all stars as the energy source.

Based on this fundamental assumption one can divide the path, along which NERNST arrived at the numerical results, into the following steps:

a) To the assumptions of EDDINGTON we have mentioned NERNST added one more, that the energy E_{ex} emitted by a star at each moment is equal to the energy E_{gen} generated by it. Hence, within a distinct period the energy emission U_{ex} is equal to the energy U_{gen} generated during this time interval: $U_{ex} = U_{gen} = U$.

b) According to the assumption the energy generation should be associated with the radioactive decay which follows an exponential law, because of this and because of assumption (a) also the energy emission will satisfy such a law: $U(t) = U_0 e^{-\nu t}$, where U_0 denotes the initial emission and t the time.

c) For a quantitative treatment of the energy or of the amount of heat U developed within the star the STEFAN-BOLTZMANN law could be used: $U(r, T) = 4\pi r^2 \sigma T^4$. Here σ is the radiation constant, T the surface or effective temperature, and r the radius of the star. If m denotes the mass of the star, assuming a spherical shape of the star one obtains the relation between r and the density δ: $\delta = 3m/(4\pi r^3)$. Then one can find the function $U(\delta, T)$ as $U(\delta, T) = (48\pi^2 m)^{2/3} \sigma \delta^{-2/3} T^4$.

d) From this relation for a star of the mass m and from the corresponding relation for the Sun $U_\odot = (48\pi^2 m_\odot)^{2/3} \sigma \delta^{-2/3} T_\odot^4$ with the assumption $m \approx m_\odot$ follows the relation $U = U_\odot (T/T_\odot)^4 (\delta_\odot/\delta)^{2/3}$, where $U_\odot$, $T_\odot$, and $\delta_\odot$ are known empirically.

e) However, NERNST could not yet indicate the function $\delta(T)$. Therefore, he used corresponding tables generated by EDDINGTON and experimental data supplied by PAUL GUTHNICK and ERNST BERNEWITZ for calculating the values of $U(\delta, T)$ from these pairs of values of δ and T in the case of stars with the mass $m_\odot$ of the Sun. For example, in this way in the case $\delta = 1.38 \cdot 10^{-3}$ g·cm^{-3} and $T = 5000$ K, i.e., for a still young star, he obtained the value $U = 57.6 \cdot 10^{34}$ J. Since HENRY N. RUSSELL had found out from the statistics of stars that giant stars have a brightness and, hence, a heat radiation U, which is nearly independent of the stage of evolution, *"with weak extrapolation"* [Nernst (1921b): 51] in the case of all stars for the initial emission NERNST took the value $U_0 = 64 \cdot 10^{34}$ J·a^{-1}.

f) In the case of the Sun for the current energy emission one had measured the value $U_\odot = 1.2 \cdot 10^{34}$ J·a^{-1}. Furthermore, from its radioactivity one had been able to determine the age of the crust of the Earth and, hence, approximately that of the Earth itself with $t_\delta = 1.5 \cdot 10^9$ a. Since according to the KANT-LAPLACE theory the origin of the planet Earth is connected with that of its central star, the age of the Sun $t_\odot$ should not be larger than that of the Earth. On the other hand, with the assumption of an approximate constancy of the mass during the whole evolutionary period, based on the Special Theory of Relativity an upper limit of $t_\odot$ with $t_{RT} = 10^{12}$ a could be indicated, such that we have $t_\delta \leq t_\odot \leq t_{RT}$. Since the planets were formed probably already soon after the generation of the Sun, NERNST estimated the age of the Sun to be close to t_δ with $t_\odot \approx 3.5 \cdot 10^9$ a.

g) Knowing the values $U_0 = U_{\odot,0}$ according to (e) as well as $t_\odot$ and $U_\odot(t_\odot)$ according to (f), by using the time dependence of the energy emission mentioned in (b) one could calculate the decay rate ν of the radioactive material supplying the energy: $\nu = t_\odot^{-1} \cdot \ln(U_{\odot,0}/U_\odot) \approx 6 \cdot 10^8$ a.

h) Since the constants U_0 and ν valid for all stars were known, according to $U(t) = U_0 e^{-\nu t}$ one could derive the energy emission of a star as a function of its age. Using the values of $U(\delta, T)$ determined according to (e) and the relation $t = \nu^{-1} \cdot \ln(U_0/U)$, NERNST determined certain values of the temperature and the density as a function of the time in order to obtain a rough picture about the evolution of stars having about the mass of the Sun and the influence of their important state parameters T and δ. NERNST emphasized [Nernst (1921b): 51.53]: *"Of course, the same path can be followed also in the case of stars with another mass, only the determination of the time becomes a little more uncertain."* PAUL GÜNTHER characterizes NERNST's accomplishments in more general terms: Because of the relation $U(t) = U_0 e^{-vt} = U(\delta, T)$ found by his teacher *"to each individual star one can ascribe an absolute age or one can calculate also these characteristic data [δ and T] for the Sun in the past and the future."* [Günther (1924): 456].

In his treatment of the novae NERNST started from the justified concept of GUTHNICK that the origin of such stars cannot have external causes such as collisions with other stars or with the cosmic cloud. The evolution of a nova consists of a very rapid process, during which the brightness increases by 10 – 15 orders of magnitude, followed by a slightly slower decrease of the brightness associated with a change in color and subsequently by an irregular fluctuation by several orders of magnitude. If finally the star becomes as bright again as before its rapid lighting up was considered likely but not certain. At any rate, by means of photographic observations one had been able to prove, that a nova does not represent a "new" star in the true sense, but instead a previously weak star becoming visible because of an extremely strong and rapid increase in brightness.

NERNST assumed that the flaring up of a nova is caused by an immense outbreak of very hot masses out of the interior of the star, where according to EDDINGTON the temperature would be several million degrees. The hot masses then would become distributed around the whole star. This rapid emission of matter to the surface would be supported by an observed strong violet-shift of some spectral lines, which could be explained by the DOPPLER effect. In this context the suggestion by NERNST to create an X-ray- and γ-astronomy is interesting: *"Up to now it has not been investigated, if very short-wave light, i.e., so-called X-rays, are emitted by the new stars, and if perhaps also, in case – as not unlikely – the masses originating from the interior of the star contain strongly radioactive substances, at the same time a very hard so-called γ-radiation is emitted by new stars. It appears advisable that at least a few observatories should be equipped with measuring instruments, in order to be prepared for examining this immensely important question during the future appearance of a new star; however, it will not be quite easy to develop such instruments of sufficient sensitivity mainly because of the disturbing absorption within our Earth's atmosphere."* [Nernst (1922a): 15].

In these enormous outbreaks of matter and, hence, also of energy NERNST saw a certain similarity to the protuberances of the Sun. Furthermore, he assumed that the strong eruptions leading to the appearance of a nova are experienced at least once by every star during its evolution. Mainly during the early stage, i.e., in the case of a giant star, they should happen with certain regularity. In the inspection of the relevant observational evidence existing up to then NERNST was assisted by FRIEDRICH PAVEL, at the time *"assistant at the observatory Neu-Babelsberg"*. Based on statistical estimates, which at the same time supported his assumptions, NERNST even arrived at the supposition that the number of the nova outbreaks roughly corresponds to that of the planets. Therefore, he considered it possible that the formation of a planet is connected with the flare-up of a nova.

NERNST's hypothesis on the existence of strongly radioactive transuranic elements within the stars and their predecessors, the cosmic clouds, was believed to be confirmed in particular by the existence of the so-called HESS' radiation, i.e., the cosmic radiation. VIKTOR FRANZ HESS had investigated the radioactivity of the atmosphere at the Radium Institute in Vienna and had found that it cannot have a terrestrial origin. In the beginning of the 1910s HESS had confirmed this observation together with WERNER KOHLHÖRSTER. In 1923 based on his measurements on the *Jungfraujoch* in Switzerland KOHLHÖRSTER was sure, that this radiation truly comes out of the cosmos, since it exceeds the observed γ-radiation on the ground by a factor of ten. Contrary to the previous assumption of its exclusive origin from the Sun, he also presumed to have demonstrated a directional effect of the radiation coming from the Milky Way. In fact today with a certain probability one considers supernova outbreaks, quasars, or pulsars to represent the main sources of this radiation, without immediately thinking of NERNST's ideas about the energy source of the stars.

PAUL GÜNTHER indicated in 1924: *"From the cosmological theory of Nernst there results an important stimulation for the chemical and physical research: The possible existence of chemical elements with a higher atomic number than that of uranium possibly in traces also still in the interior of the body of the Earth."* [Günther (1924): 456]. When in 1940 in Berkeley EDWIN M. MCMILLAN together with PHILIP H. ABELSON produced the first transuranic element, the neptunium ($_{93}$Np), by bombarding ^{238}U with neutrons, and when in 1941 GLENN T. SEABORG together with MCMILLAN, JOSEPH W. KENNEDY, and ARTHUR C. WAHL produced the second, plutonium ($_{94}$Pu), they hardly were stimulated by NERNST's prediction of the existence of these elements within the cosmos. The same applies also to ALBERT GHIORSO in Berkeley, GEORGIĬ N. FLËROV in Dubna, and the others, who continued the production of the transuranic elements. However, subsequently traces of neptunium and plutonium were actually found in uranium ores, which were not the relics from cosmic processes.

NERNST pursued his cosmological ideas developed in the beginning of the 1920s during their further course also at the Physical Institute in Berlin. In particular he held on to his theory of a stationary cosmos. In fact, during a time before in 1929 the American astronomer EDWIN P. HUBBLE saw a connection between the redshift observed in the spectra of the galaxies and an expansion of the universe, the concept of such a stationary universe was the most plausible one. So in 1922 also NERNST had declared – of course, with a remark about the *"conceptual possibility"* of a physical justification of a stationary cosmos found by him and explained above: *"Up to now nearly every scientist, who has occupied himself more deeply with the development of the sky of the fixed stars, has adopted the ground of the hypothesis of the stationary state of the sky of the fixed stars, namely of the assumption that on the average as many stars disappear by cooling down, probably are leaving even our system of the Milky Way, as new ones are generated. Without invoking this supposition in general we are hardly able to think about cosmic questions, however, on the other hand this supposition has been extremely useful qualitatively and quantitatively in particular in the area of the statistics of stars. Admittedly, the physical explanation of the possibility of that stationary state provides extraordinary difficulties, anyhow presently it appears that a conceptual possibility has been found."* [Nernst (1922a): 17].

In his paper published in 1928 in the *"Journal of the Franklin Institute"* NERNST not only repeated his definition of the stationary cosmos, but also held on to the content of his corresponding theoretical ideas: *"I may therefore hold fast to the hypothesis uttered by me that, just as the principle of the stationary condition of the cosmos demands that the radiation of the stars be absorbed by the luminiferous æther, so also finally the same thing happens with the mass, and that, conversely, strongly active elements are continually being formed from the æther, though naturally not in amounts demonstrable to us, the radio-active disintegration of which maintains correspondingly the high differences of temperature which are observed in the Universe and which at the end form the driving force of all the processes of nature in the direction demanded by the second law of thermo-dynamics. This simple hypothesis would therefore restore to us the stationary condition of the cosmos."* [Nernst (1928c): 141].

NERNST's stationary cosmos, in which *"the present fixed stars cool continually and new ones are being formed"* [Nernst (1928c): 135], represented a problem regarding the radiation existing within it, which is known as the paradox of WILHELM OLBERS. Within such an unlimited universe having eternal existence and, hence, also an eternal past, the amount of radiation must have increased continuously, such that contrary to the experience the temperature of the universe should be extremely high and the night sky should be perfectly bright. In order to solve this problem NERNST invoked his prediction of a permanent annihilation of energy corresponding to its absorption by the ether. However, also in the future there would be no possibility to check this supposition experimentally, because the amounts of energy absorbed by the ether are extremely small.

After his retirement NERNST devoted four more papers to the problem in cosmology, in which he remained faithful to his basic concepts of a stationary universe he had conceived essentially between 1916 and 1921 [Nernst (1935a), (1935b), (1937), (1938)]. As the address of the author the papers indicate *«Rittergut Zibelle b. Muskau (O. L.)»* [Nernst (1937): 661] or a similar address.

In 1935 NERNST published a paper on the evolution theory of the stars [Nernst (1935a)], in which he continued the remarks he had presented at the Academy in Berlin already in July 1931 [Nernst (1931)]. In this paper he emphasized that his theory developed for this in 1921 could have been *"only very preliminary"*, and: *"Even if I am keeping it up regarding most of the essential points, the recent physics has provided still highly important aspects, with which I want to occupy myself."* [Nernst (1935a): 511]. In this case he was assisted by the astronomer KARL PILOWSKI, and his colleagues at the University, PAUL GUTHNICK and AUGUST KOPFF, supplied him with information on astronomical questions. NERNST felt sorry that in his earlier considerations he had relied only on statistical information by EDDINGTON and had not gone back to the original literature. Otherwise he would have noticed that contrary to his earlier assumption the stars would have decreased strongly in their mass during their evolution, roughly at the ratio 100:1. This fact now well known, *"of course, stipulates a substantial change of every theory of the star evolution."* [Nernst (1935a): 511]. As a first step NERNST examined the available statistical observational material under this aspect. In this way he could derive eight classes of stars which should correspond to different evolutionary stages, because the average parameter values associated with them *"now show a perfectly regular trend."* [Nernst (1935a): 532]. In this case these and further statements referred primarily to stars having 30 – 40 Sun masses at the early stage according to the available material.

Using the known abundance values of the different classes of stars, NERNST calculated their relative lifetime and from two absolute values of age, for example, from the age of the crust of the Earth, also their absolute lifetime. In this way one could find by integration also the energy emission during the different stages, from which the energetic evolution of a star during its lifetime could be determined. It turned out that in the beginning there occurred a very strong energy emission for a relatively short time. This was *"interpreted as being connected with the agent supplying the ultra-radiation, i.e., acting according to kind of radioactive processes"*, whereas for the later stages at increasing density NERNST presumed the *"atomic fission, perhaps mainly of lithium"* acting as the energy source [Nernst (1935a): 533]. Occasionally also the gravitational work could play a role.

By holding on to his concept that the *"carriers of the ultra-radiation"* are permanently newly formed and are removed again until the vanishing of their mass, NERNST arrived at a universe in a stationary state or with constant entropy. In spite of the increase in entropy due to the decomposition processes and the unrestricted validity of the Laws of Thermodynamics, NERNST concluded that a thermal death of the universe will not happen, because *"the available energy of the universe will be supplied by a kind of Brownian motion of the zero-point energy of the light-ether."* [Nernst (1935a): 533].

Also in 1935 NERNST responded to another development concerning the state of the cosmos and connected with the redshift of spectral lines observed for cosmic objects. Already in 1916 the astronomer CARL WILHELM WIRTZ in Strasbourg had observed this effect in spectra of some so-called cosmic nebulae. Since he did not consider these objects as being extragalactic, initially his observation had no important consequences. Eight years later using the determination of the distance of the Andromeda-Nebula (M 31 or NGC 224), EDWIN P. HUBBLE was able to prove for the first time the existence of large systems of stars located far outside our Milky Way. VESTO M. SLIPHER had determined a total of 45 radial velocities of galaxies which he interpreted as a motion of our Milky Way relative to these systems of stars (drift hypothesis). Also at the Mount-Wilson-Observatory MILTON L. HUMASON known for his exact measurements had investigated the redshift and the motion of galaxies. 1929 HUBBLE presented a theory claiming a connection between the redshift interpreted as a DOPPLER effect and the distance of galaxies and, hence, implying the concept of an expanding cosmos. In particular the measurements and results by SLIPHER and HUMASON contributed to the consolidation of HUBBLE's theory.

Based on his experimental data HUBBLE found a linear relation between the redshift $z = (v_0 - v)/v$ (v observed frequency; v_0 frequency of the non-shifted spectral line) and the time t: $z = H \cdot t$. According to the DOPPLER principle, for the ratio of the "escape velocity" v and the light velocity c we have: $v/c = z = H \cdot t$, from which follows $v = Hct = Hr$. In this case r is interpreted as the distance of the cosmic object. H is the HUBBLE-constant, the current value of which is given as 72 km·s^{-1}·Mpc^{-1} (1 Mpc (Megaparsec) = $3.086 \cdot 10^{19}$ km). Assuming a constant expansion of the universe, one can look at the so-called HUBBLE-time H^{-1} as the age of the universe, where, however, this interpretation of the HUBBLE-constant is controversial.

In his second paper on the evolution of stars of 1935 [Nernst (1935b)] and in a publication in 1937 [Nernst (1937)] NERNST discussed HUBBLE's observations and results. In this case his basic interest still remained focused on the justification of his theory of the stationary state of the cosmos and its adjustment to more recent cosmological findings. On this subject he stated categorically: *"This hypothesis turned out so fruitful, that it appears unsuitable to work in astrophysics without it."* [Nernst (1937): 633].

In HUBBLE's observations NERNST saw a certain experimental confirmation of his assumption of the energy degradation. So he developed the following train of thought: Because of the reasons mentioned already, the decrease in energy follows the decay law of radioactive substances. Then in the special case of the photon energy $E = hv$ (h PLANCK's constant; v frequency) we have $dE = -HEdt$ and after integration $\ln(v_0/v) = Ht$, from which in the case of small reductions of the frequency one obtains $(v_0 - v)/v = Ht$, i.e., the expression empirically found by HUBBLE.

NERNST had formulated his law of the energy degradation probably independent of FRITZ ZWICKY, who had developed a similar idea already in 1929. Now for determining the distance r of a galaxy from the measured redshift NERNST could confront his formula $r_N = cH^{-1} \cdot \ln(v_0/v)$ with that used by HUBBLE $r_H = cH^{-1} \cdot (v_0 - v)/v$. The difference resulted from the different concepts regarding the origin of the redshift. NERNST had derived *"a simple formula fixed experimentally (in the astrophysical sense) about the nonrelativistic disappearance of the light-quanta"* [Nernst (1937): 640], whereas HUBBLE explained the redshift by the DOPPLER effect in an expanding universe. The difference between both theories would become noticeable only if the observations would be extended to objects much further away than up to then, because in the case of the latter one obtains $r_N \approx r_H$. NERNST felt quite certain about his opinion: *"Certainly, the theory of the 'exploding universe' was not plausible, we believe also due to our interpretation of the redshift of far-distant celestial objects to have replaced that theory by a scientifically useful concept."* [Nernst (1937): 640], also *"it is very remarkable, that Hubble himself ..., simultaneously with myself as I presume, according to his publications a little later* [1936], *by means of purely astronomical measurements also declares the interpretation of the redshift in terms of a Doppler effect improbable. His reasoning must be checked from the astronomical side; at any rate, his measurements showed that the decrease of the brightness of the nebulae with the distance does not happen in the way as required by a Doppler effect, but slower, as it corresponds to my new interpretation."* [Nernst (1937): 641].

In his equation $dE = -HEdt$ or $dv = -Hvdt$ NERNST saw a solution of OLBERS' paradox concerning actually only the small light intensity of the firmament together with the assumption of an infinitely large universe having a finite mass density. *"However, without any further dubious assumptions* [as, for example, the absorption by cosmic dust] *our equation provides a simple quantitative explanation."* [Nernst (1937): 644]. In such a universe an infinitely large gravitational force would be acting also upon each mass point. In order to overcome this so-called cosmological paradox, NERNST proposed a correction of NEWTON's law of gravitation $F = \kappa mMr^{-2}$ (F force; κ gravitational constant; r distance between the masses m and M) in terms of an exponential factor $F = \kappa mMr^{-2} e^{-\frac{Hr}{c}}$, where a direct experimental test of this relation between F and r would be likely not possible. NERNST had taken up this extension of the law of gravitation because of the following assumption: *"Apparently, the constant H (reciprocal of time, i.e., with the dimension of a frequency) occupies the rank of a fundamental physical constant; hence, hH has the dimension of an energy quantum. The suspicion suggests itself that not only the light-quanta disappear within these extremely small quanta, but that the same quantum is also valid in the case of the gravitational work and the kinetic energy."* [Nernst (1937): 640]. NERNST emphasized that his solution of the cosmological paradox by means of a correction of the law of gravitation would be effected in terms of the force and not of the potential, as it had been tried previously, for example, by WALTER GROTIAN and AUGUST KOPFF.

NERNST also included the neutrons discovered in 1932 by Sir CHAMES CHADWICK in his cosmological studies. So he could point out in a summarizing way: *"Now a very special theory of the cosmic radiation could be developed, based on the fact that neutrons 'von hoher lebendiger Kraft'* [with a high live force] *represent its primary component."* [Nernst (1937): 660]. Regarding the cosmic radiation NERNST had obtained suggestions and hints by ERICH REGENER. This may have included also the fact, that the presence of fast and very slow neutrons within the cosmic radiation would be confirmed. Already in 1903 for the new edition of his textbook NERNST had pointed out that the ether could consist of massless particles, to which he referred as 'neutrons': *"Apparently, the ratio of the positive and negative electrons reminds us of the optically isomeric twins. – Whether also the compound between a positive and a negative electron ($\oplus\ominus$ = neutron, electrically neutral massless molecule) really exists, apparently represents a question of high importance; we want to assume that neutrons can be present everywhere, like the light-ether, and we can add that a space filled with these molecules would have to be weightless, electrically nonconducting, but electrically* *polarizable, i.e., it must have properties as they are claimed by the physics otherwise also in the case of the light-ether."* [Nernst (1909a): 398]. After the discovery by CHADWICK it was presumed by NERNST that extremely fast neutrons are permanently generated out of the ether, which *"at their entry into our accessible world receive a quantum of rotational energy determining their mass."* [Nernst (1937): 648]. On the other hand, the permanently vanishing matter should return into the ether in the form of neutrons with small kinetic energy.

NERNST's last paper on cosmology [Nernst (1938)] dealt with the *"world temperature"*, i.e., the temperature of the intergalactic space, and it starts again from the equation $-\mathrm{d}(h\nu)/\mathrm{d}t = H \cdot (h\nu)$ explaining the redshift. NERNST justified the fact that this temperature has a definite value, although a permanent energy irradiation occurs into the intergalactic space from the galaxies uniformly distributed according to his theory, with the remark *"that, however, the world temperature remains on a certain maximum level because of the energy loss corresponding to the redshift."* [Nernst (1938): 45]. Using a few simplifying assumptions, he calculated the world temperature $T = 0.75$ K, based on the mentioned formula and the STEFAN-BOLTZMANN law. However, according to information given by REGENER this value should be 3.16 ± 0.5 K. In spite of this NERNST rated his result quite positively: *"So our value, which we cannot check presently in any other way, is not implausible."* [Nernst (1938): 46]. For him the point was particularly important, that this result had been obtained from the equation which is *"interesting also purely physically."* His paper *"perhaps provides a new contribution to the logical power of this formula."* [Nernst (1938): 47].

However, during the years after HUBBLE's fundamental paper on the cosmic redshift and the expanding cosmos the cosmological research preferred this new explanation based on the DOPPLER effect. On the one hand, this was caused by the absence of a mechanism of the energy loss assumed by NERNST, which can be tested. On the other hand, the nonstatic solutions of the equations of gravitation formulated by EINSTEIN within the General Theory of Relativity, which had been found in 1922 by ALEKSANDR A. FRIDMAN, had attracted far more attention than

NERNST's stationary theory. In contrast to EINSTEIN's stable universe still assumed at the time, FRIDMAN deduced an instable one from his results. Subsequently, he developed models of a nonlinear and retarded expansion of the cosmos starting from a singularity. Along completely different lines, between 1927 and 1933 the priest and physicist Abbé GEORGES HENRI LAMAÎTRE developed the concept of a universe evolving from a single atom, the *Uratom*. The concepts of an expanding cosmos culminated in the theory of the origin of the universe arising from a hot initial state, created in 1948 by GEORGE GAMOW, RALPH A. ALPHER, and ROBERT HERMAN.

At the same time the steady-state-theory was developed as an alternative by Sir HERMANN BONDI and THOMAS GOLD [Bondi and Gold (1948)], as well as by Sir FRED HOYLE. It had some similarity with NERNST's stationary cosmos. Actually HOYLE had coined the word 'big bang' in order to let the theory of GAMOW, ALPHER, and HERMAN, for which the term big-bang-theory had become customary, appear rather implausible. The steady-state-theory assumes a constant average density of matter in an expanding cosmos. A permanent generation of matter drives the universe. At the end of the 1950s there appeared first doubts about the theory of HOYLE, BONDI, and GOLD, when the quasars were discovered. All these objects show a strong redshift, and, therefore, in some sense represent a class of very far and very old celestial bodies. However, this feature contradicts the central idea of the steady-state-theory. Finally, a complete turn to the nonstationary big-bang-theory happened when in 1965 ARNO A. PENZIAS and ROBERT W. WILSON had discovered the cosmic background radiation.

In connection with this 2.7 K-radiation, the big-bang-theory, and with NERNST's cosmological ideas, the following modern statement is remarkable: *"What is interesting here is that Nernst, as early as 1938, had deduced that radiant flux from stars and nebulae in the region between the nebulae should heat absorbing material in this region to 2.7 K, this material then radiating like a blackbody at that temperature. In view of this prediction it is difficult to understand how Big Bang advocates can claim that 2.7 K blackbody radiation lends any support to this theory, whatever arguments Gamow et al. may have presented after the discovery."* and the general judgment *"Nernst's contributions to cosmology did not receive the attention they deserve."* [Huber and Jaakkola (1995): 57].

Actually, the thread through the research by WALTHER NERNST was thermodynamics. With a few exceptions it represented the central subject of his scientific activities during well over half a century. In 1886 his first publication was devoted to the *"Appearance of electromotic forces in metal plates which are carrying a heat current ... "* [Ettingshausen and Nernst (1886)], in 1888 his first one on physical chemistry treated the *"Heat of formation of the mercury compounds"* [Nernst (1888a)], and in 1938 his last one dealt with a research complex *"The radiation temperature of the universe"* [Nernst (1938)]. However, it appeared that NERNST's thermodynamic theory of the universe corresponding to his nature was not as far reaching as a cosmology based on EINSTEIN's General-Relativistic Theory of gravitation and its corresponding equations.

But NERNST's ideas have not been forgotten, and in the more recent cosmological research they gained again some importance. This concerns in particular his assumption, that light-energy, gravitational, and kinetic energy vanishes within the extremely small energy quanta hH (see above). Under the title *"Relevance of Nernst's Cosmology to Recent Ideas"* the British physicist P. F. BROWNE noted that for him in particular this concept of NERNST *"is of considerable interest in that precisely this hypothesis was made (1962) in order to explain the Hubble redshift. Then, the idea was that radiation from a distant galaxy lost these minute quanta of energy (gravitons) slowly to ambient radiation of the medium, which was assumed to be starlight averaged throughout the volume of the Universe."* [Browne (1995): 75]. In this way, regarding one of NERNST's last scientific conclusions *"The redshift is not a Doppler effect."* [Nernst (1937): 660], which is particularly fundamental within NERNST's cosmological research, BROWNE could state that he, *"at least, would agree with this conclusion."* [Browne (1995): 78].

Chapter 8

The Final Years (1933 – 1941)

8.1 Attitude to the Fascism

Sometimes one hears that WALTHER NERNST retired from the University in 1933, because he did not want to accept the new political situation in Germany resulting from the assumption of power by the fascists. It is correct that NERNST opposed the Nazi movement due to various reasons, and that he retired in the fall of 1933 after 44 years of activity as a university teacher. However, assuming a primary connection between both events is incorrect.

Initially, the retirement from the University had been fixed on September 30, 1932. However, since no suitable successor had been found up to this time, NERNST had been asked to continue lecturing during the winter term 1932/33 and then also during the summer term 1933 and to keep the directorship of the Physical Institute. So he entered retirement finally on October 1, 1933. Of course, after such a long period his new status was quite unfamiliar to him, which is indicated, for example, by the words *"Everything coming from the 'Alma mater' pleases the heart of the Emeritus quite particularly."* [UAHUB: II, 86], written by NERNST to the University of Berlin on June 30, 1934.

Already much earlier NERNST had publicly opposed the nationalistic and the discriminating racial activities sanctioned by the Government since 1933. During August 1920 the *"Working Group of German natural scientists for the conservation of pure science"*, with PHILIPP LENARD being one of the fathers, organized an event in the Berlin Philharmonics, in which ALBERT EINSTEIN and his theory of relativity were discriminated in the ugliest way. In particular, the *"swine Jews"* (*«Saujuden»*), one should *"really jump at the throat"* of which, were accused of theft of intellectual property, *"scientific Dadaism"*, and personal publicity obsession. As a result MAX VON LAUE wrote a statement published immediately by the press together with NERNST and HEINRICH RUBENS. In it the *"incomparably deep ideas"* of EINSTEIN have been emphasized: *"Anybody, who had the privilege to be closer acquainted with Einstein, knows that he is outdone by nobody in the respect of other people's property, in personal modesty, and aversion to publicity. It appears to be a requirement of justice to express this conviction of us without delay, even more so, since yesterday evening there had been no opportunity for this."* (quoted after [Herneck (1975): 79–80]).

NERNST's political intelligence and his political engagement and confession, but also a certain misjudgment of the approaching danger of the fascism, established already for eight years in Italy under BENITO MUSSOLINI, also in the case of Germany can be seen from a speech given by NERNST in December 1930 over the radio: *"Unfortunately it cannot be denied that wide sections of our population are burdened by heavy sorrow, and that at least in the largest part these sorrows are justified. However, we may find a consolation in the fact that our generation has lived through times, which still are in our vivid memory, were much more dangerous, and even were overcome still unexpectedly fast. In this case we do not even have to think of our collapse of 1918. Five years later, 1923, we have mastered greater dangers than may be confronting us today. In a sequence the year 23 brought to us the invasion of the Ruhr District, the communist revolt in Saxony and Thuringia, the Hitler-putsch in Bavaria, an unparalleled economical and financial crisis, and finally still the separatist movement in the Rhineland. ... Therefore, we want to commemorate with faithful thanks the two greatest leaders at the time which had died much too early,* [President] *Ebert and* [Chancellor] *Stresemann; but at the same time with the awareness, that for a nation wanting to live the principle is absolutely valid that no human being is irreplaceable. So today we do not want to lose our courage, like the devout Jews once during the building of their temple carrying the sword in one hand and the trowel in the other, and also in the coming year we must work courageously and diligently; in one hand the sword of the sharp intellect and accurate resolution, in the other the trowel as a symbol of the industrious rebuilding. However, everything, and that is most important, must happen under the lodestar of a true patriotism, and this is realized now by a large part of the people, who must turn away with horror from the permanent party quarrelling of our time. Then we can hope that finally we can cross the last stretch of the desert separating us from the Promised Land of a happy Germany."* (quoted after [Jost, W (1964): 529]).

In fact, about two years after this speech the *"party quarrelling"* was finished, however, by no means in the way imagined by NERNST. Under the Nazi dictatorship starting in 1933 truly one could not speak of a *"Promised Land of a happy Germany"*. The openly expressed opposition of NERNST to the HITLER "movement" and to the nationalistic actions of LENARD and STARK were unfavorable for the position of the *Emeritus* under the new regime similarly as his support of the Jewish physicist EINSTEIN and the fact that he was the father-in-law of two Jews.

Of course, as it is customary during a change of power, at that time all important positions in politics and also in science were filled with people loyal to the new regime. Since 1933 JOHANNES STARK directed the PTR, and the Physical Institute in Berlin was directed at first by ARTHUR WEHNELT, a member of the Nazi party. The situation was not so extreme at the Institute of Physical Chemistry. In 1936 one had expelled the disliked MAX BODENSTEIN as *Ordinarius* against his wish. However, NERNST's former student PAUL GÜNTHER was appointed as the successor. As Dean and as a follower of the Nazi politics LUDWIG BIEBERBACH directed the faculty, to which the *Emeritus* NERNST still belonged in a certain sense. This mathematician was well known as an important representative of the Geometric Theory of Functions. In 1934 he had presented a racist psychological pseudo-theory, in which he confronted «*artfremde*» (foreign), mainly Jewish and French, with «*arteigene*» (species-specific), i.e., German types and styles of mathematical creations [Bieberbach (1934)]. In this way to the "German Physics" of LENARD a similarly racial discriminating "German Mathematics" had been added.

"From the protocol of the plenary session [of the Berlin Academy] *of April 27* [1933] *one can see that also Nernst protested against the pro-fascist statements by Bieberbach and Heymann."* [Hoffmann and Schlicker (1987): 523]. The background for this public statement by NERNST was the resignation of ALBERT EINSTEIN from the Academy on March 28. On his trip to the USA the latter had publicly expressed his concern about the assumption of power by the Nazis. As a result, on April 1 ERNST HEYMANN, Professor of Law, had published on his own in the press a statement by the Academy, in which he declared that the Academy would have no occasion to feel sorry about the resignation of EINSTEIN. In particular MAX VON LAUE but also the botanist GOTTLIEB JOHANN FRIEDRICH HABERLANDT had protested against the previous measures of the management of the Academy against EINSTEIN.

NERNST's attitude regarding the old and the new rulers can also be illustrated with the example of his reactions to congratulations on his 70th birthday. On the one hand, he considered the congratulations of President PAUL VON HINDENBURG, occupying his office since 1925, as being honorable and very important. On the other hand, his daughter EDITH reported: *"In principle, it was quite irrelevant to my father, whether the Physikal. techn. Reichsanstalt congratulated him to his 70th birthday. When it did not do it, with its Nazi attitude, my father said: This is Stark! ... – it is inappropriate to write in this case as Mr. Eggert did, this is 'stark'* [German adjective = rude, sassy, ...], *since my father felt more honored to have received no congratulations from that side."* [Zanthier (1964a)]. Actually, in the printed version other than in the manuscript JOHN EGGERT had NERNST say regarding this point, distorting the meaning: *"'Furthermore, all friends and Institutes have congratulated, with one exception: the Reichsanstalt. Isn't that 'stark'* [German adjective, see above]*?'"* [Eggert (1964): 453].

Furthermore, NERNST's attitude regarding the Nazis can be seen from documents in his personal file at the University of Berlin [Bartel (1990)]. During the years after the emigration of his daughters HILDE and AN-

GELA and their families NERNST developed a busy traveling activity abroad, which was caused also by this sad circumstance in addition to his research activities. So in 1935 BERNHARD RUST, the Imperial and Prussian Secretary of Science, School and Public Education, had *"approved lectures in the Urania in Prague and at the Technische Hochschule in Brünn, as well as at the Chemical Society located there"* [UAHUB: II, 18]. RUST had also approved to participate *"at the end of August of this year in the meeting of the foundation board of trustees of the research station Jungfrauenjoch in Switzerland"* [UAHUB: II, 17]. On the instructions by RUST, THEODOR VAHLEN, being in charge of the Department of Science, allowed *"that during January of 1936 Professor Dr. Nernst gives lectures at the Chemical-Physical Society in Vienna on the 'Establishment of Some New Fundamental Physical Laws Based on Astronomical Measurements' and in the Urania-Vienna and the Urania-Graz on the 'Origin of the Fixed stars'."* [UAHUB: II, 12].

A trip to the University of Oxford and to his favorite former student FREDERICK A. LINDEMANN, who as "Prof" was established and was also Director of a Low-Temperature Laboratory in the style of his teacher in Berlin, certainly had the primary goal to see again his emigrated daughters, sons-in-law, and grandchildren. On December 6, 1935 BIEBERBACH forwarded NERNST's application regarding this matter *"with approval"* [UAHUB: III, 31] to the Secretary. On January 8, 1936 RUST approved, *"that Professor Dr. Walter Nernst gives a lecture at the Science Faculty of the University of Oxford."* [UAHUB: II, 9].

However, in the meantime something had happened which was quite unfavorable for NERNST's departure to Oxford. On December 15, 1935 the «cand. math.» OTTO RICHTER had approached the Dean BIEBERBACH in a letter asking for his help at the recommendation of the State Ministry of Propaganda. As part of his *"final university examination thesis on the Nobel Laureates of Chemistry"* he had *"performed detailed racial-biological investigations in order to verify different basic laws (Galton etc.) in the case of the Nobel Laureates"* [UAHUB: III, 29], and in this context he had sent questionnaires to the corresponding scientists or their relatives. RICHTER had approached BIEBERBACH, since *"unfortunately*

Herr Geheimrat Nernst has not filled out completely his questionnaire; according to his words he had more important things to do" [UAHUB: III, 29].

In fact, at the bottom of the questionnaire NERNST had written: *"N. B. I hope that also all other Nobel Laureates have more important things to do, than to answer all your questions!!!"* [UAHUB: III, 26]. However, this hope did not come true, since on January 4, 1936 RICHTER would proudly report, *"that 28 questionnaires had been answered already from America, France, Scandinavia, England, and Germany."* [UAHUB: III, 22; Bartel (1990): 131-133]. Among those, of whom RICHTER had obtained their genealogical table, were IRVING LANGMUIR, HAROLD C. UREY, MARIE CURIE, IRÈNE and FRÉDÉRIC JOLIOT-CURIE, ERNEST RUTHERFORD, Sir WILLIAM RAMSAY, SVANTE ARRHENIUS, THE SVEDBERG, EMIL FISCHER, CARL BOSCH, FRIEDRICH BERGIUS and HANS FISCHER (complete list: [Bartel (1990): 133]). Different to his foreign and German colleagues or their relatives or in the case of MARIE CURIE even to the Archiepiscopal Ordinariate in Warsaw, NERNST had apparently understood that RICHTER's questionnaires had an anti-Semitic and *"racial-biological"* nature and wanted to demonstrate the superiority of a master race above lower races in a physical and an intellectual sense, roughly like the doctrine of FRANCIS GALTON of the improvement of the human race (eugenics). He could have read clear indications of this kind from LENARD or could have heard them from BIEBERBACH in 1934 in a meeting of the Academy. Certainly he had also connected the defamation of EINSTEIN and of other Jewish intellectuals with RICHTER's intentions.

Initially RICHTER had asked BIEBERBACH only to arrange for him access to the genealogical documents of NERNST existing possibly in the University files. Since such documents did not exist, BIEBERBACH at first suggested other sources. However, on December 30, 1935 he wrote to RICHTER: *"Regarding the matter of Geheimrat Prof. Dr. Nernst I would like to report to my superior Ministry ..., perhaps, that the required information can be obtained in this way."* [UAHUB: III, 27]. On January 7, 1936 BIEBERBACH informed Secretary RUST about RICHTER's project and its questionnaires and indicated: *"Except for the Jewish Laureates*

Wallach and Willstaetter, only Herr Geheimrat Nernst presented difficulties. ... The reason of my report is mainly the fact, that I believe that from the scantiness of the information I must conclude that Geheimrat Nernst, who earlier has confirmed officially that he would be of Aryan descent, has only astonishingly little knowledge about his ancestors. Therefore, I would consider it appropriate and I am asking for it, to be authorized to ask Herrn Geheimrat Nernst officially, to fill out for me the so-called large questionnaire (form II, 64/65) for the files. I am asking for this request also, since from the comment made by Herr Geheimrat Nernst in connection with the non-answering of the questionnaire of the candidate Richter it emerges that Herr Geheimrat Nernst still does not appear to have the correct idea about the meaning and the importance of the relevant concepts of the new Reich." [UAHUB: III, 21]. As a result on February 17 the Secretary ordered the University administration *"to have the customary questionnaires for proving the Aryan descent filled out for Prof. Dr. Nernst and his wife and to present them to me again at the largest possible speed."* [UAHUB: III, 19].

Since primarily NERNST did not want to ruin for himself the possibility to be allowed to travel to Oxford because of the reason we have mentioned, he gave in: *"z. Z. Rittergut [manor] Oberzibelle $^{25}/_2$ 36. Dear Herr Dekan [Dean]! Following your letter from the 20th of the month, which I found here having just returned from my lecture trip to Vienna, I have immediately filled out the attached questionnaires, a few, likely quite irrelevant gaps I could still provide, which, however, would require some time. Most sincerely yours W. Nernst."* [UAHUB: III, 17]. It is interesting that the "State Office of Clan Research" (*Reichsstelle für Sippenforschung*) asked the University *"for the speedy return of the proof of descent (questionnaire) of Prof. Dr. Walter Nernst for inspection"* and its *"immediate transmission"* [UAHUB: I, 92].

The trip to Oxford did not happen any more in 1936. The following year was that of NERNST's 50th anniversary of his PhD. Perhaps LINDEMANN wanted to push for the visit of his admired teacher by applying for an honorary PhD of NERNST at the University of Oxford in connection with this memorable event. It was not so much this recognition, but instead the personal and family reasons we had mentioned, which now caused NERNST to adopt means, which did not correspond to his attitude, but rather to his capability to express for pragmatic reasons and using ironic phrases exactly what truly was not meant at all. In this case NERNST did not even shy away from using the so-called *«deutschen Gruß»* (German greeting) in official letters. This behavior can be documented by three letters, which he addressed to the Dean BIEBERBACH relevant for him, in order to win him over directly and indirectly for his trip to Oxford. So on May 25, 1937 he wrote from Bad Elster: *"Dear Colleague! By postponing for a few days the 'official' thanks to you as Dean, being on an automobile trip, already today I want to thank you quite personally most cordially for your words which were humanely as well as scientifically so impressive for me!"* [UAHUB: III, 2–3]. The announced thanks came on May 31: *"To your Spectability I want to express my most sincere thanks for your extremely kind congratulation to the 50th anniversary of my oral PhD examination in Wurzburg. I have admired quite particularly, that the Herr Dekan* [Dean] *is so well informed about a former member of his Faculty! Heil Hitler! Dr. Walther Nernst"* [UAHUB: III, 1]. On June 2, 1937 BIEBERBACH received the news: *"Herewith I inform you most devotedly, as happened already today with you and the Herr Prorector, that the University of Oxford has invited me to accept the Honorary PhD of Sciences in a ceremonial session on Wednesday, June 23. Herewith I apply also in writing with nine copies to effect the approval of the Mr. Secretary for accepting the above distinction; upon inquiry at the Ministry the Herr Prorector was informed already today that there would be no problems. Heil Hitler! Prof. Dr. W. Nernst."* [UAHUB: III, 41].

NERNST's tactics were successful. On June 4 BIEBERBACH forwarded the application *"with approval in eight copies"* [UAHUB: II/III, 4/40], and on June 15 the following letter was sent to NERNST: *"Upon your application from June 2, 1937 I approve that on June 23, 1937 you travel to a session in Oxford, in order to accept the Honorary PhD of Sciences. ... Furthermore, during your stay abroad if possible you have to contact the local foreign organization of the NSDAP* [Nazi party]*, which will be* *informed about your trip. ... The Rector* [H. KRÜGER] *p.p. Hoppe"* [UA-HUB: II, 1].

Fig. 8.1   FREDERICK A. LINDEMANN and WALTHER NERNST 1937 in Oxford.

NERNST's actual position in Germany during the time of the fascism starting in 1933 can be seen from the statement of MAX BODENSTEIN, at the time remarkably courageous and discussed already in Section 4.4, in connection with NERNST's daughters, *"two of which together with their families are living abroad, because of the difficulties of the present times almost inaccessible, ..., a heavy fate, which he also felt as such – although he had never complained about this."* [Bodenstein (1942a): 81]. Within this context also the memory of NERNST's former student EMIL ABEL is revealing: *"On New Year's Day in 1939 I received in Vienna an open postcard from Professor Nernst, ignoring the possible dangers for the writer and the recipient: 'Hold out and wait for better times.' These few words tell much about the mental attitude of Nernst."* [Abel (1954): 156].

8.2   An Attempt to Participate in the War-Related Research during the Second World War

"Unfortunately I must take it easy after a heart attack 1½ years ago." WALTHER NERNST had written to WALTER OSTWALD on September 14, 1940 (see Fig. 5.11). In spite of his bad health, his opposition to the Nazi regime, and his painful experiences during the First World War, one year before his death for completely inexplicable reasons the scholar put himself at the disposal for research relating to the Second World War. PAUL

GÜNTHER reported about this very last research project of his former teacher [Günther (1951): 558] (see also [Bartel (1992): 43]). Since 1936 GÜNTHER directed the Physical-Chemical Institute, in which existed a Section since the beginning of the war, where war-related research was carried out. It was reported that a classified Section of the Institute had worked for the Office "Foreign Countries/Counter-Intelligence" of the High Command of the Army directed by WILHELM CANARIS. Also MAX BODENSTEIN, GÜNTHER's predecessor, who had been expelled from his *Ordinariat* against his wish, had returned to the *Bunsenstraße*. He had been called back in 1939 to take up teaching duties again in order to provide the current Institute Director with freedom for war-related projects.

On the side of the enemy without NERNST's knowledge his favorite former student FREDERICK A. LINDEMANN was actively involved in war-related research in a high position [Bartel (1992): 43–44]. Because of the acquaintance of LINDEMANN with Sir WINSTON CHURCHILL starting in 1921 and developing later into a friendship, it happened that during the Second World War *"Churchill appointed him his personal adviser."* [Birkenhead (1962): 220]. Together with CHURCHILL's political career also LINDEMANN's position went up: in 1939 he started the S[tatistical] Branch, and he also became its Director when in 1940 it changed into CHURCHILL's Prime Minister's Statistical Section. Already as Viscount CHERWELL in 1942, LINDEMANN was elevated to the rank of Pay Master General. Similar to NERNST during the First World War on the German side, now he had a strong influence on the British conduct of the war. LINDEMANN invented several technical procedures connected with the air warfare. A proposal by LINDEMANN, the English physicist most closely connected with the German people and their culture prior to the fascist period, had terrible consequences. This was expressed by MAX BORN, an emigrant himself on the British Isle, with the words: *"It was his idea to break the fighting spirit of the German people by means of air attacks upon the centers of the large cities."* [Born (1972): 101].

As reported by GÜNTHER, 1940 in Berlin NERNST put himself at the disposal of the Navy, in order to improve the compressed-air drive of torpedoes. For this purpose he wanted to apply the slowly burning propulsion elements he had developed during the First World War for trench-mortars. The work was carried out in a room in the basement of his former Institute in the *Bunsenstraße*, where he was assisted by a laboratory technician. Because NERNST had been only poorly informed by the Navy, he got for himself literature on naval warfare. Also the vessel provided by this client, in which the experiments were carried out for some time, turned out to be completely unsuitable. It was *"torn up"* because of *"an unwanted, but not at all unexpected explosion."* [Günther (1951): 558]. A few days after this accident NERNST returned once again to the Institute, in order to inform himself about the repair of the damage and to announce that he did not continue to work on the project and, hence, would not return any more to the Institute. *"When he went out of the door he pointed with his finger to the opposite wall, which bordered a room in the basement of the Physical Institute located within the same building, and smiling he quoted the line of Platen: 'More than half of this world was mine.'"* [Günther (1951): 558].

8.3 The End in the Village of Zibelle

We have pointed out already in Section 5.8 that after leaving his Chair at the University of Berlin in 1933, WALTHER NERNST retired to his manor Oberzibelle. This location became now his proper place of residence. However, occasionally the *Emeritus* also went back to Berlin. So for example, on December 24, 1935 it was stated in a "political certification": *"Regarding the Volksgenosse* [national comrade] *Dr. phil. Walther Nernst ... [,] residing in Berlin NW 40, Hindersinstr. 5[,] here in political matters nothing detrimental has become known."* [UAHUB: II, 8].

These stays in Berlin were caused by the fact that NERNST further engaged himself scientifically and in the organization and the politics of science also after he had been released from his teaching and management duties. His research effort in the field of cosmology and astrophysics has been discussed above. He continued to attend events organized by the Physical Society and by the Academy of Berlin. NERNST had left the Senate of the Kaiser-Wilhelm-Society (KWG), however, at the request of the Society he remained a member of the Board of the Physics Institute and of the Research Station Jungfraujoch. On June 7, 1933 he had indicated to MAX PLANCK, at the time President of the Society: *"I am quite happy to offer with my modest means my cooperation with the corresponding Boards, further serving the KWG, however, with the reservation that I am allowed to resign any time if I run into certain difficulties."* [AGMPG; Zott (1996): 223].

After the *"late disappointment"* as it was called by PAUL GÜNTHER, which *"struck him in his rank as a scientist"* [Günther (1951): 558], which we have discussed in the last Section, the scholar now being quite ill retired nearly completely to his manor Oberzibelle together with his wife. Upon the increasing deterioration of NERNST's health his daughter EDITH visited Zibelle more frequently together with her family. On November 15, 1941 she was called urgently to the sick-bed of her father, whom she met, however, only unconscious. A few days later, on November 18 a quarter before three o'clock in the morning WALTHER NERNST passed into the final eternal rest within the rustic environment which he had loved since his childhood.

"And now after this rich life Walther Nernst had passed away. As I was told by Mrs. Nernst, it was one of his last words, 'I have always striven for the truth.' And this is truly what we recognize in all his words and works. He was a divinely gifted searcher of the truth, but he was also a divinely gifted finder of the truth, and in both qualities he will always live in our memory and far beyond in the history of our science." [Bodenstein (1942a): 104]. MAX BODENSTEIN finished his obituary with these words.

JOHANNA ZIEGLER, who was employed at the time in the household of NERNST similar to her sister HILDEGARD HÖGEL, reported: *"We have also witnessed the funeral, when he passed by us on a gun-carriage."* [Muche (19091): 17]. In Berlin on November 21 WILLY HOPPE, Rector of the Friedrich-Wilhelm-University, announced: *"Hereby I make the sad announcement that the Full Professor in the Faculty of Mathematics and the Natural Sciences of the University of Berlin, Geheimer Regierungsrat Dr. phil., Dr. phil. h.c., Dr. med. h.c., Dr.-Ing. e.h., Dr. sc.*

h.c. Walther Nernst died during the night from the 17th to the 18th of the month. The cremation takes place on Tuesday, the 25th of the month at 14.30 o'clock at the crematorium Berlin-Wilmersdorf." [UAHUB: III, 95]. As the representative of the Faculty of Mathematics and the Natural Sciences PAUL GÜNTHER spoke during the memorial service in the crematorium at the coffin of his teacher: *"And if now very many students in our homeland and in the whole world take their leave from him, Walther Nernst, who himself during his long life has seen much death and leave taking, then for the German science there remains a high and lasting legacy, for every achievement at its location the valid standard. And for everybody, who is sensitive to the greatness of his work, there remains a consolation: it is the consolation which even during difficult times of war in 1917 he himself felt in his work as he testified in the introduction to one of his books, the consolation which results from each truly deeper insight into the given law of the world."* [UAHUB: II, 39; Schultze (1992): 53].

Fig. 8.2 Grave of WALTHER NERNST at the Central Cemetery in Göttingen.

NERNST had decreed not to be buried in Göttingen, since the *Georgia Augusta* had turned into a stronghold of the Nazis. So his urn was buried at the St. Thomas Cemetery in Berlin-Neukölln. Ten years later under the date of November 1, 1951 the «*Göttinger Chronik*» noted: *"Today in a simple ceremony the urns of Professor Walther Nernst and his wife were buried at the Central Cemetery."* [SAG]. In 1949 EMMA NERNST had died and was buried in London. Today the couple is laid to rest together with their daughters HILDE and EDITH at the cemetery of the town (Fig. 8.2), in which they were married, their children were born, and in which the great scholar had started his academic career. Very nearby MAX PLANCK, MAX VON LAUE, and OTTO HAHN found their last resting-place.

Chapter 9

Honors and Memorials

The importance of WALTHER NERNST and the distinctive features of his personality can be seen also from the many obituaries written by his former students and colleagues at many locations in the world. In the previous Chapters we have quoted several times from those of MAX BODENSTEIN [Bodenstein (1942a)] and of ALBERT EINSTEIN [Einstein (1942)]. Here standing in for many others we want to quote the concluding remarks from the obituaries from NERNST's former students FREDERICK A. LINDEMANN and FRANZ SIMON and from the physical chemist ANTON SKRABAL of Vienna: *"On those who knew him, Nernst made an unforgettable impression. His quickness to seize a new idea, his profundity in apprehending its application, his clarity in presenting the most intricate trains of thought, marked him out amongst the scientists of his time. Though he did not suffer fools gladly he was an excellent friend of those who were able to appreciate him, and his pupils who remember his kindness and sense of humour, his generosity and devotion to their interests, will for ever gratefully treasure his memory."* [Cherwell and Simon (1942): 106] – *"However, how many well-known scholars had to wait many years – if they experienced it at all – until their accomplishments were recognized and appreciated by the professional community. Quite different in the case of Nernst! The problems studied and solved by him were always relevant at the time, and so his research results immediately found interest and acceptance with the contemporaries. ... Now with the passing away of Walther Nernst an era has finally come to an end, which the historians of our science will call that of the 'classical physical chemistry'."* [Skrabal (1942): 199].

The recognition, which NERNST could experience during his life, found its expression in many awards and honors. The Nobel Prize in Chemistry, the honorary membership and the Memorial Medal of the German Bunsen Society, the Franklin Medal of the Franklin Institute, and the membership in the Academy of the Sciences of Berlin and in the Ordre pour le Mérite have been mentioned already. Also the German Society of Technical Physics had made NERNST an honorary member. In addition to the Academies of Berlin, Göttingen, and Munich, as time went on NERNST became honored by the membership in the Academies of Turin, Modena, Venice, Budapest, Oslo, Stockholm, Vienna, and Leningrad (now Saint Petersburg), as well as in the Royal Society of London. Several Universities awarded the great scientist an honorary doctorate. So NERNST had become Dr. phil. h.c. in Graz, Dr. med. h.c. in Erlangen and Göttingen, Dr.-Ing. e.h. in Danzig and Munich, and Doctor of Science in Oxford.

Also the many articles dedicated to NERNST on the occasions of his anniversaries have to be mentioned among the honors received by him. In this case it is interesting that his 50th birthday has been celebrated already in this way [Bugge (1914)]. On the occasion of the 25th anniversary of his PhD his former students prepared a commemorative volume with 487 pages [FS Nernst (1912)]. Also after his death, on the occasion of anniversaries and commemoration days connected with him, articles have been published which indicate that the person and the work of WALTHER NERNST cannot be forgotten.

The same also applies to the colloquia and similar events organized because of such reasons. So in 1964 on the occasion of NERNST's 100th birthday at the University of Göttingen a memorial ceremony of the Academy of the Sciences and of the German Bunsen Society during its meeting, and the Walther-Nernst-Memorial-Symposium at the Humboldt University in Berlin lasting several days were organized. Similar to the meeting of the Bunsen Society in Berlin in 1964, at the latter, JOHN EGGERT had covered the life and the activities of his teacher in a memorial lecture. During the memorial symposium the lecture hall, in which in 1905 NERNST had formulated for the first time the Third Law of Thermodynamics, was given his name, as is indicated by the bronze-plaque unveiled at this location (Fig. 5.7). In addition to many other personalities, PAUL GÜNTHER, KURT MENDELSSOHN, and NERNST's daughters ANGELA and EDITH had participated in the symposium. The latter summarized the impression of the sisters with the words *"We obtained such an extensive picture of the activity of my father as it could not have been more magnificent."* [Zanthier (1964b)].

In 1983 in connection with a meeting on the occasion of the 100th anniversary of the existence of the research laboratory in the *Bunsenstraße* in Berlin, of course, in the historic lecture by FRIEDRICH HERNECK [Herneck (1985)] also NERNST's activities from 1905 until 1922 were covered. Furthermore, a memorial plaque for NERNST and BODENSTEIN was attached to the historic building. In June of 1996 in Göttingen the hundred-year ceremony of the opening of the Institute of Physical Chemistry and Electrochemistry founded by NERNST took place. Since 1999 at the location where the former Physical Institute of the University of Berlin existed and where today the *Arbeitsgemeinschaft der Rundfunkanstalten Deutschlands* (ARD) (Working Group of the Radio Stations of Germany) resides, a plaque reminds us of the activity of NERNST at this former important research place of physics.

On the 50th anniversary of the death of the scholar, in November 1991 colloquia were organized at the Humboldt University of Berlin and at the Göttingen University. On this occasion already in June a memorial plaque had been fixed at the birthplace of NERNST in Wąbrzeźno/Briesen following an initiative supported by the German-Polish Society, the University of Göttingen, and the University of Torun, among others (Fig. 9.1) [Niedzielska *et al.* (1991)]. The memorial plaque unveiled in 1992 in Niwica/Zibelle serves similarly the honoring memory of WALTHER NERNST and the international understanding (Fig. 9.1). *"These memorial plaques for a great scientist from the German-Polish frontier region indicate the cultural bonds between both nations."* [WH (1993)].

In the year 2005 the event *"A Century of the Third Law of Thermodynamics – Memorial Colloquium in Honor of Walther Nernst"* was organized at the Humboldt University of Berlin.

W TYM DOMU URODZIŁ SIĘ LAUREAT NAGRODY NOBLA W DZIEDZINIE CHEMII **WALTHER NERNST** 1864 – 1941 IN DIESEM HAUS WURDE NOBELPREISTRÄGER IN CHEMIE GEBOREN MIESZKAŃCY WĄBRZEŹNA 1991	IN MEMORIAM Hier lebte / żył tudaj von / od 1922 r. – 1941 r. Prof. Dr. Walther H. Nernst NOBEL-Preis Chemie 1920 r. Präs. / Prez. PTR 1922 r. – 1924 r. * Briesen / Westpreußen 25.6.1864 r. Wąbrzeźno Prusy Zachodnie † Zibelle / Schlesien 18.11.1941 r. Niwica Śląska NOVEMBER / LISTOPAD 1992 r.

Fig. 9.1 Memorial plaques for WALTHER NERNST in Wąbrzeźno/Briesen (left) and in Niwica/Zibelle (right).

Translations
(left) *"In this house the Nobel Prize Laureate in Chemistry Walther Nernst (1864–1941) was born. / The citizens of Wąbrzeźno 1991"*
(right) *"In memoriam / Here, Prof. Dr. Walther H. Nernst (Nobel Prize in Chemistry 1920, President of the PTR 1922 – 1924, born Briesen/West Prussia 25/6/1864, died Zibelle/Silesia 18/11/1941) lived from 1922 – 1941. / November 1992"*

The naming of streets and roads in towns are also included in to the honors serving for keeping alive the memory of NERNST. So, in 1950 the *Schulstraße* in Hamburg-Ottensen was renamed *Nernstweg* and a little later (1951) the *Warburgstraße* in Leipzig-Möckern *Nernststraße*. 1957 Göttingen got a *Walther-Nernst-Weg*, and since 1974 there is also a *Nernstweg* in the Berlin District of Neukölln, where NERNST's final resting place was located for about a decade. On the ground of the former Academy of the Sciences of the German Democratic Republic (GDR) in Berlin-Adlershof within East-Berlin at the time a street was named in memory of the great scientist. It still exists now on the present Science Campus, officially since 1998. The lecture building of the Institutes for Chemistry and for Physics on the *Campus Adlershof* is named *Walther Nernst-Haus*.

As in the case of many other personalities of the sciences, also a crater on the moon was named after NERNST. It has the coordinates 35.3° latitude north and 94.8° longitude east. Its diameter is 116 km.

On the occasion of the 250th anniversary of the founding of the Academy of the Sciences in Berlin in 1950 the Post Office of the GDR issued a set of stamps showing the famous members. The 20-penny-stamp shows WALTHER NERNST (Fig. 9.2). Thirty years later in Sweden there were issued four 2-crown-stamps showing the Nobel Laureates of 1920, one of the stamps displaying NERNST and his lamp.

Fig. 9.2 WALTHER NERNST on a stamp of the German Democratic Republic of 1950.

Every external memory and all honors are exceeded by the fact, however, that the name of WALTHER NERNST is connected with several laws of nature, the importance of which cannot be overestimated. Furthermore, the statement by ANTON SKRABAL remains indelibly valid *"At the cradle of 'physical chemistry' as an independent field of knowledge there stood four men, whose names cannot be excluded from history of science: the propagandistic 'Feuergeist'* [fire-spirit, salamander] *of Wilhelm Ostwald, the prophet and finder Jacobus Henricus van't Hoff, the reformer Svante Arrhenius, whose theory of the ions exceeded every-*

thing in terms of boldness, and the youngest among them, the quietly thinking and reflecting Walther Nernst." [Skrabal (1942): 194]. So all, who seriously deal with the natural science inevitably will run into the work of the great scientist and human being. Hardly anybody can accomplish more than WALTHER NERNST, who dedicated his life to science and its advance.

References

Abel (1954)
E. Abel: Zur Erinnerung an Walther Nernst – Anläßlich seines 90. Geburtstages, 25. Juni 1954. *Österreichische Chemiker-Zeitung* **55** (1954), 11/12, 151–156.

AGMPG
Archiv zur Geschichte der Max-Planck-Gesellschaft, Akte Nernst, Abt. I, Rep. 1A, Nr. 3001.

Arrhenius (1887)
S. Arrhenius: Über die Dissociation der in Wasser gelösten Stoffe. *Zeitschrift für physikalische Chemie* **1** (1887), 631–648.

Arrhenius (1901a)
S. Arrhenius: Zur Berechnungsweise des Dissociationsgrades starker Elektrolyte. *Zeitschrift für physikalische Chemie* **36** (1901), 28–40.

Arrhenius (1901b)
S. Arrhenius: Zur Berechnungsweise des Dissociationsgrades starker Elektrolyte II. *Zeitschrift für physikalische Chemie* **37** (1901), 315–322.

Arrhenius (1903)
S. Arrhenius: *Lehrbuch der kosmischen Physik.* (Two tomes), S. Hirzel: Leipzig 1903.

Asen (1955)
J. Asen: *Gesamtverzeichnis des Lehrkörpers der Universität Berlin, Band I: 1810 – 1945.* O. Harrossowitz: Leipzig 1955.

Assche (1980)
P. van Assche: The Ignored Discovery of the Element Z = 43. *Nuclear Physics* A**480** (1980), 205–214.

BAK
Bundesarchiv Koblenz, Dienststelle Potsdam. Reichserziehungsministerium (REM) Nr. N 43 (Personalakte W. Nernst), Blatt 6.

Barkan (1995)

D. K. Barkan: *Theory, Practice, and a Perspectival View of Nature: The Work of Walther Nernst from Electrochemistry to Solid State Physics.* California Institute of Technology: Pasadena 1995.

Barkan (1999)

D. K. Barkan: *Walther Nernst and the Transition to Modern Physical Science.* Cambridge University Press: Cambridge, New York, Melbourne 1999.

Bartel (1988)

H.-G. Bartel: Theoretische Chemie im Wandel. *Spectrum* **19** (1988) 10, 22–24.

Bartel (1989)

H.-G. Bartel: *Walther Nernst* (Biographien hervorragender Naturwissenschaftler, Techniker und Mediziner, Band 90). B.G. Teubner Verlagsgesellschaft: Leipzig 1989.

Bartel (1990)

H.-G. Bartel: Die Stellung Walther Nernsts zum Faschismus. *Internationale Studien – Leipziger Hefte zur Friedensforschung: Ärzte und Chemiker in der Konfrontation mit Krieg und Faschismus.* Heft Nr. 8, Karl-Marx-Universität Leipzig 1990, 124–141.

Bartel (1992)

H.-G. Bartel: Walther Nernst und Frederick Alexander Lindemann als militärische Forscher und Berater – Anmerkung für eine Analyse ihres Verhaltens in den Weltkriegen. *Wissenschaftliche Zeitschrift der Humboldt-Universität zu Berlin, Reihe Mathematik/Naturwissenschaften* **41** (1992) 4, 41–44.

Bartel (1996)

H.-G. Bartel: *Mathematische Methoden in der Chemie.* Spektrum: Heidelberg, Berlin, Oxford 1996.

Bartel et al. (1983)

H.-G. Bartel, G. Scholz, F. Scholz: Die Nernst-Lampe und ihr Erfinder. *Zeitschrift für Chemie* **23** (1983), 277–287.

Bechstein (1931)

C. Bechstein Pianofortefabrik, Berlin: *Die Resonanz der Presse über den neuen Bechstein-Siemens-Nernst Flügel.* 1931.

Bennewitz (1909)

K. Bennewitz: Beiträge zur Frage der Zersetzungsspannung. *Inaugural-Dissertation*, Friedrich-Wilhelms-Universität zu Berlin 1909; *Zeitschrift für physikalische Chemie* **72** (1910), 202–224.

Berthelot (1879)

M. Berthelot: *Essai de mécanique chimique fondée sur la thermochimie, tome 2.* Dunod: Paris 1879.

Bieberbach (1934)
L. Bieberbach: Stilarten mathematischen Schaffens. *Sitzungsberichte der Preußischen Akademie der Wissenschaften zu Berlin, Physikalisch-mathematische Klasse* **1934**, 351–360.

Birchall et al. (1973)
J.D. Birchall, J. Park (Imp. Chem. Ltd., London): GB Patent 1 449 510 (11/27/1973); L.H. Cadoff (Westinghouse Electric Corp, Pittsburgh): US Patent 4 016 446 (1/31/1975).

Birkenhead (1962)
F. [W. F. Furneaux Smith] 2nd Earl of Birkenhead: *The Professor and the Prime Minister: The Official Life of Professor F.A. Lindemann, Viscount Cherwell*. Riverside Press: Cambridge, Boston 1962.

BIZ (1915)
Die Wissenschaft und der Krieg. *Berliner Illustrirte Zeitung* **24** (1915)-08-29=35, 475–476.

Bloehm (1916)
W. Bloem: Der Geist im Heere. In: Bund deutscher Gelehrter und Künstler (Kulturbund) (ed.): *Deutsche Volkskraft nach zwei Kriegsjahren – Vier Vorträge*. B.G. Teubner: Leipzig, Berlin 1916, 24–34.

Blücher (1906)
H. Blücher (ed.): *Auskunftsbuch für die Chemische Industrie (V./VI. Jahrgang 1906/1907)*. G. Ziemsen: Berlin 1906.

Bodenstein (1934)
M. Bodenstein: Walther Nernst zum siebzigsten Geburtstage. *Die Naturwissenschaften* **22** (1934), 437–439.

Bodenstein (1942a)
M. Bodenstein: Walther Nernst (25.6.1864 – 18.11.1941). *Berichte der Deutschen Chemischen Gesellschaft* **75** (1942) 6A, 79–104.

Bodenstein (1942b)
M. Bodenstein: Gedächtnisrede auf Walther Nernst. *Jahrbuch der Preußischen Akademie der Wissenschaften zu Berlin* 1942, 140–142.

Bodenstein (1942c)
M. Bodenstein: Hundert Jahre Photochemie des Chlorknallgases. *Berichte der Deutschen Chemischen Gesellschaft* **75A** (1942), 119–136.

Böhme (1975)
K. Böhme (ed.): *Aufrufe und Reden deutscher Professoren im Ersten Weltkrieg*. (Reclams Universal-Bibliothek Nr. 9787). Reclam: Stuttgart 1975, 47–49.

Bondi and Gold (1948)
H. Bondi, T. Gold: The Steady-State Theory of the Expanding Universe. *Monthly Notices of The Royal Astronomical Society* **1948**, 252–270.

Bonhoeffer (1922)
  K.F. Bonhoeffer: Photochemische Sensibilisierung und Einstein'sches Äquivalentgesetz. *Inaugural-Dissertation*, Universität Berlin 1922.
Bonhoeffer (1943)
  K.F. Bonhoeffer: Zur Theorie des elektrischen Reizes. *Die Naturwissenschaften* **31** (1943), 270–275.
Born (1972)
  M. Born: *Albert Einstein, Hedwig und Max Born, Briefwechsel 1916–1955*. Rowohlt: Reinbek bei Hamburg 1972.
Brill (1905)
  O. Brill: Ueber einige Erfahrung beim Gebrauch der Mikrowaage für Analysen. *Berichte der Deutschen Chemischen Gesellschaft* **38** (1905), 140–146.
Browne (1995)
  P.F. Browne: The Cosmological Views of Nernst: an Appraisal. *Apeiron* **2** (1995), 72–78.
Bugge (1914)
  G. Bugge: Walter Nernst – Zum 50. Geburtstag am 25. Juni. *Reclams Universum-Jahrbuch* **30** (1914), 257–259.
Busch (1920)
  H. Busch: Widerstände mit rückfallender Charakteristik. *Physikalische Zeitschrift* **21** (1920), 632–634; Über die Erwärmung von Drähten in verdünnten Gasen durch den elektrischen Strom. *Annalen der Physik* **64** (1921), 401–450.
Callen (1960)
  H.B. Callen: *Thermodynamics – An Introduction to the Physical Theories of Equilibrium Thermostatics and Irreversible Thermodynamics*. J. Wiley & Sons: New York 1960.
Casimir (1964)
  H.B.G. Casimir: Walther Nernst und die Quantentheorie der Materie. *Berichte der Bunsengesellschaft* **68** (1964), 530–534.
Cassel (1916)
  H. Cassel: Über Entflammung und Verbrennung von Sauerstoff-Wasserstoff-Gemischen. *Annalen der Physik* 4. Folge **51** (1916), 685–704.
Cherwell and Simon (1942)
  [Viscount] Cherwell, F. Simon: Walther Nernst 1864–1941. *Obituary Notices of Fellows of the Royal Society of London* **4** (1942), 11, 101–112.
Clusius (1943)
  K. Clusius: Dem Andenken an Walther Nernst: Spezifische Wärmen von Festkörpern. *Die Naturwissenschaften* **31** (1943), 397–400.

Cremer (1971)
E. Cremer: Max Bodenstein in memoriam. *Berichte der Bunsen-Gesellschaft für Physikalische Chemie* **75** (1971), 964–967.
Cremer (1987)
E. Cremer: Walther Nernst und Max Bodenstein. In: *Berlinische Lebensbilder „Naturwissenschaftler"*. Colloquium Verlag: Berlin 1987, 183–202.
CZ (1917)
Chemiker-Zeitung **41** (1917).
Czapski (1884)
S. Czapski: Ueber die thermische Veränderlichkeit der electromotorischen Kraft galvanischer Elemente und ihrer Beziehung zur freien Energie derselben. [Wiedemanns] *Annalen der Physik und Chemie*, Neue Folge **21** (1884), 209–243.
Diestel (1896)
G. Diestel: Hochschulen. In: *Berlin und seine Bauten, Teil II (Der Hochbau / Öffentliche Bauten)*. Ernst & Sohn: Berlin 1896, 257 *et sqq.*
Dobel (1968)
R. Dobel (ed.): *Lexikon der Goethe-Zitate*. Artemis: Zürich, Stuttgart 1968, c. 1065, 25–27.
Drude and Nernst (1890)
P. Drude, W. Nernst: Einfluß der Temperatur und des Aggregatzustandes auf das Verhalten des Wismuths im Magnetfelde. *Nachrichten von der Königlichen Gesellschaft der Wissenschaften und der Georg-Augusts-Universität zu Göttingen* **1890**, 471–481; *Annalen der Physik und Chemie* 3. Folge **42** (1891), 568–580.
Drude and Nernst (1891)
P. Drude, W. Nernst: Über die Fluorescenzwirkung stehender Lichtwellen. *Nachrichten von der Königlichen Gesellschaft der Wissenschaften und der Georg-Augusts-Universität zu Göttingen* **1891**, 346–358; Annalen der Physik und Chemie 3 (1892), 460–474.
Drude and Nernst (1894)
P. Drude, W. Nernst: Über Elektrostriktion durch freie Ionen. *Zeitschrift für physikalische Chemie* **15** (1894), 79–85.
Ebert (1943)
L. Ebert: Dem Andenken an Walther Nernst: Dielektrizitätskonstante, Ionengleichgewichte und Ionenkräfte in Lösungen. *Die Naturwissenschaften* **31** (1943), 263–265.
Eder (1952)
F.X. Eder: *Moderne Meßmethoden in der Physik, Teil I: Mechanik, Akustik*. Deutscher Verlag der Wissenschaften: Berlin 1952.
Eggert (1918)
J. Eggert: Über Acetylensilber. *Zeitschrift für Elektrochemie* **24** (1918), 150–154.

Eggert (1926)
 J. Eggert: *Lehrbuch der physikalischen Chemie in elementarer Darstellung.* S. Hirzel: Leipzig 1926.
Eggert (1941)
 J. Eggert, L. Hock: *Lehrbuch der Physikalischen Chemie in elementarer Darstellung.* S. Hirzel: Leipzig ^{5}1941.
Eggert (1943a)
 J. Eggert: Walther Nernsts Lehrbuch. *Die Naturwissenschaften* **31** (1943), 412–415.
Eggert (1943b)
 J. Eggert: Erinnerungen an Walther Nernst. *Zeitschrift für den physikalischen und chemischen Unterricht* **56** (1943), 43–50.
Eggert (1964)
 J. Eggert: Walther Nernst – Zur hundertsten Wiederkehr seines Geburtstages am 25. Juni 1964. *Angewandte Chemie* **76** (1964), 445–455.
Eggert and Schimank (1917)
 J. Eggert, H. Schimank: Einige Vorlesungsversuche zur Theorie der Explosivstoffe. *Zeitschrift für Elektrochemie* **23** (1917), 189–192; *Chemiker-Zeitung* **41** (1917), 11.
Eggert and Schimank (1918)
 J. Eggert, H. Schimank: Über einige Vorlesungsversuche mit Acetylensilber. *Berichte der Deutschen Chemischen Gesellschaft* **51** (1918), 454–456.
Einstein (1907)
 A. Einstein: Die Plancksche Theorie der Strahlung und die Theorie der spezifischen Wärme. *Annalen der Physik* **22** (1907), 180–190.
Einstein (1920)
 A. Einstein: *Äther und Relativitätstheorie – Rede gehalten am 5. Mai 1920 an der Reichs-Universität zu Leiden.* J. Springer: Berlin 1920.
Einstein (1924)
 A. Einstein: Geleitwort. In: H. Diels: *T. Lucretius Carus. De rerum natura, lateinisch und deutsch II* [translation]. Wiedemannsche Verlagsbuchhandlung: Berlin 1924, IV a–b.
Einstein (1942)
 A. Einstein: The Work and Personality of Walther Nernst. *Scientific Monthly* **54** (1942), 195–196.
Einstein, Alf (1931)
 A. Einstein: Das Universalklavier. *Berliner Tageblatt, Morgenausgabe für Berlin* **60** (1931)-08-27=402, 2.
Enders (1988)
 B. Enders: *Lexikon der Musikelektronik.* Deutscher Verlag für Musik: Leipzig 1988.

Engels (1968)
F. Engels: *Dialektik der Natur.* In: K. Marx, F. Engels: *Werke, Band 20.* Dietz: Berlin 1968.
Ettingshausen and Nernst (1886)
A. v. Ettingshausen, W. Nernst: Über das Auftreten elektromotorischer Kräfte in Metallplatten, welche von einem Wärmestrom durchflossen werden und sich im magnetischen Felde befinden. [Wiedemanns] *Annalen der Physik und Chemie,* 3. Folge **29** (1886), 343–347; *Anzeiger der Kaiserlichen Akademie der Wissenschaften in Wien, Mathematisch-Naturwissenschaftliche Classe* **23** (1886), Nr. 13, 114–118; *Zeitschrift für Elektrotechnik* <Wien> **4** (1886), 549–551. Über das Hall'sche Phänomen. [Exners] *Repertorium der Physik* **23** (1887), 93–136; *Sitzungsberichte der Mathematisch-naturwissenschaftlichen Classe der Kaiserlichen Akademie* <Wien> **94** (1887), II. Abth., 560–610. Über das thermische und galvanische Verhalten einiger Wismuth-Zinn-Legierungen im magnetischen Felde. [Wiedemanns] *Annalen der Physik und Chemie,* 3. Folge **33** (1888), 474–492; *Sitzungsberichte der Mathematisch-naturwissenschaftlichen Classe der Kaiserlichen Akademie* <Wien> **96** (1888), II. Abth., 787–806.
Eucken (1907)
A. Eucken: Über den stationären Zustand zwischen polarisierten Wasserstoffelektroden. *Inaugural-Dissertation,* Friedrich-Wilhelms-Universität zu Berlin 1907; *Zeitschrift für physikalische Chemie* **59** (1907), 72–117.
Eucken (1909)
A. Eucken: Über die Bestimmung spezifischer Wärmen bei tiefen Temperaturen. *Physikalische Zeitschrift* **10** (1909), 586–589.
Eucken (1922)
A. Eucken: *Grundriss der physikalischen Chemie für Studierende der Chemie und verwandter Fächer.* Akademische Verlagsgesellschaft: Leipzig 1922.
Eucken (1930)
A. Eucken: *Lehrbuch der chemischen Physik.* Akademische Verlagsgesellschaft: Leipzig 1930.
Eucken (1943)
A. Eucken: Rückblicke auf die Entwicklung unserer Kenntnisse über die Molwärme der Gase. *Die Naturwissenschaften* **31** (1943), 314–322.
FS Nernst (1912)
Festschrift – W. Nernst zu seinem fünfundzwanzigsten Doktorjubiläum gewidmet von seinen Schülern. W. Knapp: Halle/S. 1912.
Gans (1906)
L. Gans: Gegen die chemische Reichsanstalt! *Die chemische Industrie* **29** (1906), 589–593.

Goethe (1893)
Goethes Werke, herausgegeben im Auftrage der Großherzogin Sophie von Sachsen. II. Abtheilung: Goethes Naturwissenschaftliche Schriften, 3. Band. Zur Farbenlehre – Historischer Theil I. H. Böhlau: Weimar 1893.

Großklaus and Heitzsch (1985)
S. Grossklaus, O. Heitzsch: *Walter Nernst – Leben und Schaffen im Deutschland der Jahrhundertwende.* (Jugendobjekt „Philosophische Probleme in den Naturwissenschaften"). Karl-Marx-Universität Leipzig 1985.

Günther (1924)
P. Günther: Die kosmologischen Betrachtungen von Nernst. *Zeitschrift für angewandte Chemie* **37** (1924), 454–457.

Günther (1951)
P. Günther: Zum 10. Todestag von Walther Nernst. *Physikalische Blätter* **7** (1951), 556–558.

Haag (1984)
H. Haag: Les sondage de paix de W. Nernst auprès de F. Philippson (1915–1917). *Bulletin de la Commission royale d'Histoire (Académie Royale de Belgique)* **CL** (1984), 328–356.

Haber (1905)
F. Haber: *Thermodynamik technischer Gasreaktionen, 7 Vorlesungen.* R. Oldenbourgh: München, Berlin 1905.

Haberditzl (1960)
W. Haberditzl: Walther Nernst und die Traditionen der physikalischen Chemie an der Berliner Universität. In: *Festschrift zur 150-Jahr-Feier der Humboldt-Universität zu Berlin, Band I.* Deutscher Verlag der Wissenschaften: Berlin 1960, 401–416.

Handbuch Physik (1926)
Handbuch der Physik, Band IX: Theorien der Wärme. Springer, Berlin 1926.

Herneck (1975)
F. Herneck: *Albert Einstein.* (Biographien hervorragender Naturwissenschaftler, Techniker und Mediziner, Band 14). B.G. Teubner Verlagsgesellschaft: Leipzig 1975.

Herneck (1985)
F. Herneck: Zur Geschichte der Physikalischen Chemie an der Berliner Universität. *Wissenschaftliche Zeitschrift der Humboldt-Universität zu Berlin, Mathematisch-Naturwissenschaftliche Reihe* **36** (1985) 1, 6–13.

Herrmann (1972)
D.B. Herrmann: Walther Nernst und sein Neo-Bechstein-Flügel – Eine Episode aus der Geschichte der elektronischen Musik. *Schriftenreihe für Geschichte der Naturwissenschaften, Technik und Medizin (NTM)* **9** (1972), 40–48.

Herz (1912)
W. Herz: *Leitfaden der theoretischen Chemie. Als Einführung in das Gebiet für Studierende der Chemie, Pharmazie und Naturwissenschaften, Ärzte und Techniker.* F. Enke: Stuttgart 1912, ²1920, ³1923, ⁴1930.

Hoechst (1966)
Farbwerke Hoechst Aktiengesellschaft (ed.): Vertrag mit Prof. W. Nernst. *Dokumente aus Hoechster Archiven* **18** (1966), 19–21.

Hoffmann (1990)
D. Hoffmann: Walter Nernst und die Physikalisch-Technische Reichsanstalt – Zum 125. Geburtstag des Gelehrten. *PTB-Mitteilungen: Forschen + Prüfen. Amts- und Mitteilungsblatt der Physikalisch-Technischen Bundesanstalt Braunschweig-Berlin* **100** (1990), 40–45.

Hoffmann (1992)
D. Hoffmann: Refugien eines Vielbeschäftigten. *Wissenschaftliche Zeitschrift der Humboldt-Universität zu Berlin, Reihe Mathematik/Naturwissenschaften* **41** (1992) 4, 37–39.

Hoffmann (1999)
D. Hoffmann: Nernst: architect of physical revolution. *Physics World* September 1999, 53.

Hoffmann and Schlicker (1987)
D. Hoffmann, W. Schlicker: Wissenschaft unter dem braunen Stiefel 1933 bis 1945. In: H. Laitko (ed.) *Wissenschaft in Berlin – Von den Anfängen bis zum Neubeginn nach 1945.* Dietz: Berlin 1987, 502–591.

Hohenester (1992)
A. Hohenester: Walther Nernst und die Grazer Physik. *Wissenschaftliche Zeitschrift der Humboldt-Universität zu Berlin, Reihe Mathematik/Naturwissenschaften* **41** (1992) 4, 13–22.

Huber and Jaakkola (1995)
P. Huber, T. Jaakkola: The Static Universe of Walther Nernst. *Apeiron* **2** (1995), 53–57.

Huebener (2001)
R.P. Huebener: *Magnetic Flux Structures in Superconductors.* J. Springer: Berlin ²2001.

Hund (1967)
F. Hund: *Geschichte der Quantentheorie.* Bibliographisches Institut: Mannheim 1967.

IMWKT (1914)
Walther Nernst in Argentinien. – Zur argentinischen Kultur; ein deutsches Institut in Buenos Aires. *Internationale Monatsschrift für Wissenschaft, Kunst und Technik* **8** (1914), col. 1285–1288.

Jaenicke (1996)
W. Jaenicke: *100 Jahre Bunsen-Gesellschaft 1894–1994.* Steinkopff: Darmstadt 1996.
Jahn (1882)
H. Jahn: *Die Grundzüge der Thermochemie und ihre Bedeutung für die theoretische Chemie.* Hölder: Wien 1882.
Jahn (1892)
H. Jahn: *Die Grundzüge der Thermochemie und ihre Bedeutung für die theoretische Chemie.* A. Hölder, Wien ^{2}1892.
Jahn (1895)
H. Jahn: *Grundriss der Elektrochemie.* A. Hölder: Wien 1895.
Jahn (1901)
H. Jahn: Ueber die Nernst'schen Formeln zur Berechnung der elektromotorischen Kraft der Concentrationselemente – Eine Erwiderung an Herrn Arrhenius. *Zeitschrift für physikalische Chemie* **36** (1901), 453–460.
Joffe (1967)
A.F. Joffe: *Begegnungen mit Physikern.* Teubner, Leipzig ^{2}1967.
Jost, F (1908)
F. Jost: Über die Verwendung eines elektrischen Druckofens bei Behandlung chemischer Gleichgewichte. 1. Über die Einwirkung von Wasserstoff auf Kohle bei hohen Temperaturen. 2. Über das Ammoniakgleichgewicht. *Inaugural-Dissertation,* Friedrich-Wilhelms-Universität zu Berlin 1908.
Jost, W (1964)
W. Jost: Zum 100. Geburtstag von Walther Nernst. *Berichte der Bunsengesellschaft für Physikalische Chemie* **68** (1964), 525–534.
Jost, W (1966)
W. Jost: The First 45 Years of Physical Chemistry in Germany. *Annual Review of Phyical Chemistry* **17** (1966), 1–14.
Kant (1786)
I. Kant: *Metaphysische Anfangsgründe der Naturwissenschaft.* J.F. Hartknoch: Riga 1786.
Kohlrausch and Maltby (1899)
F. Kohlrausch, M.E. Maltby: Das elektrische Leitvermögen wässriger Lösungen von Alkalichloriden und Nitraten. *Berichte der Königlichen Preußischen Akademie der Wissenschaften zu Berlin* **1899**, 665–671; *Wissenschaftliche Abhandlungen der Physikalisch-Technischen Reichsanstalt* **3** (1900), 156–227; *Zeitschrift für physikalische Chemie* **36** (1901), 750–752.
Körber (1969)
H.-G. Körber (ed.): *Aus dem wissenschaftlichen Briefwechsel Wilhelm Ostwalds, II. Teil.* Akademie-Verlag: Berlin 1969.

Körber and Ostwald (1961)
H.G. Körber; G. Ostwald (eds.): *Aus dem wissenschaftlichen Briefwechsel Wilhelm Ostwalds, I. Teil: Briefwechsel mit Ludwig Boltzmann, Max Planck, Georg Helm und Josiah Willard Gibbs.* Akademie-Verlag: Berlin 1961.
Kretschmann (1883)
H. Kretschmann: Kgl. evangel. Gymnasium zu Graudenz. XVII. Jahresbericht über das Schuljahr Ostern 1882 bis Ostern 1883. Graudenz 1883.
Krüger (1903)
F. Krüger: Über Polarisationskapazität. *Zeitschrift für physikalische Chemie* **45** (1903), 1–74.
Krüger (1939)
F. Krüger: 50 Jahre seit dem Erscheinen von W. Nernsts Arbeit über „Die elektromotorische Wirksamkeit der Ionen. *Die Naturwissenschaften* **27** (1939), 553–555.
Lepsius (1964)
R. Lepsius: Zur hundertsten Wiederkehr des Geburtstages von Walther Nernst. *Chemiker-Zeitung / Chemische Apparatur* **83** (1964), 603–606.
Loeb and Nernst (1888)
M. Loeb, W. Nernst: Zur Kinetik der in Lösung befindlichen Körper. 2. Überführungszahlen und Leitvermögen einiger Silbersalze. *Zeitschrift für physikalische Chemie* **2** (1888), 948–963.
Lomonosov (1961)
M.V. Lomonosov: *Die Elemente der mathematischen Chemie.* In: M.W. Lomonossow: Ausgewählte *Schriften in zwei Bänden, Band I.* Akademie-Verlag: Berlin 1961, S. 68–77.
Magnus (1906)
A. Magnus: Ein neues Widerstandsgefäfs zur Bestimmung des Leitvermögens von Flüssigkeiten. *Verhandlungen der Deutschen Physikalischen Gesellschaft* **8** (1906), 1–8.
Maltby (1895)
M.E. Maltby: Methode zur Bestimmung grosser elektrolytischer Widerstände. *Zeitschrift für physikalische Chemie* **18** (1895), 133–158.
Maltby (1897)
M.E. Maltby: Methode zur Bestimmung der Periode electrischer Schwingungen. *Wiedemanns Annalen der Physik* **61** (1897), 553–577.
Martius (1906)
C.A. von Martius: Eine Chemische Reichsanstalt? *Die chemische Industrie* **29** (1906), 135–139.
Mendelssohn (1973)
K. Mendelssohn: *The World of Walther Nernst – The Rise and Fall of German Science.* MacMillan: London, Basingstoke 1973; *Walther Nernst und seine Zeit – Auf-*

stieg und Niedergang der deutschen Naturwissenschaften. Physik Verlag: Weinheim 1976.
Mittasch (1951)
A. Mittasch: *Geschichte der Ammoniaksynthese.* Verlag Chemie, Weinheim 1951.
Muche (1989)
E.E. Muche: Zibelle und seine berühmten Mitbürger. I. Professor Dr. Walther Nernst. *Oberlausitzer Rundschau* 1989, Nr. 1, 17–18.
Muche (1991)
E.E. Muche: Zibelle und seine berühmten Mitbürger. IV. Teil. *Oberlausitzer Rundschau* 1991, Nr. 10, 17–18.
Nagel et al. (1991)
A. von Nagel et al.: *Stickstoff – Die Chemie stellt die Ernährung sicher.* Schriftenreihe des Unternehmensarchivs der BASF Aktiengesellschaft ^{2}1991.
Naumann (1984)
M. Naumann (ed.): *Artikel aus Diderots Enzyklopädie.* Reclam: Leipzig 1984.
Nernst (1887)
W. Nernst: *Über die elektromagnetischen Kräfte welche durch den Magnetismus in von einem Wärmestrome durchflossenen Metallplatten geweckt werden.* Metzger & Wittig: Leipzig 1887; [Wiedemanns] *Annalen der Physik und Chemie,* 3. Folge **31** (1887), 760–789, 1048.
Nernst (1888a)
W. Nernst: Über die Bildungswärme der Quecksilberverbindungen. *Zeitschrift für physikalische Chemie* **2** (1888), 23–28.
Nernst (1888b)
W. Nernst: Zur Kinetik der in Lösung befindlichen Körper. 1. Theorie der Diffusion. *Zeitschrift für physikalische Chemie* **2** (1888), 613–637.
Nernst (1889a)
W. Nernst: Die elektromotorische Wirksamkeit der Jonen. *Zeitschrift für physikalische Chemie* **4** (1889), 129–181.
Nernst (1889b)
W. Nernst: *Die elektromotorische Wirksamkeit der Jonen.* W. Engelmann: Leipzig 1889.
Nernst (1889c)
W. Nernst: Zur Theorie umkehrbarer galvanischer Elemente. *Sitzungsberichte der Königlich Preußischen Akademie der Wissenschaften zu Berlin* **1889**, 83–95.
Nernst (1890a)
W. Nernst: Über ein neues Prinzip der Molekulargewichtsbestimmung. *Nachrichten von der Königlichen Gesellschaft der Wissenschaften und der Georg-Augusts-Universität zu Göttingen* **1890**, 57–66; *Zeitschrift für physikalische Chemie* **6** (1890),

16–36; Über eine neue Verwendung des Gefrierapparates zur Molekulargewichtsbestimmung. *Zeitschrift für physikalische Chemie* **6** (1890), 573–577.
Nernst (1890b)
W. Nernst: Über die Verteilung eines Stoffes zwischen zwei Lösungsmitteln. *Nachrichten von der Königlichen Gesellschaft der Wissenschaften und der Georg-Augusts-Universität zu Göttingen* **1890**, 401–416; Verteilung eines Stoffes zwischen zwei Lösungsmitteln und zwischen Lösungsmittel und Dampfraum. *Zeitschrift für physikalische Chemie* **8** (1891), 110–139.
Nernst (1891)
W. Nernst: Physikalische Chemie. In: R. Meyer (ed.): *Jahrbuch der Chemie* **1** (1891), 1–66.
Nernst (1892a)
W. Nernst: Über die mit der Vermischung k(c)onzentrierter Lösungen verbundene Änderung der freien Energie. *Nachrichten von der Königlichen Gesellschaft der Wissenschaften und der Georg-Augusts-Universität zu Göttingen* **1892**, 428–438; *Annalen der Physik und Chemie* **53** (1894), 57–68.
Nernst (1892b)
W. Nernst: Allgemeiner Teil. In: O. Dammer (ed.): *Handbuch der anorganischen Chemie, Band I*. F. Enke: Stuttgart 1892, p. 1–358.
Nernst (1893a)
W. Nernst: Dielektric(z)itätskonstante und chemisches Gleichgewicht. *Nachrichten von der Königlichen Gesellschaft der Wissenschaften und der Georg-Augusts-Universität zu Göttingen* **1893**, 491–496; *Zeitschrift für physikalische Chemie* **13** (1894), 531–536; Methode zur Bestimmung von Dielektric(z)itätskonstanten. *Nachrichten von der Königlichen Gesellschaft der Wissenschaften und der Georg-Augusts-Universität zu Göttingen* **1893**, 762–774; *Zeitschrift für physikalische Chemie* **14** (1894), 622–663.
Nernst (1893b)
W. Nernst: *Theoretische Chemie vom Standpunkte der Avogadroschen Regel und der Thermodynamik*. F. Enke: Stuttgart 1893 (= 2nd edition of [Nernst (1892b)]).
Nernst (1894/95)
W. Nernst: Über Flüssigkeitsketten. *Zeitschrift für Elektrotechnik und Elektrochemie* **1** (1894/95), 153–155; Über die Auflösung von Metallen in galvanischen Elementen. *Zeitschrift für Elektrotechnik und Elektrochemie* **1** (1894/95), 243–246.
Nernst (1895)
W. Nernst: *Theoretical Chemistry from the Standpoint of Avogadro's Rule and Thermodynamics*. Macmillan: London, New York 1895.

Nernst (1896a)
W. Nernst: *Das Institut für Physikalische Chemie und besonders Elektrochemie an der Universität Göttingen. Festschrift zur Einweihungsfeier am 2. Juni 1896.* W. Knapp: Halle/S. 1896.

Nernst (1896b)
W. Nernst: *Die Ziele der physikalischen Chemie. Festrede gehalten am 2. Juni 1896 zur Einweihung des Instituts für physikalische Chemie und Elektrochemie der Georgia Augusta zu Göttingen.* Vandenhoeck & Ruprecht: Göttingen 1896.

Nernst (1896c)
W. Nernst: Zur elektrochemischen Messkunde. *Zeitschrift für Elektrochemie* **3** (1896), 52–54.

Nernst (1897a)
W. Nernst: Ueber das chemische Gleichgewicht, elektromotorische Wirksamkeit und elektrolytische Abscheidung von Metallgemischen. *Zeitschrift für physikalische Chemie* **22** (1897), 539–542.

Nernst (1897b)
W. Nernst: Verfahren zur Erzeugung von elektrischem Glühlicht. *Deutsches Reichspatent* Nr. 104872 (6.7.1897).

Nernst (1897c)
W. Nernst: Vorrichtung zum Erhitzen Nernstscher Glühkörper. *Deutsches Reichspatent* Nr. 107533 (2.10.1897).

Nernst (1899)
W. Nernst: Zur Theorie der elektrischen Reizung. *Nachrichten von der Königlichen Gesellschaft der Wissenschaften zu Göttingen, Mathematisch-physikalische Klasse* **1899**, 1, 104–108.

Nernst (1899/1900)
W. Nernst: Über die elektrolytische Leitung fester Körper bei sehr hohen Temperaturen. *Zeitschrift für Elektrochemie* **6** (1899/1900), 41–43.

Nernst (1900/01)
W. Nernst: Über Elektrodenpotentiale (Nach Versuchen und Berechnungen von Wilsmore). *Zeitschrift für Elektrochemie* **7** (1900/01), 253–257.

Nernst (1901a)
W. Nernst: Erwiderung auf einige Bemerkungen der Herren Arrhenius, Kohnstamm, Cohen und Noyes. *Zeitschrift für physikalische Chemie* **36** (1901), 596–604.

Nernst (1901b)
W. Nernst: Zur Theorie der Lösungen. *Zeitschrift für physikalische Chemie* **38** (1901), 487–500.

Nernst (1903)
W. Nernst: Über Molekulargewichtsbestimmungen bei sehr hohen Temperaturen. *Zeitschrift für Elektrochemie* **9** (1903), 622–627.

Nernst (1904)
W. Nernst: Theorie der Reaktionsgeschwindigkeit in heterogenen Systemen. *Zeitschrift für physikalische Chemie* **47** (1904), 52–55.
Nernst (1905)
W. Nernst: Physikalisch-chemische Betrachtungen über den Verbrennungsprozeß in den Gasmotoren. *Zeitschrift des Vereins deutscher Ingenieure* **1905**, 1–36.
Nernst (1906a)
W. Nernst: Über die Helligkeit glühender schwarzer Körper und über ein einfaches Pyrometer. *Physikalische Zeitschrift* **7** (1906), 380–383.
Nernst (1906b)
W. Nernst: Über die Beziehungen zwischen Wärmeentwicklung und maximaler Arbeit bei kondensierten Systemen. *Sitzungsberichte der Königlich Preußischen Akademie der Wissenschaften zu Berlin* **1906**, 933–940 (ausgegeben am 10.1.1907).
Nernst (1906c)
W. Nernst: Ueber die Berechnung chemischer Gleichgewichte aus thermischen Messungen. *Nachrichten von der Königlichen Gesellschaft der Wissenschaften zu Göttingen, Mathematisch-physikalische Klasse* **1906**, 1, 1–40.
Nernst (1907a)
W. Nernst: Experimental and Theoretical Application of Thermodynamics to Chemistry. Yale University Press: New York – New Haven 1907 (= Yale University Mrs. Hepsa Ely Silliman Memorial Lecture 4).
Nernst (1907b)
W. Nernst: Über das Ammoniakgleichgewicht (Nach Versuchen des Herrn F. Jost). *Zeitschrift für Elektrochemie* **13** (1907), 521–524.
Nernst (1908)
W. Nernst: Zur Theorie des elektrischen Reizes. *Pflügers Archiv für die gesamte Physiologie des Menschen und der Tiere* **122** (1908), 275–314.
Nernst (1909a)
W. Nernst: *Theoretische Chemie vom Standpunkte der Avogadroschen Regel und der Thermodynamik.* F. Enke: Stuttgart ^{6}1909.
Nernst (1909b)
W. Nernst: Über die Berechnung elektromotorischer Kräfte aus thermischen Größen. *Sitzungsberichte der Königlich Preußischen Akademie der Wissenschaften zu Berlin* **1909**, 247–267.
Nernst (1910a)
W. Nernst: Letter to Otto Wiener, October 27, 1910. Universität Leipzig, Handschriftenabteilung der Universitätsbibliothek.
Nernst (1910b)
W. Nernst: Letter to Otto Wiener,-November 27, 1910. Universität Leipzig, Handschriftenabteilung der Universitätsbibliothek.

Nernst (1911a)
W. Nernst: Letter to Otto Wiener, January 22, 1911. Universität Leipzig, Handschriftenabteilung der Universitätsbibliothek.
Nernst (1911b)
W. Nernst: Über einen Apparat zur Verflüssigung von Wasserstoff. *Zeitschrift für Elektrochemie* **17** (1911), 735–737.
Nernst (1911c)
W. Nernst: Zur Theorie der spezifischen Wärme und über die Anwendung der Lehre von den Energiequanten auf physikalisch-chemische Fragen überhaupt. *Zeitschrift für Elektrochemie* **17** (1911), 265–275.
Nernst (1911d)
W. Nernst: Über die Verträglichkeit des von mir aufgestellten Wärmetheorems mit der Gleichung von van der Waals bei sehr tiefen Temperaturen. *Koninklijke Akademie van Wetenschappen te Amsterdam, Verslag van de Gewone Vergaderingen der Wis- en Natuurkundige Afdeeling* **20** (1911), 64–67.
Nernst (1912a)
W. Nernst: Thermodynamik und spezifische Wärme. *Sitzungsberichte der Königlich Preußischen Akademie der Wissenschaften zu Berlin, physikalisch-mathematische Classe* **1912**, 134–140.
Nernst (1912b)
W. Nernst: Zur neueren Entwicklung der Thermodynamik (Verhandlungen der Gesellschaft Deutscher Naturforscher und Ärzte). A. Pries: Leipzig 1912.
Nernst (1913a)
W. Nernst: Elektrische Lampe, bei der als Hauptleiter Metalldampf, z. B. Quecksilberdampf, dient und bei der als färbende Bestandteile Salzdämpfe verwendet werden. *Deutsches Reichspatent* 288229 (19.8.1913).
Nernst (1913b)
W. Nernst: *Die Bedeutung des Stickstoffs für das Leben – Fest-Vortrag aus Anlaß der 10. Jahresversammlung gehalten in München am 1. Oktober 1913.* (= Deutsches Museum, Vorträge und Berichte 13). Deutsches Museum: München 1913.
Nernst (1913c)
W. Nernst: Zur neueren Entwicklung der Thermodynamik. *Verhandlungen der Gesellschaft Deutscher Naturforscher und Ärzte* **84** (1913), 100–116.
Nernst (1914a)
W. Nernst: Anwendung der Quantentheorie auf eine Reihe physikalisch-chemischer Probleme. In: A. Eucken (ed.): *Die Theorie der Strahlung und der Quanten. Verhandlungen einer von E. Solvay einberufenen Zusammenkunft (30. Oktober bis 3. November 1911).* (= W. Nernst (ed.): *Abhandlungen der Deutschen Bunsen-Gesellschaft für angewandte physikalische Chemie* 7). W. Knapp: Halle/S. 1914, 208–233 (Discussion 234–244).

Nernst (1914b)
W. Nernst: [Letter to Max Le Blanc, April 23, 1914]. *Zeitschrift für Elektrochemie* **20** (1914), 357.
Nernst (1914c)
W. Nernst: Über die Anwendung des neuen Wärmesatzes auf Gase. *Zeitschrift für Elektrochemie* **20** (1914), 357–360.
Nernst (1915)
W. Nernst: Zur Registrierung schnell verlaufender Druckänderungen. *Sitzungsberichte der Königlich Preußischen Akademie der Wissenschaften* **1915**, 896–901.
Nernst (1916a)
W. Nernst: Der Krieg und die deutsche Industrie. In: Bund deutscher Gelehrter und Künstler (Kulturbund) (ed.): *Deutsche Volkskraft nach zwei Kriegsjahren – Vier Vorträge*. B.G. Teubner: Leipzig, Berlin 1916, 12–23.
Nernst (1916b)
W. Nernst: Über einen Versuch, von quantentheoretischen Betrachtungen zur Annahme stetiger Energieänderungen zurückzukehren. *Verhandlungen der Deutschen Physikalischen Gesellschaft* **18** (1916), 83–116.
Nernst (1916c)
W. Nernst: Über die experimentelle Bestimmung chemischer Konstanten. *Zeitschrift für Elektrochemie* **22** (1916), 185–194.
Nernst (1917)
W. Nernst: Innere und äussere Ballistik der Minenwerfer. *Neue Leipziger Illustrierte Zeitung* (1917)-11-30=3882.
Nernst (1918a)
W. Nernst: *Die theoretischen und experimentellen Grundlagen des neuen Wärmesatzes*. W. Knapp: Halle/S. 1918.
Nernst (1918b)
W. Nernst: Zur Anwendung des Einsteinschen photochemischen Äquivalentgesetzes I. *Zeitschrift für Elektrochemie* **24** (1918), 335–336.
Nernst (1920)
W. Nernst: Zur Konstitution der Hydride (Nach Versuchen von Herrn K. Moers). *Zeitschrift für Elektrochemie* **26** (1920), 323–324.
Nernst (1921a)
W. Nernst: *Zum Gültigkeitsbereich der Naturgesetze – Rede zum Antritt des Rektorats der Friedrich-Wilhelms-Universität in Berlin, gehalten in der Aula am 15. Oktober 1921*. Norddeutsche Buchdruckerei und Verlagsanstalt: Berlin 1921.
Nernst (1921b)
W. Nernst: *Das Weltgebäude im Lichte der neueren Forschung*. J. Springer: Berlin 1921.

Nernst (1922a)

W. Nernst: *Ueber das Auftreten neuer Sterne – Rede zur Gedächtnisfeier des Stifters der Berliner Universität König Friedrich Wilhelms III in der Aula am 3. August 1922.* Norddeutsche Buchdruckerei und Verlagsanstalt: Berlin 1922.

Nernst (1922b)

W. Nernst: Letter to Svante Arrhenius, January 6, 1922 (p. 4). *Kungliga Vetenskapsakademien, Stockholm/Sverige, Nobelarkivet*, collection of manuscripts: Arrhenius, Svante.

Nernst (1923)

W. Nernst: Studien zur chemischen Thermodynamik – Nobel-Vortrag den 12 December [1921] in Stockholm gehalten. In: *Les Prix Nobel en 1921–1922.* Norstedt & fils: Stockholm 1923.

Nernst (1924a)

W. Nernst: *Die theoretischen und experimentellen Grundlagen des neuen Wärmesatzes.* W. Knapp: Halle/S. ^{2}1924; *The New Heat Theorem, its Foundations in Theory and Experiment.* Dutton: New York 1926, Methuen: London 1926.

Nernst (1924b)

W. Nernst: T. Lucretius Carus. De rerum natura, lateinisch und deutsch von Hermann Diels. Bd. II: Lukrez, Von der Natur, übersetzt von Hermann Diels. (Buchbesprechung [book review]). *Deutsche Literaturzeitung für Kritik der internationalen Wissenschaft* **45**/Neue Folge **1** (1924), col. 1741–1743.

Nernst (1927)

W. Nernst: Zur Theorie der elektrischen Dissoziation. *Zeitschrift für Elektrochemie* **33** (1927), 428–431.

Nernst (1928a)

W. Nernst: Über die Berechnung der elektrolytischen Dissoziation aus der elektrischen Leitfähigkeit. *Sitzungsberichte der Preußischen Akademie der Wissenschaften zu Berlin, Physikalisch-mathematische Klasse* **1928**, 4–8.

Nernst (1928b)

W. Nernst: Zur Theorie der elektrischen Dissoziation. *Zeitschrift für physikalische Chemie* **135** (1928), 237–250.

Nernst (1928c)

W. Nernst: Physico-chemical Considerations in Astrophysics. *Journal of the Franklin-Institute* **206** (1928), 135–142.

Nernst (1929)

W. Nernst: Persönliche Erinnerung an Oberst Max Bauer. *Deutsche Allgemeine Zeitung (Ausgabe für Groß-Berlin)* **68** (1929)-05-12=216, 1.2; *(Reichs-Ausgabe)* **68** (1929)-05-14=216–217, 3.

Nernst (1930)
W. Nernst: Albert v. Ettingshausen. (Eine Erinnerung an meine Grazer Studienzeit.). *Elektrotechnik und Maschinenbau* **48** (1930), 279–281.
Nernst (1931)
W. Nernst: Kritische physikalische Bemerkungen zu neueren astrophysikalischen Theorien (Zusammenfassung). *Sitzungsberichte der Preußischen Akademie der Wissenschaften zu Berlin, Physikalisch-mathematische Klasse* **1931**, 430.
Nernst (1935a)
W. Nernst: Physikalische Betrachtungen zur Entwicklungstheorie der Sterne. *Zeitschrift für Physik* **97** (1935), 511–534.
Nernst (1935b)
W. Nernst: Einige Anwendungen der Physik auf die Sternentwicklung. *Sitzungsberichte der Preußischen Akademie der Wissenschaften zu Berlin, Physikalisch-mathematische Klasse* **1935**, 473–479.
Nernst (1937)
W. Nernst: Weitere Prüfung der Annahme eines stationären Zustandes im Weltall. *Zeitschrift für Physik* **106** (1937), 633–661.
Nernst (1938)
W. Nernst: Die Strahlungstemperatur des Universums. *Annalen der Physik* **5**, Folge 32 (1938), 44–48.
Nernst (1939)
W. Nernst: Zur Erinnerung an den hundertsten Geburtstag von Willard Gibbs. *Die Naturwissenschaften* **27** (1939), 393–394.
Nernst and Abegg (1894)
W. Nernst, R. Abegg: Über den Gefrierpunkt verdünnter Lösungen. *Nachrichten von der Königlichen Gesellschaft der Wissenschaften zu Göttingen, mathematisch-physikalische Klasse* **1894**, 141–153; *Zeitschrift für physikalische Chemie* **15** (1894), 681–693; **18** (1895), 658–661; R. Abegg, W. Nernst: On the Freezing-points of Dilute Solutions. *The London, Edinburgh, and Dublin Philosophical Magazine and Journal of Science* **5**, ser. 41 (1896), 196–199.
Nernst and Barratt (1904)
W. Nernst, J.O.W. Barratt: Über elektrische Nervenreizung durch Wechselströme. *Zeitschrift für Elektrochemie* **10** (1904), 664–668.
Nernst and Heffter (1922)
Rektorwechsel an der Friedrich-Wilhelms-Universität zu Berlin am 15. Oktober 1922. I. Bericht des abtretenden Rektors W. Nernst über das Amtsjahr 1921/22 – II. Rede des antretenden Rektors A. Heffter. Norddeutsche Buchdruckerei und Verlagsanstalt: Berlin 1922.
Nernst and Lange (1928)
W. Nernst, F. Lange: Kurt Urban †. *Die Naturwissenschaften* **16** (1928), 796.

Nernst and Lieben (1900/01)
W. Nernst, R. von Lieben: Über ein neues phonographisches Prinzip. *Zeitschrift für Elektrochemie* **7** (1900/1901), 533–534; R. von Lieben: Einige Beobachtungen am „Elektrochemischen Phonographen". *Zeitschrift für Elektrochemie* **7** (1900/1901), 534–538.

Nernst and Lindemann (1911)
W. Nernst, F.A. Lindemann: Spezifische Wärme und Quantentheorie. *Zeitschrift für Elektrochemie* **17** (1911), 817–827.

Nernst and Noddack (1923)
W. Nernst, W. Noddack: Zur Theorie photochemischer Vorgänge. *Sitzungsberichte der Preußischen Akademie der Wissenschaften, Physikalisch-mathematische Klasse* **1923**, 110–115.

Nernst and Orthmann (1926)
W. Nernst, W. Orthmann: Die Verdünnungswärme von Salzen bei sehr kleinen Konzentrationen. *Sitzungsberichte der Preußischen Akademie der Wissenschaften zu Berlin, Physikalisch-mathematische Klasse* **1926**, 51–56; **1927**, 136–141.

Nernst and Orthmann (1928)
W. Nernst, W. Orthmann: Die Verdünnungswärme von Salzen bei sehr kleinen Konzentrationen. *Zeitschrift für physikalische Chemie* **135** (1928), 199–208.

Nernst and Pauli (1892)
W. Nernst, R. Pauli: Weiteres zur electromotorischen Wirksamkeit der Ionen. *Annalen der Physik und Chemie* **45** (1892), 353–359; W. Nernst: Über die Potentialdifferenz verdünnter Lösungen. *Annalen der Physik und Chemie* **45** (1892), 360–369.

Nernst and Riesefeld (1903)
W. Nernst, E.H. Riesenfeld: Ueber quantitative Gewichtsanalyse mit sehr kleinen Gewichtsmengen. *Berichte der Deutschen Chemischen Gesellschaft* **36** (1903), 2086–2093.

Nernst and Sand (1909)
W. Nernst, J. Sand: Das physikalisch-chemische Institut an der Universität Berlin. *Zeitschrift für Elektrochemie* **15** (1909), 229–232.

Nernst and Sand (1910)
W. Nernst, J. Sand: Das physikalisch-chemische Institut. In: M. Lenz (ed.): *Geschichte der Königlichen Friedrich-Wilhelms-Universität zu Berlin, Band III*. Verlag der Buchhandlung des Waisenhauses: Halle/S. 1910, 306–310.

Nernst and Schoenflies (1895)
W. Nernst, A. Schoenflies: *Einführung in die mathematische Behandlung der Naturwissenschaften – Kurzgefaßtes Lehrbuch der Differential- und Integralrechnung mit besonderer Berücksichtigung der Chemie*. Wolff: München, Leipzig 1895.

Nernst and Warburg (1957)
W. Nernst, L. Warburg: Zwischen Raum und Zeit (Ein Märchen). *Physikalische Blätter* **13** (1957), 564–565.
Nernst and Wartenberg (1906)
W. Nernst, H. v. Wartenberg: Einige Bemerkungen zum Gebrauch des Wannerpyrometers. *Verhandlungen der Deutschen Physikalischen Gesellschaft* **8** (1906), 146–150.
Nernst and Wild (1900/01)
W. Nernst, W. Wild: Einiges über das Verhalten elektrolytischer Glühkörper. *Zeitschrift für Elektrochemie* **7** (1900/1901), 373–376.
Nernst et al. (1910)
W. Nernst, F. Koref, F.A. Lindemann: Untersuchungen über die spezifische Wärme bei tiefen Temperaturen, I, II. *Sitzungsberichte der Königlich Preußischen Akademie der Wissenschaften zu Berlin* **1910**, 247–261, 262–282.
Niedzielska et al. (1991)
M. Niedzielska, Z. Gardzielewski, H. Appel, K. Antonowicz. *Walther Nernst 1864 – 1941*. Toruńskie Towarzystwo Kultury: Wąbrzeźno, Grudziądz, Toruń 1991.
Ogilvie (2000)
M.B. Ogilvie: Maltby, Margaret Eliza. In: M. Ogilvie, J. Harvey: *The Biographical Dictionary of Women in Science – Pioneering Lives from Ancient Times to the Mid-20th Century, Volume 2.* Routledge: New York, London 2000.
Ostwald (1888)
W. Ostwald: Über die Dissociationstheorie der Elektrolyte. *Zeitschrift für physikalische Chemie* **2** (1888), 270–283.
Ostwald (1892)
W. Ostwald: „Handbuch der anorganischen Chemie", Stuttgart 1892. *Zeitschrift für physikalische Chemie* **9** (1892), 774–775; „Jahrbuch der Chemie. Frankfurt/M. 1892". *Zeitschrift für physikalische Chemie* **9** (1892), 776.
Ostwald (1906a)
W. Ostwald: Die Chemische Reichsanstalt. *Zeitschrift für angewandte Chemie* **19** (1906), 1025–1027.
Ostwald (1906b)
W. Ostwald: Für die chemische Reichsanstalt! *Die chemische Industrie* **29** (1906), 645–647.
Ostwald (1906c)
W. Ostwald: *Die chemische Reichsanstalt.* Akademische Verlagsgesellschaft: Leipzig 1906.
Ostwald (1927)
W. Ostwald: *Lebenslinien – Eine Selbstbiographie, Zweiter Teil.* Klasing & Co.: Berlin 1927.

Ostwald and Nernst (1889)
W. Ostwald, W. Nernst: Über freie Jonen. *Zeitschrift für physikalische Chemie* **3** (1889), 120–130.
Partington (1953)
J.R. Partington: Hermann Walther Nernst. The Nernst Memorial Lecture, delivered before the Chemical Society at Burlington House on March 19th, 1953. *Journal of the Chemical Society (London)* **1953**, 2853–2872.
Pauli (1893)
R. Pauli: *Bestimmung der Empfindlichkeitskonstanten eines Galvanometers mit astatischem Nadelpaar und aperiodischer Dämpfung.* Dissertation, Universität Göttingen 1893.
Pfleiderer (1909)
G. Pfleiderer: Die Sauerstoffentwicklung bei der Salzsäureelektrolyse mit Platinanode. Beiträge zur Frage der Zersetzungsspannung. *Inaugural-Dissertation*, Friedrich-Wilhelms-Universität zu Berlin 1909; *Zeitschrift für physikalische Chemie* **68** (1910), 49–82.
Pinner (1918)
F. Pinner: *Emil Rathenau und das elektrische Zeitalter.* Akademische Verlagsgesellschaft: Leipzig 1918.
Planck (1887)
M. Planck: Über die molekulare Konstitution verdünnter Lösungen. *Zeitschrift für physikalische Chemie* **1** (1887), 577–582.
Planck (1927)
M. Planck: *Vorlesungen über Thermodynamik.* De Gruyter: Berlin, Leipzig ^{8}1927.
Planck (1967)
M. Planck: *Wissenschaftliche Selbstbiographie.* (= Lebensdarstellungen deutscher Naturforscher Nr. 5). J.A. Barth: Leipzig ^{4}1967.
Pusch (1918)
L. Pusch: Zur Anwendung des Einsteinschen photochemischen Äquivalentgesetzes II. *Zeitschrift für Elektrochemie* **24** (1918), 336–339.
Remme (1926)
K. Remme (ed.): *Die Hochschulen Deutschlands – Ein Führer durch Geschichte, Landschaft, Studium (Ausgabe für Ausländer).* Verlag des Akademischen Auskunftsamts: Berlin 1926.
Riecke (1895)
E. Riecke: *Lehrbuch der Physik zu eigenem Studium und zum Gebrauche bei Vorlesungen.* Veit & Co.: Leipzig 1895.
Riesenfeld (1924)
E.H. Riesenfeld: Walter Nernst zu seinem sechzigsten Geburtstag. *Zeitschrift für angewandte Chemie* **37** (1924), 437–439.

Riesenfeld (1930)
  E.H. Riesenfeld: Svante Arrhenius (1859–1927). *Berichte der Deutschen Chemischen Gesellschaft* **63** (1930), 1–40.
Ritter (1798)
  J.W. Ritter: *Beweis, dass ein beständiger Galvanismus den Lebensprocess in dem Thierreich begleite: Nebst neuen Versuchen und Bemerkungen über den Galvanismus.* Industrie-Comptoir: Weimar 1798.
Ritter (1810)
  J.W. Ritter: *Fragmente aus dem Nachlasse eines jungen Physikers Ein Taschenbuch für Freunde der Natur.* Mohr und Zimmer: Heidelberg 1810.
Rolland (1966)
  R. Rolland: Über dem Getümmel (Au-dessus de la mêlée). In: *Gesammelte Werke in Einzelbänden: Der freie Geist (L'esprit libre).* Rütten & Loening: Berlin 1966, 92–97.
Roloff (1894)
  M. Roloff: Beiträge zur Kenntnis der photochemischen Wirkung in Lösungen. *Zeitschrift für physikalische Chemie* **13** (1894), 327–365.
Rösler (1999)
  W. Rösler: Hermann Diels und Albert Einstein: Die Lukrez-Ausgabe von 1923/24. In: *Hermann Diels (1848–1922) et la Science de l'Antiquité.* (= Entretiens sur l'Antiquité Classique, tome XLV). Fondation Hardt: Vandœvres-Genève 1999.
Roth (1949)
  W.A. Roth: Aus den Erinnerungen eines alten Thermochemikers. *Die Naturwissenschaften* **36** (1949), 225–229.
Rubens (1910)
  H. Rubens: Das physikalische Institut. In: M. Lenz (ed.): *Geschichte der königlichen Friedrich-Wilhelms-Universität zu Berlin, Band III.* Verlag der Buchhandlung des Waisenhauses: Halle/S. 1910, 278–296.
Rubner (1916)
  M. Rubner: Unsere Ernährung. In: Bund deutscher Gelehrter und Künstler (Kulturbund) (ed.): *Deutsche Volkskraft nach zwei Kriegsjahren – Vier Vorträge.* B.G. Teubner: Leipzig, Berlin 1916, 2–12.
SAG
  http://www.stadtarchiv.goettingen.de/chronik/1951_11.
Sawade (1933)
  S. Sawade: Beiträge zur Theorie alter und neuer Klavierformen. *Zeitschrift für technische Physik* **14** (1933), 353–362.
Schimank (1930)
  H. Schimank: *Epochen der Naturforschung – Leonardo / Kepler / Faraday.* Wegweiser-Verlag: Berlin 1930.

Schultze (1992)

W. Schultze: Quellen zum Wirken von Walther Nernst an der Berliner Universität im Archiv der Humboldt-Universität zu Berlin. *Wissenschaftliche Zeitschrift der Humboldt-Universität zu Berlin, Reihe Mathematik/Naturwissenschaften* **41** (1992) 4, 45–53.

Seeger (1966)

H. Seeger (ed.): *Musiklexikon, Erster Band.* Deutscher Verlag für Musik: Leipzig 1966.

Skrabal (1942)

A. Skrabal: Walther Nernst (Gestorben am 18. November 1941). *Almanach der Akademie der Wissenschaften in Wien* **92** (1942), 193–199.

Solvay (1914)

E. Solvay: Ansprache zur Eröffnung des Kongresses. In: A. Eucken (ed.): *Die Theorie der Strahlung und der Quanten. Verhandlungen einer von E. Solvay einberufenen Zusammenkunft (30. Oktober bis 3. November 1911).* (= W. Nernst (ed.): *Abhandlungen der Deutschen Bunsen-Gesellschaft für angewandte physikalische Chemie* 7). W. Knapp: Halle/S. 1914, 1–4.

Stark (1987)

J. Stark: *Erinnerungen eines deutschen Naturforschers.* Bionomia-Verlag: Mannheim 1987.

Stenzel (1976)

R. Stenzel: Begründung für die Verschmelzung der Reichsanstalt für Maß und Gewicht mit der Physikalisch-Technischen Reichsanstalt in Berlin im Jahre 1923. *Annals of Science* **33** (1976), 289–306.

Szöllösi-Janze (1998)

M. Szöllösi-Janze: *Fritz Haber 1868 – 1934, Eine Biographie.* Beck: München 1998.

Tammann and Nernst (1891)

G. Tammann, W. Nernst: Über die Maximaltension, mit welcher Wasserstoff aus Lösungen durch Metalle in Freiheit gesetzt wird. *Nachrichten von der Königlichen Gesellschaft der Wissenschaften und der Georg-Augusts-Universität zu Göttingen* **1891**, 201–212; *Zeitschrift für physikalische Chemie* **9** (1892), 1–11.

UAG

Universitätsarchiv Göttingen, Dekanatsakten der Philosophischen Fakultät Nr. 175 a.

UAHUB PhF

Universitätsarchiv der Humboldt-Universität zu Berlin, Philosophische Fakultät, Nr. 1470.

UAHUB *volume, sheet(s)*

Universitätsarchiv der Humboldt-Universität zu Berlin, Universitätskurator Personalia N 21.

Van't Hoff (1896)
J. H. van't Hoff: Antrittsrede. *Sitzungsberichte der Königlich Preußischen Akademie der Wissenschaften zu Berlin* 1896^{II}, 745–747.

Vierling (1936)
O. Vierling: *Das elektroakustische Klavier.* VDI-Verlag: Berlin 1936.

Wang et al. (2001)
Y. Wang, Z. A. Xu, T. Kakeshita, S. Uchida, S. Ono, Y. Ando, N. P. Ong: Onset of the vortexlike Nernst signal above T_c in $La_{2-x}Sr_xCuO_4$ and $Bi_2Sr_{2-y}La_yCuO_6$. *Physical Review* **B 64**, 224519 (2001).

Wazek (2005)
M. Wazek: „Einstein auf der Mordliste!"Die Angriffe auf Einstein und die Relativitätstheorie 1922. In: J. Renn (ed.): *Albert Einstein – Ingenieur des Universums – Hundert Autoren für Einstein.* Wiley-VCH: Weinheim 2005.

WH (1993)
W. H.: Gedenktafel für Walther Nernst. *PTB-Mitteilungen: Forschen + Prüfen. Amts- und Mitteilungsblatt der Physikalisch-Technischen Bundesanstalt Braunschweig-Berlin* **103** (1993), 112.

Wiechert (1921)
E. Wiechert: *Der Äther im Weltbild der Physik.* Weidmannsche Buchhandlung: Berlin 1921.

Wiener (1910)
O. Wiener: Letter to Walther Nernst, December 12, 1910. Universität Leipzig, Handschriftenabteilung der Universitätsbibliothek.

Willstätter (1958)
R. Willstätter: *Aus meinem Leben – Von Arbeit, Muße und Freunden.* Verlag Chemie: Weinheim ^{2}1958.

Winckel (1931)
F. W. Winckel: Das Radio-Klavier von Bechstein-Siemens-Nernst – Klangfarben auf Bestellung. *Die Umschau: Forschung, Entwicklung, Technologie* **35** (1931), 840–843.

Young and Linebarger (1900)
J. W. A. Young, C. E. Linebarger: *The Elements of the Differential and Integral Calculus, Based on Kurzgefaßtes Lehrbuch der Differential- und Integralrechnung von W. Nernst und A. Schönflies.* Appleton: New York 1900.

Zanthier (1964a)
E. von Zanthier: Letter to Werner Haberditzl, September 25, 1964 (unpublished).

Zanthier (1964b)
E. von Zanthier: Letter to Werner Haberditzl, October 12, 1964 (unpublished).

Zanthier (1973a)
E. von Zanthier: Letter to Friedrich Herneck, February 21, 1973 (unpublished).

Zanthier (1973b)
  E. von Zanthier: Letter to Friedrich Herneck, September 23, 1973 (unpublished).
Zanthier (1977)
  E. von Zanthier: Letter to Friedrich Herneck, January 1977 (unpublished).
Zanthier (1978)
  E. von Zanthier: Letter to Friedrich Herneck, September 15, 1978 (unpublished).
ZfE (1914)
  Zeitschrift für Elektrochemie **20** (1914).
ZfE (1915)
  Zeitschrift für Elektrochemie **21** (1915).
ZfE (1917)
  Zeitschrift für Elektrochemie **23** (1917).
Ziman (1960)
  J.M. Ziman: *Electrons and Phonons – The Theory of Transport Phenomena in Solids.* Clarendon Press: Oxford 1960.
Zott (1996)
  R. Zott (ed.): *Wilhelm Ostwald und Walther Nernst in ihren Briefen sowie in denen einiger Zeitgenossen.* Verlag für Wissenschafts- und Regionalgeschichte Dr. Michael Engel: Berlin 1996.
ZSAM
  Zentrales Staatsarchiv Merseburg, Nachlaß Althoff, Rep. 92, Althoff C, Nr. 20.
ZSAP
  Zentrales Staatsarchiv Potsdam, RMdI 27079.
ZSAP-N
  Zentrales Staatsarchiv Potsdam, REM Nr. N[ernst] 43.
ZSAP-P
  Zentrales Staatsarchiv Potsdam, REM Nr. P[aschen] 44.

Name Index

Abderhalden, Emil (1877–1950)   148
Abegg, Richard Wilhelm Heinrich (1869–1910)   96, 97, 100, 101, 216, 217
Abel, Emil (1875–1958)   335
Abelson, Philip Hange (b. 1913)   317
Abrikosov (Абрикосов), Alekseĭ Alekseevich (Алексей Алексеевич) (b. 1928)   30
Achard, Franz Carl (1753–1821)   134
Albert I (1875–1934, King of the Belgians 1909–1934)   255
Albrecht, Wilhelm Eduard   (1800–1876) 59
Alembert, see: d'Alembert
Alpher, Ralph Asher (b. 1921)   325
Althoff, Friedrich Theodor (1839–1908)   86, 87, 89, 92, 140, 208, 275
Ampère, André Marie (1775–1836)   5, 270
Arco, Georg Wilhelm Alexander Graf von (1869–1940)   219
Aristotle ('Αριστοτέλης) (384–322 BC)   4, 36, 89
Arrhenius, Svante August (1859–1927)   6, 8, 23, 25, 26, 33, 35, 38, 39, 41–43, 50, 52–54, 80, 82, 83, 92, 97, 103, 140, 223, 252, 253, 256, 257, 260, 262, 266, 267, 273, 275, 287–290, 306, 308, 332, 345
Askenasy, Paul (1869–1936)   217
Auer von Welsbach, Carl Freiherr (1858–1929)   105, 107, 111
Aulinger, Eduard Franz Karl (1854–?)   23
Avogadro, Lorenzo Romano Amadeo Carlo A. Conte di Quaregna e Ceretto (1776–1856)   5, 42, 73, 74, 85

Baade, Walter (1893–1960)   307
Baeyer, Adolf Johann Friedrich Wilhelm Ritter von (1835–1917)   83, 252
Bar, Carl Ludwig von (1836–1913)   92
Barkan, Diana Kormos   vi
Bauer, Max Hermann (1869–1929)   238–240
Bechstein, Carl junior (1860–1931)   299
Bechstein, Edwin (1859–1934)   299
Bechstein, Friedrich Wilhelm Carl senior (1826–1900)   298, 299
Bechstein, Helene (née Capito)   299

Bechstein, Johannes (1863–1905) 299
Beckmann, Ernst Otto (1853–1923) 40, 51, 80, 82, 92, 207–210
Beethoven, Ludwig van (1770–1827) 251, 302
Behn, Ulrich Andreas Richard (1868–1908) 170
Behring, Emil Adolph von (1854–1917) 252
Beilstein, Friedrich Konrad (1838–1906) 74
Benndorf, Hans (1870–1953) 23
Bennewitz, Gustav David Kurt (1886–1964) 198, 201, 274
Berg, Otto (1874–?) 278
Bergius, Friedrich Carl Rudolf (1884–1949) 332
Bernewitz, Ernst (1892–1921) 314
Bernstein, Elmer (1922–2004) 296
Bernthsen, August Heinrich (1855–1931) 215
Berthelot, Marcelin Pierre Eugène (1827–1907) 64, 156, 197
Berzelius, Jöns (Joens) Jakob (Jacob) (1779–1848) 6
Bethmann Hollweg, Theobald Theodor Friedrich Alfred von (1856–1921) 255
Beuth, Christian Peter Wilhelm (1781–1853) 138
Bieberbach, Ludwig Georg Elias Moses (1886–1982) 329–334
Biedermann, Karl (1812–1901) 11
Biwald, Leopold (1731–1805) 20
Bjerrum, Niels Janniksen (1879–1958) 178, 191, 289, 291, 292
Bloem (pseudonym B. Walter), Walter Julius Gustav (1868–1951) 245, 248
Blücher, Gebhard Leberecht Fürst B. von Wahlstatt (1742–1819) 9
Blümner, Hugo (1844–1919) 11
Bode, Harald (1909–1987) 297
Bodenstein, Max Ernst August (1871–1942) 49, 73, 96, 120, 145, 165, 168, 183, 189, 191, 199, 204, 205, 215–217, 219, 232, 237, 243, 260, 273, 274, 284, 297, 303, 329, 335, 336, 338, 341, 343
Boeckh (Böckh), Philipp August (1785–1867) 136, 225
Boerhaave, Herman (1668–1738) 2, 3
Boettinger, Henry Theodore von (1848–1920) 86, 89
Bohr, Niels Henrik David (1885–1962) 180, 266
Bois-Reymond, see: du Bois-Reymond
Boltzmann, Ludwig Eduard (1844–1905) 21–26, 29, 32, 35, 37, 70, 75, 83, 84, 107, 126, 290, 306, 307, 313, 324
Bondi, Sir Hermann (1919–2005) 325
Bonhoeffer, Karl Friedrich Otto Hans (1899–1957) 115, 117, 205, 216, 217, 274
Bopp, Franz (1791–1867) 136
Borchers, Johann Albert Wilhelm (1856–1925) 92, 213, 216, 217
Born, Fritz Karl Theodor (1896–1955) 274
Born, Max (1882–1970) 175, 176, 336
Börnstein, Richard Leopold (1852–1913) 18, 137, 140
Bosch, Carl (1874–1940) 167, 212, 248, 265, 332

Bosse, Julius Robert (1832–1901) 84, 87, 89, 92
Bothe, Walther Wilhelm Georg Franz (1891–1957) 272, 279
Böttinger, Henry (Heinrich) Theodor(e) von (1848–1920) 214, 215
Boyle, Robert (1627–1691) 2, 4, 46
Brandes, Heinrich Wilhelm (1777–1834) 36
Braun, Karl Ferdinand (1850–1918) 218
Braune, Hermann Konrad Ludwig (1886–1977) 190, 191
Bredig, Georg (1868–1944) 52, 223, 273
Brill, Otto (1881–?) 120
Brillouin, Marcel Louis (1854–1948) 179, 180
Broglie, Maurice de (1875–1960) 180
Brønsted, Johannes Nicolaus (1879–1947) 163
Brosig, Max (1853–?) 13
Brown, Robert (1773–1858) 139, 320
Browne, P.F. 326
Brühl, Julius Wilhelm (1850–1911) 57, 58
Brunner, Erich (1878–1934) 102
Bruns, Ernst Heinrich (1848–1919) 37
Buchner, Eduard (1860–1917) 137, 140, 252
Bülow, Hans Guido Freiherr von (1830–1894) 298
Bunsen, Robert Wilhelm Eberhard (1811–1899) 7, 57, 105, 144, 203, 213
Busch, Hans Walter Hugo (1884–1973) 110
Butenandt, Adolf Friedrich Johann (1903–1995) 212

Cahn, Heinz 71, 72, 331
Callen, Herbert B. 27
Canaris, Wilhelm Franz (1887–1945) 336
Carnot, Nicolas Léonard Sadi (1796–1832) 5, 151–153, 157, 185, 186
Cartesius, see: Descartes
Casimir, Hendrik Brugt Gerhard (1909–2000) 182, 183
Caspari, William Augustus (1877–1951) 100
Cassel, Hans Maurice (Moritz) (1891–1981) 242
Cato (Maior), Marcus Porcius (Censorius) (234–149 BC) 11
Chadwick, Sir James (1891–1974) 272, 323, 324
Chisholm, Grace, see: Young
Chladni (Chladny, Chladenius), Ernst Friedrich Florenz (1756–1827) 293, 302
Churchill, Sir Winston Leonard Spencer (1874–1965) 336
Claisen, Ludwig (1851–1930) 148
Clapeyron, Benoît Pierre Émile (1799–1864) 161
Clark, Josiah Latimer (1822–1898) 163
Clausius, Rudolf Julius Emanuel (1822–1888) 5, 6, 21, 52, 152, 153, 161, 187, 188, 197, 307
Clebsch, Rudolf Friedrich Alfred (1833–1872) 60, 61, 63
Clement, John Kay (1880–?) 261

Clusius, Klaus Paul Alfred (1903–1963) 175, 177
Cohen, Ernst Julius (1869–1944) 288
Colding, Ludwig August (1815–1889) 151
Cooper-Hewitt, Peter (1861–1921) 112
Cottrell, Frederick Gardner (1877–1948) 135
Coulomb, Charles Augustin de (1736–1806) 290
Cowell, Henry Dixon (1897–1965) 295
Cremer, Erika (1900–1996) 16, 273
Crookes, Sir William (1832–1919) 230
Curie, Marie (née Skłodowska, Marya) (1867–1934) 148, 179, 180, 199, 332
Curie, Pierre (1859–1906) 148, 199
Czapski, Siegfried (1861–1907) 41
Czerny, Marianus (1896–1985) 287

D'Albert, Eugène (Eugen) Francis Charles (1864–1932) 304
d'Alembert, Jean Baptiste Le Rond (1717–1783) 127, 128
D'Ans, Jean (1881–1969) 135
Dahlmann, Christoph (1785–1860) 59
Dalton, John (1766–1844) 4, 128
Dammer, Otto (1839–1916) 74, 77
Dan(n)eel, Heinrich Ludwig Julius August Maria (1867–1942) 98, 99, 102, 216, 217
Daniell, John Frederic (1790–1845) 7
Davy, Sir Humphry (wrongly Humphrey) (1778–1829) 6, 105
de Geer, Gerard Jacob Freiherr (1858–1943) 193
de Haas, Wander Johannes (1878–1960) 270, 272
Debye (Debije), Peter (Petrus) Joseph(us) Wilhelm(us) (1884–1966) 36, 173, 176, 178, 180, 289–291
Delbrück, Ludwig (1860–1913) 210
Delbrück, Max Emil Julius (1850–1919) 137
Democritus (Δημόκριτος) (circa 460–370 BC) 4, 259
Des Coudres, Theodor (1862–1926) 37, 96, 97, 101
Descartes (Cartesius), René (Renatus) (1596–1650) 4, 127
Dewar, Sir James (1842–1923) 125, 170
Diderot, Denis (1713–1784) 127, 128
Diels, Hermann Alexander (1848–1922) 259
Diels, Otto Paul Hermann (1876–1954) 148
Dirichlet, Peter Gustav Lejeune (1805–1859) 60
Dolezalek, Friedrich (Fritz) (1873–1920) 96, 98, 100, 272
Donau, Julius (1877–1960) 121
Doppler, Christian Johann (1803–1853) 315, 321, 322, 324, 326
Dove, Heinrich Wilhelm (1803–1879) 136, 225
Drägert, Willy (1890–?) 169, 191
Driescher, Hans 297–300, 302
Drude, Paul Karl Ludwig (1863–1906) 29, 37, 63, 64, 67, 139, 148, 284

Name Index

Drummond, Thomas (1797–1840) 105
du Bois-Reymond, Emil Heinrich (1818–1892) 115, 137, 226, 268, 285
Duisberg, Friedrich Carl (1861–1935) 238
Dulong, Pierre Louis (1785–1838) 17, 171
Dunant, George 167, 191

Ebert, Friedrich (1871–1925) 279, 328
Ebert, Hermann (1861–1913) 37, 67
Ebert, Johannes Ludwig (1894–1956) 292
Eddington, Sir Arthur Stanley (1882–1944) 309, 312–315, 319
Edison, Thomas Alva (1847–1931) 105, 113
Eggert, John Emil Max (1891–1973) 75, 79, 112, 194, 205, 231, 232, 240–243, 274, 275, 330, 342
Ehrenfest, Paul (1880–1933) 180
Ehrlich, Paul (1854–1915) 252
Eigen, Manfred (b. 1927) 34
Einstein, Albert (1879–1955) 9, 14, 15, 19, 139, 159, 169–173, 175, 176–180, 182, 184, 203, 204, 211, 212, 223, 250, 253, 256–259, 266, 270, 272, 282, 302, 310, 311, 324–326, 327–330, 332, 341
Einstein, Alfred (1880–1952) 303, 304
Elbs, Karl (1858–1933) 215
Emich, Friedrich (1860–1940) 121
Encke, Johann Franz (1791–1865) 136, 225
Engelmann, Wilhelm (1808–1878) 11
Engelmann, Wilhelm Friedrich (1785–1823) 41
Engels, Friedrich (1820–1895) 4
Enke, Ferdinand (1810–1869) 79
Epicure(s) ('Επίκουρος) (341–271 BC) 259
Erdmann, Hugo (1862–1910) 78
Erler, Heinrich Wilhelm (1820–1896) 12
Erman, Paul (1764–1851) 136, 284
Ernst August II (1771–1851, Duke of Cumberland since 1799, Duke of Brunswig-Luneburg and King of Hanover 1837–1851) 59
Erzberger, Matthias (1875–1921) 257
Esau, Abraham (1884–1955) 269, 272
Ettingshausen, Albert Freiherr von (1850–1932) 23–30, 32, 34, 35, 63, 69, 76
Ettingshausen, Andreas Freiherr von (1796–1878) 24
Ettingshausen, Constantin Freiherr von (1826–1897) 24
Eucken, Arnold Thomas (1884–1950) 34, 79, 80, 96, 124, 172, 175, 176, 178, 179, 182, 189, 191, 199, 200, 202, 215, 246
Eucken, Rudolf Christoph (1846–1926) 246, 252
Euler, Leonhard (1707–1783) 127
Euler-Chelpin, Hans Karl August Simon von (1873–1964) 135
Ewald, Georg Heinrich August von (1803–1875) 59

Fajans, Kasimir (1887–1975) 273
Falkenhayn, Erich Georg Anton Sebastian von (1861–1922) 238, 239, 246, 254
Faraday, Michael (1791–1867) 6, 7, 51, 52, 55, 99, 204, 263
Fechner, Gustav Theodor (1801–1887) 36
Fehling, Hermann Christian von (1812–1885) 74
Feit, Wilhelm Friedrich August (1867–1956) 278
Fichte, Johann Gottlieb (1762–1814) 135, 136
Fick, Adolf Eugen (1829–1901) 43, 44, 102
Fischer, Alfred (1894–1917) 232
Fischer, Emil Hermann (1852–1919) 31, 32, 71, 86, 100, 129,139–142, 148, 174, 198, 207–210, 232, 238, 244, 252, 275, 332
Fischer, Hans (1881–1945) 332
Fischer, Walter (1891–1916) 232
Flërov (Флёров), Georgiï Nikolaevich (Георгий Николаевич) (1913–1990) 317
Foerster, Fritz (Friedrich) (1866–1931) 215, 273
Foerster, Wilhelm Julius (1832–1921) 18, 139, 226, 252, 268
Fowler, Sir Ralph Howard (1889–1944) 153
Franck, James (1882–1964) 195
Franklin, Benjamin (1706–1790) 222, 265, 293
Frankó 297, 298
Franz II (I) (1768–1835, Holy German Emperor F. II 1792–1806, Austrian Emperor F. I 1804–1835) 20
Frenkel' (Френкель), Jakov Il'ich (Яков Ильич) (1894–1952) 114
Frick, Wilhelm (1877–1946) 282
Fridman (Friedmann, Фридман), Aleksandr Aleksandrovich (Александр Александрович) (1888–1925) 325
Friedrich I, «der Streitbare» (the Warlike) (1370–1428, Margrave of Meissen since 1381, Elector of Saxony 1423–1428) 36
Friedrich II, "the Great" (1712–1786, King of Prussia 1740–1786) 134, 245
Friedrich III (crown prince Friedrich Wilhelm) (1831–1888, German Emperor and King of Prussia 1888) 268
Friedrich III/I (1657–1713, F. III: Elector of Brandenburg since 1688, F. I: King in Prussia 1701–1713) 133
Friedrich Wilhelm I, the «Soldatenkönig» (1688–1740, King of Prussia 1713–1740) 133
Friedrich Wilhelm III (1770–1840, King of Prussia 1797–1840) 10, 135, 237, 258
Friedrich Wilhelm IV (1795–1861, King of Prussia 1840–1861) 245
Frisch, Christoph Albert (1840–1918) 186
Fulda, Ludwig Anton Salomon (1862–1939) 251

Gabriel, Sigismund (1851–1924) 18
Gaede, Wolfgang (1878–1945) 123–125
Galilei, Galileo (1564–1642) 311
Galton, Sir Francis (1822–1911) 331, 332
Galvani, Luigi (1737–1798) 5, 6

Gamow (Gamov, Гамов), George Anthony (Georgiĭ Antonovich, Георгий Антонович) (1904–1968) 325, 326
Gans(-Landau), Leo Ludwig (1843–1935) 207
Gassendi (Gassend), Pierre (Petrus) (1592–1655) 4
Gauss, Carl Friedrich (1777–1855) 12, 31, 58, 60
Geer, see: de Geer
Gehrcke, Ernst Johannes Ludwig (1878–1960) 272, 280, 283
Geiger, Hans (Johannes) Wilhelm (1882–1945) 270, 272
Geissler, Heinrich Johann Wilhelm (1814–1879) 125
Georg II August (1683–1760, Elector of Hanover and as George II King of Great Britain and Ireland 1727–1760) 58
George, Benjamin (1712–1771) 143
Gervinus, Georg Gottfried (1805–1871) 59
Ghiorso, Albert (b. 1915) 317
Gibbs, Josiah Willard (1839–1903) 5, 7, 48, 152, 155, 156, 159, 160, 196, 261
Goebel (Göbel), Johann Heinrich (Henry) Christoph (1818–1893) 106
Goethe (Göthe), Johann Wolfgang von (1749–1832) 1, 36, 251, 264
Gold, Thomas (1920–2004) 325
Goldschmidt, Hans (Johannes) Wilhelm (1861–1923) 215
Goldschmidt, Robert Benedict (1877–1935) 180, 254
Gordon, Clarence McCheyne (1870–1962) 99
Graefe, Carl Ferdinand von (1787–1840) 70
Graefe, Friedrich Wilhelm Ernst Albrecht von (1828–1870) 70
Gren, Friedrich Albrecht Carl (1760–1798) 6
Grimm, Hans August Georg (1887–1958) 215, 216
Grimm, Jacob Ludwig Carl (1785–1863) 59, 137
Grimm, Wilhelm Carl (1786–1859) 59, 137
Grotian, Walter (1890–1955) 323
Grotthuss (Grothuss), Freiherr Christian Johann Dietrich Theodor (Teodors) von (1785 to 1822) 6
Grove, Sir William Robert (1811–1896) 7
Gruess, Peter (b. 1949) 213
Grüneisen, Eduard August (1877–1949) 148, 175, 181, 272, 273
Guericke, Otto von (1602–1686) 4
Guillaume, Charles Édouard (1861–1938) 195
Gullstrand, Allvar (1862–1930) 194
Günther, Paul (1892–1969) 22, 191, 215–217, 274, 307, 312, 315, 317, 329, 336, 338, 339, 343
Guthnick, Paul (1879–1947) 308, 314, 315, 319

Haagn, Ernst (1875–1929) 98
Haas, see: de Haas
Hába, Alois (1893–1973) 304

Haber, Fritz (1868–1934)   111, 165, 167, 168, 195, 196, 202, 209, 210, 212, 217, 220, 221, 223, 230, 238, 239, 244, 248, 252, 273, 284
Haberlandt, Gottlieb Johann Friedrich (1854–1945)   330
Haeussermann, Carl (1853–1918)   74
Hahn, Albert   71, 72, 331
Hahn, Otto (1879–1968)   212, 273, 340
Hall, Edwin Herbert (1855–1938)   25, 27–29
Halske, Johann Georg (1814–1890)   111, 219
Hammond, Laurens (1895–1973)   296
Haniel von Haimhausen, Edgar Karl Alfons (1870–1935)   260
Hankel, Wilhelm Gottlieb (1814–1899)   36, 37, 104
Harding, Warren Gamaliel (1865–1923)   260
Harnack, Karl Gustav Adolf von (1851–1930)   208–210, 212, 252
Hasenöhrl, Friedrich (1874–1915)   179, 180, 182
Hauptmann, Gerhart Johann Robert   (1862–1946) 252
Hausen, Christian August (1693–1743)   36
Hausmanninger, V.   23
Heffter, Arthur Carl Wilhelm (1859–1925)   226
Hefner-Alteneck, Friedrich Franz Heinrich Philipp von (1845–1904)   105, 106, 111, 112
Hegel, Georg Wilhelm Friedrich (1770–1831)   128, 136
Heim, Carl Ludwig (1858–1924)   92
Heisenberg, Werner Karl (1901–1976)   37
Hell, Carl Magnus von (1849–1926)   74
Hellwig, Moritz (1841–1912)   143, 286
Helmholtz, Ellen von, see: Siemens
Helmholtz, Hermann Ludwig Ferdinand von (1821–1894)   5, 7, 19, 21, 41, 42, 48, 54, 88, 94, 95, 101, 111, 139, 143, 151, 155, 156, 159, 160, 196, 226, 253, 268, 269, 274, 277, 284–286, 289, 306
Hempel, Adolph Friedrich (1767–1834)   70
Heraeus, Wilhelm Carl (1827–1904)   118, 120
Herman, Robert (1914–1997)   325
Hermbstaedt, Sigismund Friedrich (1760–1833)   136
Herneck, Friedrich (1909–1993)   249, 343
Herrmann, Dieter B. (b. 1939)   302
Hertling, Georg Friedrich Graf von (1843–1919)   256
Hertz, Gustav Ludwig (1887–1975)   138
Hertz, Heinrich Rudolf (1857–1894)   66, 88, 94
Herz, Walter Georg (1875–1930)   77
Herzen, Édouard   180
Herzfeld, Karl Ferdinand (1892–1978)   198
Hess, Viktor Franz (1883–1964)   316
Hesse, Albert (1866–1924)   65
Hettner, Georg Gerhard (1892–1968)   287

Hettner, Hermann Georg (1854–1914) 18, 287
Heydweiller, Adolf (1856–1926) 32, 33
Heymann, Ernst (1870–1946) 330
Hiecke, Richard (1864–1948) 23, 24
Hilbert, David (1862–1943) 62, 243
Hill, Archibald Virian (1886–1977) 117
Himstedt, Franz Wilhelm Adolph Albert (1852–1933) 80
Hindemith, Paul (1895–1963) 296
Hindenburg, Paul Ludwig Hans Anton von Beneckendorf und von (1847–1934) 256, 330
Hindersin, Gustav Eduard von (1804–1872) 337
Hitler (Hiedler), Adolf (1889–1945) 299, 328, 329
Hittorf, Johann Wilhelm (1824–1914) 6, 42, 45, 52, 288
Hock, Lothar Erich Kurt (1890–1978) 79
Hoffmann, Dieter (b. 1948) vi
Hoffmann, E.T.A. (Ernst Theodor Amadeus), really: Ernst Theodor Wilhelm (1776–1822) 303
Hoffmann, Gerhard (1880–1945) 36
Hofmann, August Wilhelm von (1818–1892) 19, 97, 139, 143, 226
Högel, Hildegard (née Kulka) 235, 338
Hohmann, C. 65
Höhnow (Hönow, A.) 173, 174
Holborn, Ludwig Friedrich Christian (1860–1926) 280
Honegger, Arthur (1892–1955) 296
Höpfner, Ernst (1836–1915) 84, 89, 92
Hoppe, Willy (1884–1960) 335, 338
Horak, Franz (1880–?) 168, 191
Horstmann, August Friedrich (1842–1929) 7, 8, 42, 51, 75
Hostelet, G. 180
Hoyle, Sir Fred (1915–2001) 325
Hubble, Edwin Powell (1889–1953) 317, 320–322, 324, 326
Hückel, Erich Armand Arthur Joseph (1896–1980) 289–291
Humason, Milton Lasell (La Salle) (1891–1972) 321
Humboldt, Friedrich Wilhelm Christian Carl Ferdinand Freiherr von (1767–1835) 135, 136, 206
Humboldt, Friedrich Wilhelm Heinrich Alexander Freiherr von (1769–1859) 135–137
Hund, Friedrich (1896–1997) 37, 180, 183
Hutten, Ulrich von (1488–1523) 36

Ihne, Ernst Eberhard von (1848–1917) 210, 245
Inghen, Marsilius von (circa 1340–1398) 57
Ioffe (Иоффе), Abram Fëdorovich (Абрам Фёдорович) (1880–1960) 114, 223, 307

Jablokoff, Pierre, see: Yablochkov
Jahn, Hans (Johannes) Max (1853–1906)   vii, 8, 25, 26, 129, 130, 140, 145, 148, 156, 157,196, 197, 201, 288, 289
Jarres, Karl (1874–1951) 279, 280
Jeans, Sir James Hopwood (1877–1946) 107, 179, 180
Jedlik (Jedlík), Ányos (Anián, Anton) István (Štefan) (1800–1895) 7
Jellinek, Karl (1882–1971) 165, 167
Joachim, Joseph (1831–1907) 69
Joffe, see: Ioffe
Johannsen, Heinrich (1902–?) 284
Joliot-Curie, Irène (1897–1956) 332
Joliot-Curie, Jean Frédéric (1900–1957) 332
Jones, Harry Clary (1865–1916) 97
Jordis, Eduard Friedrich Alexander (1868–1917) 256
Jordis, W. 256
Joseph II (1741–1790, Holy German Emperor 1765–1790) 20
Jost, Fritz (1885–?) 166, 167, 191, 222
Jost, Wilhelm Friedrich (1903–1988) 8, 96, 114, 183, 265
Joule, James Prescott (1818–1889) 5, 107, 151
Julius Echter von Mespelbrunn (1545–1617) 30

Kahlbaum, Georg Wilhelm August (1853–1905) 92
Kalinin (Калинин), Mikhail Ivanovich (Михаил Иванович) (1875–1946) 127
Kamerlingh Onnes, Heike (1853–1926) 92, 170, 173, 179, 180, 223
Kant (Cant), Immanuel (1724–1804) 1, 126–128, 251, 306, 312, 314
Kapp, Wolfgang (1858–1922) 228
Karl II (1540–1590, Austrian archduke 1564–1590) 19
Kármán, Theodore von (1881–1963) 176
Keating, William Hypolitus (1799–1844) 265
Keil, Carl Friedrich (1807–1888) 11
Kelvin, see: Thomson, Sir William
Kennedy, Joseph William (1916–1957) 317
Kepler, Friedrich Johannes (1571–1630) 20
Kirchhoff, Adolf Johann Wilhelm (1826–1908) 18, 159, 160
Kirchhoff, Gustav Robert (1824–1887) 7, 19, 57, 139, 226
Klaproth, Martin Heinrich (1743–1817) 134, 136
Klein, Christian Felix (1849–1925) 61, 62, 84, 86, 87, 89, 103, 252
Klemenčič, Ignaz (Ignacij) (1853–1901) 23, 26
Klingenberg, Georg (1870–1925) 246
Klitzing, Klaus(-Olaf) von (b. 1943) 29
Kluck, Alexander Heinrich Rudolf von (1846–1934) 237
Knapp, Wilhelm (1840–1908) 213
Knop, Johann August Ludwig Wilhelm (1817–1891) 40
Knudsen, Martin Hans Christian (1871–1921) 179, 180, 223

Koeth, Joseph (1870–1936) 246
Köhler, Fritz 263
Kohlrausch, Friedrich Wilhelm Georg (1840–1910) 30–34,42, 45, 66, 67, 104, 139, 263, 269, 274
Kohnstamm, Philipp Abraham (1875–1951) 197, 288
Kolhörster (Kohlhörster), Werner Heinrich Julius Gustav (1887–1946) 272, 316
Kopff, August Adalbert (1882–1960) 319, 323
Kopp, Hermann Franz Moritz (1817–1892) 61, 169, 171
Koppel, Leopold (1854–1933) 209–211
Koref, Fritz (1884–1969) 122, 190, 191
Kossel, Albrecht Ludwig Karl Martin Leopold (1853–1927) 252
Köster, Adolf (1883–1930) 229
Krazer, Carl Adolf Joseph (1858–1926) 32
Kretschmann, H. 11, 12
Kries, Johannes von (1853–1928) 116
Krogh, Schack August Steenberg (1874–1949) 195
Kronecker, Leopold (1823–1891) 140
Kroth, Ewald (1880–1952) 247
Krüger, Friedrich August Heinrich (1877–1940) 50, 100
Krüger, Wilhelm (1898–1977) 335
Krupp von Bohlen und Halbach, Gustav (1870–1950) 209
Krüss, Hugo Andres (1879–1945) 209, 210
Kuhlmann, Wilhelm H.F. 121
Kummer, Ernst Eduard (1810–1893) 136, 140, 226
Kundt, August Adolf Eduard Eberhard (1839–1894) 54, 86, 88, 139, 283, 284
Kurlbaum, Ferdinand (1857–1927) 138, 272
Küster, Friedrich Wilhelm (1861–1917) 92, 97

Laar, Johannes Jacobus van (1860–1938) 165
Lamaître, Georges Henri (1894–1966) 325
Lampadius, Wilhelm August Eberhard (1772–1842) 105
Lamprecht, Karl Gotthard (1856–1915) 206
Landolt, Hans Heinrich (1831–1910) 8, 18, 32, 88, 92, 137, 139–142, 144, 145,148, 268, 272
Langenbeck, Bernhard Rudolph Conrad von (1810–1887) 70
Langenbeck, Conrad Martin Johann (1776–1851) 69, 70
Langevin, Paule (1872–1946) 179, 180
Langl, Josef (1843–1916) 11
Langmuir, Irving (1881–1957) 112, 261, 266, 332
Laplace, Pierre Simon Marquis de (1749–1827) 306, 314
Laue, Max Theodor Felix von (1879–1960) 212, 272, 279, 282, 283, 328, 330, 340
Lavoisier, Antoine Laurent de (1743–1794) 3, 4
Le Blanc, Max Julius Louis (1865–1943) 92, 100, 141, 214, 215, 242, 263, 264
Le Chatelier, Henry Louis (1850–1936) 155, 159

Le Rossignol, Robert (1884–1976)   167
Leduc, Sylvestre Anatole (1856–1937)   27
Lehmann, Christian (1675–1739)   36
Leibniz, Gottfried Wilhelm Freiherr von (1646–1716)   127, 128, 133
Lejeune-Dirichlet, Peter Gustav (1805–1859)   136, 140
Lemberg, Johann Theodor (1842–1902)   38
Lémery, Nicolas (1645–1715)   5
Lenard, Philipp Eduard Anton (1862–1947)   252, 283, 327, 329, 330, 332
Lepsius, Bernhard Richard Alexander (1854–1934)   221, 241
Lepsius, Carl Richard (1810–1884)   136
Lepsius, Carl Richard Reinhold (1885–1969)   236, 241
Lermontova (Лермонтова), Yulia (Julia) Vsevolodovna (Юлия Всеволодовна) (1847–1919)   103
Leuckart, Rudolf Karl Georg Friedrich (1822–1898)   37
Libavius (Libau, Basilius de Varna), Andreas (circa 1550–1616)   2
Lichtenberg, Georg Christoph (1742–1799)   58
Lie, Marius Sophus (1842–1899)   37
Lieben, Robert von (1878–1913)   218, 219
Liebermann, Karl Theodor (1842–1914)   19
Liebermann, Max (1847–1935)   186, 187, 245, 252
Liebig, Justus Freiherr von (1803–1873)   17, 81
Lilienfeld, Julius Edgar (1882–1963)   174
Linde, Carl Paul Gottfried von (1842–1934)   90
Lindemann, Charles Lionel (1885–1970)   181, 189, 191
Lindemann, Frederick Alexander (Viscount Cherwell) (1886–1957)   122, 175–177, 180, 181, 190, 191, 249, 331, 333, 336, 341
Lippmann, Gabriel (1845–1921)   101
Lipschitz, Rudolf (1832–1903)   12
Liszt, Franz (Ferenc) von (1811–1886)   298
Lodygin (Лодыгин), Aleksandr Nikolaevich (Александр Николаевич) (1847–1923)   111
Loeb, Jaques (1859–1924)   262
Loeb, Morris (1863–1912)   43, 261
Loewenherz (Löwenherz), Ludwig (1847–1892)   272
Lohmeyer, Karl Ferdinand (1826–1911)   69
Lohmeyer, Meta (1867–?)   70
Lohmeyer, Minna (née Heyne) (1839–1874)   70
Lomonosov (Ломоносов), Mikhail Vasil'evich (Михаил Васильевич) (1711–1765)   3, 127, 128
Lorentz, Hendrik Antoon (1853–1928)   22, 28–30, 179, 180, 182, 184
Lorenz, Lothar   77
Lorenz, Otto Ferdinand (1838–1896)   286
Lorenz, Richard (1863–1929)   96, 97

Loschmidt, Johann Joseph (1821–1895) 6, 24, 171, 290
Löwenstein, Leo (1879–1956) 121
Ludendorff, Erich Friedrich Wilhelm (1865–1937) 256, 265
Ludwig V (1577–1626, Landgrave of Hesse-Darmstadt 1596–1626) 81
Ludwig, Carl Friedrich Wilhelm (1816–1895) 37
Luise Auguste Princess of Prussia (1808–1870) 110
Łukasiewicz, Ignacy (1822–1882) 105
Lukrez (*Titus Lucretius Carus*) (94–55 BC) 259
Lummer, Otto Richard (1860–1925) 272, 273
Lüst, Reimar (b. 1923) 213

Mach, Ernst (1838–1916) 5, 22, 75
Mager, Jörg Georg Adam (1880–1939) 295, 297
Magnus, Alfred (1880–1960) 199
Magnus, Heinrich Gustav (1802–1870) 136, 226, 284, 285
Maltby, Margaret Eliza (1860–1944) 98, 102–104, 227, 261, 272
Marckwald, Willy (1864–1942) 148, 199, 241, 273, 274
Marggraf, Andreas Sigismund (1709–1782) 134
Mariotte, Edme Seigneur de Chazeuil (circa 1620–1684) 2, 46
Markl, Hubert (b. 1938) 213
Marquardt, Joachim Karl (1812–1882) 11
Marquart, Paul (1849–1917) 215
Martenot, Maurice (1898–1980) 296
Martinů, Bohuslav (1890–1959) 295
Martius, Carl Alexander von (1838–1920) 207
Maupertuis, Pierre Louis Moreau de (1698–1759) 134
Maximilian II (1527–1576, Holy German Emperor 1564–1576) 30
Maxwell, James Clerk (1831–1879) 21, 94, 170
Mayer, Christian Gustav Adolph (1839–1908) 37
Mayer, Julius Robert von (1814–1878) 5, 151
McMillan, Edwin Mattison (1907–1991) 317
Meißner, Fritz Walther (1882–1974) 271, 272, 279
Melanchthon (originally Schwarzerd[t]), Philipp (1497–1560) 57
Mendelssohn, Franz von (1835–1935) 210
Mendelssohn, Kurt Alfred Georg (1906–1980) v, vi, 70, 72, 232, 236, 343
Merriam, Edmund Sawyer (1880 – after 1933) 200, 202, 261
Merrick, Samuel (Richard) Vaughan (1801–1870) 265
Merz, Victor (1839–1893) 17, 32
Meyer, Arnold (1844–1896) 17
Meyer, Julius Lothar (1830–1895) 88
Meyer, Richard Emil (1846–1926) 74
Meyer, Victor (1848–1897) 57, 88, 118, 120
Meyerhoffer, Wilhelm (1864–1906) 52, 134
Michaelis, Georg (1857–1936) 256

Micheles, Theodor L. A. (1872–?) 240
Minkowski, Hermann (1864–1909) 62
Miquel, Johannes von (1829–1901) 89
Mises, Richard Martin Edler von (1883–1953) 284
Mitscherlich, Eilhard (1794–1863) 136, 225
Mittasch, Paul Alwin (1869–1953) 167, 215
Mitterbacher, Elise Magdalena (1801–1828) 10
Moellendorff, Wichard Georg Otto von (1881–1937) 246
Moers, Kurt 202, 203
Moltke, Helmuth Carl Bernhard Graf von (1800–1891) 268
Mommsen, Christian Mathias Theodor (1817–1903) 11, 136
Mond, Ludwig (1839–1909) 208
Moog, Robert (Bob) (1934–2005) 295
Mozart, Johannes Chrysostomus Wolfgang(us) Amadeus (Theophilius, Gottlieb, Amade) (1756–1791) 293
Müller-Breslau, Heinrich Franz Bernhard (1851–1925) 244
Mussolini, Benito (1883–1945) 328

Nahrath, Max 302
Napoleon I (original Bonaparte (Buonaparte), Napoléon) (1769–1821, French Emperor 1804–1814/15) 135
Naumann, Alexander (1837–1922) 80
Neesen, Friedrich (1849–1923) 149
Nerger, Anna 10–12
Nerger, Edith 12
Nerger, Frida 12
Nerger, Karl August (1800–1886) 10, 236
Nerger, Ottilie, see: Nernst
Nerger, Rudolf 10, 11, 70, 71, 229
Nernst, Angela (married Hahn) (1903–?) 11, 70–73, 330, 331, 335, 343
Nernst, Christian (1721–?) 9
Nernst, Edith Primula (married von Zanthier) (1900–1980) vi, 70–73, 249, 252, 256, 299, 303, 330, 338, 340, 343
Nernst, Emma (née Lohmeyer) (1871–1949) 12, 68–71, 73, 225, 233, 239, 257, 273, 302, 333, 338, 340
Nernst, Gustav junior (1896–1917) 12, 70, 71, 73, 193, 232, 233, 257
Nernst, Gustav senior (1827–1888) 10, 71
Nernst, Hermann (1789–1848) 9
Nernst, Hilde(gard) Elektra (married Cahn) (1894–1955) 12, 70–73, 86, 330, 331, 335, 340
Nernst, Johann David (1759–?) 9
Nernst, Johannes Christian (1769–?) 9
Nernst, Ottilie (née Nerger) (1833–1876) 10
Nernst, Philipp (1792–1844) 9, 10

Name Index

Nernst, Rudolf (1893–1914)   11, 12, 70, 71, 73, 193, 232, 233, 256, 257, 264
Nernst, Walther Hermann (1864–1941)   *passim*
Neumann, Carl Gottfried (1832–1925)   59–61
Neumann, Franz Ernst (1798–1895)   60, 63, 169, 171
Newton, Sir Isaac (1643–1727)   20, 127, 128, 311, 322
Nippoldt, Wilhelm August (1843–1904)   31
Nobel, Alfred Bernhard (1833–1896)   253, 289
Noddack, Ida Eva (née Tacke) (1896–1978)   273, 278
Noddack, Walter Karl Friedrich (1893–1960)   205, 272, 274, 278
Noyes, Arthur Amos (1866–1936)   52, 103, 261, 262, 288, 290
Nussbaumer, Otto (1876–1930)   24

Ochsenfeld, Robert (1901–1993)   271, 272
Oettingen, Arthur Joachim von (1836–1920)   38, 67
Ohm, Georg Simon (1789–1854)   32
Olbers, Heinrich Wilhelm Matthias (1758–1840)   318, 322
Onsager, Lars (1903–1976)   29, 292
Oordt, Gabriel van   165
Ornstein, Leonard Salomon (1880–1941)   197
Orthmann, Wilhelm (1901–1945)   131, 287, 291, 292
Ostwald, Helene (née von Reyher) (1854–1946)   68
Ostwald, Walter Karl Wilhelm (1886–1958)   164, 335
Ostwald, Wilhelm Friedrich (1853–1932)   5, 8, 22, 25, 33, 35–41, 50–54, 63–68, 74, 75, 78, 80, 82, 83, 86–88, 89, 91, 92, 97, 101–103, 129, 130, 135, 140, 141, 153, 164, 206–210, 213–216, 220, 221, 227, 231, 235, 252, 253, 261, 262, 288, 289, 345

Paalzow, Carl Adolf (1823–1908)   138, 139, 268
Palazzo, Luigi (1861–1933)   33
Palmaer, Knut Wilhelm (1868–1942)   228
Palmer, Charles Skeele (1858–1939)   77, 288
Paracelsus (properly Hohenheim, Philippus Aureolus Theophrastus Bombastus von) (1493–1541)   4
Partington, James Riddick (1886–1965)   42, 190, 191, 287
Paschen, Friedrich Louis Carl Heinrich (1865–1947)   212, 269, 271, 280–282
Pauli, Robert (1866–?)   67, 68
Pauli, Wolfgang Ernst (1900–1958)   182
Pavel, Friedrich Wilhelm Herbert (1889–1954)   316
Pebal, Leopold von (1826–1887)   25
Peltier, Jean Charles Athanase (1785–1845)   27, 29, 30
Penzias, Arno Allan (b. 1933)   325
Perrier, Carlo (1886–1948)   278
Perrin, Jean Baptiste (1870–1942)   179, 180
Petit, Alexis Thérèse (1791–1820)   17, 171
Pfeffer, Wilhelm Friedrich Philipp (1845–1920)   37, 42

Pfleiderer, Georg (1886–?) 201
Pflüger, Eduard (1829–1910) 116
Philippson, Franz Moses (1851–1929) 254–257
Pichler, Franz (1866–1919) 24
Pier, Mathias (1881–1965) 168, 189, 191, 217, 241
Pilowski, Karl (1905–1991) 319
Pinner, Felix (1880–1942) 5
Planck, Max Karl Ernst Ludwig (1858–1947) 19, 48, 49, 52–54, 107, 129, 139–142, 148, 152, 153, 165, 169, 170, 173, 178–180, 182, 187–189, 210–212, 223, 226, 245, 252, 258, 266, 273, 280, 283, 288, 302, 309, 321, 338, 340
Planté, Raymond Louis Gaston (1834–1889) 7, 100
Platen-Hallermünde, Karl August Georg Maximilian Graf von (1796–1835) 337
Plücker, Julius (1801–1868) 61
Pohl, Robert Wichard (1884–1976) 272
Poincaré, Jules Henri (1854–1912) 179, 180
Polanyi, Michael (1891–1976) 182
Pollitzer, Franz (1885–1942) 124, 168, 174, 189, 192
Pregl, Fritz (1869–1930) 121
Pringsheim, Ernst Georg (1859–1917) 148, 272
Pringsheim, Peter (1881–1963) 287
Proust, Joseph Louis (1754–1826) 3
Prym, Friedrich Emil (1841–1915) 32
Pusch, see: Volmer

Quenstedt, Friedrich August (1809–1889) 12

Rammelsberg, Carl Friedrich (1813–1899) 139, 143, 144
Ramsay, Sir William (1852–1916) 332
Ranke, Leopold Franz von (1795–1886) 11, 136
Rasch, Gustav (1863–1939) 33
Rathenau, Emil (1838–1915) 106, 110, 219
Rathenau, Walther (1867–1922) 214, 245, 246, 254, 258, 259
Ratzel, Friedrich (1844–1904) 37
Rayleigh, Lord John William (Strutt, John William), Third Baron (1842–1919) 107, 179
Regener, Erich Rudolph Alexander (1881–1955) 323, 324
Reicke, Georg (1863–1923) 251
Reil, Johann Christian (1759–1813) 6
Reisz, Eugen 218
Reithoffer, Max (1864–1945) 219
Remme, Karl (1881–1947) 226
Reuss, Eduard (1804–1891) 11
Richards, Theodore William (1868–1928) 198, 223, 261
Richter, Jeremias Benjamin (1762–1807) 3, 128
Richter, Otto 331–333

Name Index

Riecke, Carl Victor Eduard (1845–1915) 58–60, 62–64, 68, 80, 81, 89, 103, 104
Riedler, Alois (1850–1936) 244
Riemann, Bernhard Georg Friedrich (1826–1866) 60
Riesenfeld, Ernst Hermann (1877–1957) 33, 35, 54, 100, 120, 149, 150, 216, 273, 274, 306
Righi, Augusto (1850–1920) 27
Ritter, Johann Wilhelm (1776–1810) 6, 105, 114, 115
Röhl, Johann Christian Friedrich 16
Rohmer, M. 256
Rohn, Wilhelm Julius Paul (1887–1943) 256
Rolland, Romain (1866–1944) 253
Roloff, Friedrich Max (1870–1915) 96, 97
Roloff, Max (1871–?) 68
Röntgen, Wilhelm Conrad von (1845–1923) 31, 81, 111, 220, 252
Roon, Albrecht Theodor Emil Graf von (1803–1879) 285
Roscoe, Sir Henry Enfield (1833–1913) 203
Rose, Gustav (1798–1873) 136
Rose, Heinrich (1795–1864) 136
Rosen, Friedrich (1856–1935) 260
Rosvænge (Rosvaenge, Roswaenge), Helge (1897–1972) 304
Roth, Walter Adolf (1873–1950) 18, 129, 145
Rubens, Heinrich Leopold (1865–1922) 138, 139, 179, 180, 210, 252, 283, 284, 287, 328
Rubner, Max (1854–1932) 245, 248, 252
Ruprecht I (1309–1390, Elector Palatine since 1353) 57
Russell, Henry Norris (1877–1957) 314
Russolo, Luigi (1885–1947) 295
Rust, Bernhard (1883–1945) 331, 332
Rutherford, Ernest, Lord of Nelson and Cambridge (1871–1937) 179, 180, 182, 332

Saha, Meghnad (মেঘনাদ সাহা) (1893–1956) 194
Sala, Oskar (1910–2002) 296
Sand, Julius Wilhelm (1878–1917) 145
Savigny, Friedrich Carl von (1779–1861) 136
Sawade, Siegfried (b. 1908) 305
Schellbach, Karl Heinrich (1805–1892) 267, 268
Schenck, Friedrich Rudolf (1870–1965) 215, 273
Schiller, Johann Christoph Friedrich von (1759–1805) vi
Schillings, Max von (1868–1933) 304
Schimank, Hans Friedrich Wilhelm Erich (1888–1979) 190, 192, 242, 243
Schlegel, August Wilhelm von (1767–1845) 106
Schlegel, Karl Wilhelm Friedrich von (1772–1829) 106
Schleiermacher, Friedrich Daniel Ernst (1768–1834) 135, 136
Schlenk, Wilhelm Johann (1879–1943) 284

Schlieffen, Alfred Graf von (1833–1913) 237
Schmalz, Theodor Anton Heinrich (1760–1831) 225
Schmidt, Carl Ernst Heinrich (1822–1894) 38
Schmidt, Erhard Oswald Johann (1876–1959) 226, 284
Schmidt, Leopold (1824–1892) 11
Schmidt-Ott, Friedrich (1860–1956) 209
Schönflies (Schoenflies), Arthur Moritz (1853–1928) 129–131
Schonrock, Otto Paul Hermann (1870–1946) 272
Schottky (Schottki), Walter Hans (1886–1976) 114
Schottky, Hermann Viktor (1885–?) 122
Schrödinger, Erwin Rudolf Josef Alexander (1887–1961) 223
Schüler, Edmund (?– after 1930) 260, 261
Schwab, Georg-Maria (1899–1984) 216
Schwarz, Hermann Amandus (1843–1921) 61, 140
Schwers, Frédéric 125, 176, 192
Scriabin (Скрябин), Aleksandr Nikolaevich (Александр Николаевич) (1872–1915) 295
Seaborg, Glenn Theodore (1912–1999) 317
Seckel, Paul Georg Emil (1864–1924) 226
Seebeck, Thomas Johannes (1770–1831) 27–30
Segrè, Emilio Gino (1905–1989) 278
Seibt, Elisabeth (?–1992) 233
Selling, Eduard (1834–1920) 32
Sheldon, Samuel (1862–1920) 33
Siemens, Arnold Wilhelm von (1853–1918) 269
Siemens, Ellen von (née von Helmholtz) (1864–1941) 269
Siemens, Ernst Werner von (1816–1892) 7, 90, 105, 111, 139, 219, 258, 268, 269, 299
Siemens, George Wilhelm von (1855–1919) 258
Siggel, Alfred (1884–?) 169, 192
Silliman, Augustus Ely (1807–1884) 263
Silliman, Benjamin senior (1779–1864) 105
Silliman, Hepsa Ely (?–1883) 163, 169, 172, 198, 263
Simon, Franz (Sir Francis) Eugen (Eugene) (1893–1956) 177, 193, 249, 274, 341
Simson, Clara von (1896–1983) 274
Singer, George John (1786–1817) 6
Sinsteden, Wilhelm Josef (1803–1891) 7
Skrabal, Anton (1877–1957) 341, 345
Slaby, Adolph Karl Heinrich (1849–1913) 219
Slipher, Vesto Melvin (1875–1969) 320, 321
Solvay, Ernest Gaston Joseph (1838–1922) 179, 180, 182, 254, 255
Sommerfeld, Arnold Johannes Wilhelm (1868–1951) 79, 179, 180, 182, 244
Speckpeter, Heinrich (1873–1933) 215
Sperling, Auguste (married Nerger) (1810–1886) 10, 236

Name Index

Spieker, Paul Emmanuel (1826–1896) 286
Staab, Heinz August (b. 1929) 213
Stackelberg, Nikolai Mark Otto August Freiherr von (1896–1971) 217
Stahl, Georg Ernst (1660–1734) 133
Stark, Johannes Nicolaus (1874–1957) 203, 269, 271, 282, 283, 329, 330
Stefan (Štefan), Joseph (Jožef) (1835–1893) 21, 22, 24, 107, 313, 324
Stock, Alfred Eduard (1876–1946) 284
Stock, Franz (?–1939) 211
Strauss, Richard (1864–1949) 296
Strauss, Siegmund (1875–1942) 218
Streintz, Franz (1855–1902) 23, 26
Streintz, Heinrich (1848–1892) 23, 26
Stresemann, Gustav (1878–1929) 328
Stroof, Ignaz (1838–1920) 263
Stumpf, Carl (1848–1936) 302
Sudermann, Hermann (1857–1928) 245, 251
Svedberg, The (Theodor) (1884–1971) 332
Swan, Sir Joseph Wilson (1828–1914) 106
Sylvester, James Joseph (1814–1897) 128

Tacke, Ida, see: Noddack
Tafel, Julius (1862–1918) 100, 201
Tait, Peter Guthrie (1831–1901) 197
Tammann, Gustav Heinrich Johann Apollon (1861–1938) 52, 65, 96, 141, 197, 215, 256, 262
Tangl, Michael (1861–1921) 223
Tappen, Dietrich Gerhard Emil Theodor (1866–1953) 238
Tappen, Hans (1879–?) 238
Termen (Термен, Theremin), Lev (Лев, Leon) Sergeevich (Сергеевич) (1896–1993) 295, 296
Thaer, Albrecht Daniel (1752–1828) 137
Thiessen, Peter Adolf (1899–1990) 215
Thomson, Sir Joseph John (1856–1940) 66
Thomson, Julius (1826–1909) 41, 42, 156
Thomson, Sir William (Lord Kelvin of Largs) (1824–1907) 82, 151, 153, 187, 306
Thür, Georg (1846–1924) 145
Tiemann, Johann Karl Wilhelm Ferdinand (1848–1899) 18
Tilden, William Augustus (1842–1926) 170
Toepler, August Joseph Ignaz (1836–1912) 20, 24
Tralles, Johann Georg (1763–1822) 136
Trautwein, Friedrich Adolf (1888–1956) 295
Trott zu Solz, August Bodo Wilhelm Klemens Paul von (1855–1938) 210
Tudor, Henri Owen (1859–1928) 7

Ullmann, Max (1865–1941)   193
Urban, Kurt (1904–1928)   222
Urey, Harold Clayton (1893–1981)   332

Vahlen, Theodor Karl (1869–1945)   331
van der Waals, Johannes Diderik (1837–1923)   179, 197
van't Hoff, Jacobus Henricus (1852–1911)   8, 17, 37, 39, 41–45, 47, 48, 50, 75, 86, 92, 134, 135, 140, 142, 148, 158, 160, 163, 198, 211, 215, 262, 345
Varèse (Varesse), Edgar (Edgard Victor Achille Charles) (1883–1965)   295, 296
Vasyukhnova (Васюхнова, Wasjuchnowa), Maria Mitrofanovna (Мариа Митрофанова) (1882–?)   168, 192
Vierling, Oskar Walther (1904–1986)   297, 298, 305
Virchow, Rudolf Ludwig Carl (1821–1902)   136
Vogel, Friedrich Adolf (1856–1907)   213
Voght, Johann Caspar Reichsfreiherr (Baron) von (1752–1839)   1
Vögler (Voegler), Albert (1877–1945)   212
Voigt, Woldemar (1850–1919)   63
Volmer, Lotte (née Pusch) (1890–?)   204
Volmer, Max (1885–1965)   273
Volta, Alessandro Giuseppe Antonio Anastasio conte di (1745–1827)   6, 7
Voss, Christian Friedrich (1722–1795)   88, 92

Waals, see: van der Waals
Wagner, Carl (1901–1977)   114
Wagner, Julius Eugen (1857–1922 or 1924)   40, 52, 216
Wahl, Arthur Charles (b. 1917)   317
Walden, Paul (1863–1957)   141
Waldeyer-Hartz, Heinrich Gottfried Wilhelm von (1836–1921)   245, 252
Walker, Sir James (1863–1935)   52
Wallach, Otto Hermann Theodor Gustav (1847–1931)   89, 223, 224, 252, 333
Walter, Bruno, really: Schlesinger, Bruno Walter (1876–1962)   304
Wanner, H.   122
Warburg, Emil Gabriel (1846–1931)   14, 101, 116, 139, 140, 142, 148, 179, 180, 210, 212, 228, 267, 269, 270, 275, 284
Warburg, Lotte (married Meyer Viol) (1884–1948)   14
Warnstedt, Adolf von (1813–1897)   90
Wartenberg, Hans Joachim von (1880–1960)   122, 167, 168, 189, 192, 217, 264, 273
Wasjuchnowa, see: Vasyukhnova
Watson, Eric   214
Weber, Heinrich (1842–1913)   61
Weber, Heinrich Friedrich (1843–1912)   17, 170
Weber, Wilhelm Eduard (1804–1891)   31, 34, 36, 59, 60
Wegschneider, Rudolf Franz Johann (1859–1935)   78
Wehnelt, Arthur Rudolph Berthold (1871–1944)   218, 283, 287, 329

Weierstrass, Karl Theodor Wilhelm (1815–1897) 140, 226
Weinhold, Adolf Ferdinand (1841–1917) 12
Weiss, Christian Samuel (1780–1856) 136, 225
Wenzel, Carl Friedrich (1740–1793) 3
Westinghouse, George (1846–1914) 262
Wheatstone, Sir Charles (1802–1875) 65, 288
Wichelhaus, Carl Hermann (1842–1927) 143, 145, 148
Wicke, Ewald (1914–2000) 79, 92
Wiechert, Johann Emil (1861–1928) 310
Wiedemann, Eilhard Ernst Gustav (1852–1928) 53, 54
Wiedemann, Gustav Heinrich (1826–1899) 36, 37, 40, 41, 48, 53
Wien, Max Carl Werner (1866–1938) 101, 280
Wien, Wilhelm Karl Werner Otto Fritz Franz (1864–1928) 107, 179, 180, 182, 252, 272, 273, 280
Wiener, Otto Heinrich (1862–1927) 36, 64, 173, 174, 207
Wilhelm II (born Friedrich Wilhelm Albert Viktor) (1859–1941, German Emperor and King of Prussia 1888–1918) 88, 92, 141, 174, 209–211, 237, 244, 249, 265
Wilke, Arthur (1853–1913) 213, 216
Willstätter, Richard Martin (1872–1842) 66, 252, 333
Wilsmore, Norman Thomas Mortimer (1868–1940) 99
Wilson, Robert Woodrow (b. 1936) 325
Winckel, Fritz Wilhelm (1907–2000) 298
Winkler, Clemens Alexander (1838–1904) 214
Winkler, Johann Heinrich (1703–1770) 36
Wirth, Joseph Karl (1879–1956) 260
Wirtz, Carl Wilhelm (1876–1939) 320
Wislicenus, Johannes (1835–1902) 32, 37, 40
Wöhler, Friedrich (1800–1882) 138
Wolf, Christian Wilhelm Friedrich August (1759–1824) 136
Wollaston, William Hyde (1766–1828) 3
Wundt, Wilhelm Maximilian (1832–1920) 37
Wüst, Fritz (Wilhelm Friedrich) (1860–1938) 244

Yablochkov (Яблочков), Pavel Nikolaevich (Павел Николаевич) (1847–1894) 105, 106, 111
Yamaha, Torakusu (山葉寅楠) (1851–1916) 304
Young, Grace (née Chisholm) (1868–1944) 103

Zacher, Hans F. (b. 1928) 213
Zaeslein-Benda, Vicky 141, 142
Zanthier, Angela von (married Klingmüller) (1923–1987) 73
Zanthier, Edith von, see: Nernst
Zanthier, Rudolf Ernst von (1889–?) 72, 73
Zastrau, Karl Albert Fritz (1837–1899) 143, 286

Zenneck, Jonathan Adolf Wilhelm (1871–1959) 280
Zeppelin, Ferdinand Adolf August Heinrich Graf von (1838–1917) 244, 245
Ziegler, Johanna (née Kulka) 235, 338
Zirkel, Ferdinand (1838–1912) 37
Zwicky, Fritz (1898–1974) 307, 322